ULRYK

AF606573

ULRYK

BIOLOGICAL AND MEDICAL PHYSICS, BIOMEDICAL ENGINEERING

Springer
Berlin
Heidelberg
New York
Hong Kong
London
Milan
Paris
Tokyo

BIOLOGICAL AND MEDICAL PHYSICS, BIOMEDICAL ENGINEERING

The fields of biological and medical physics and biomedical engineering are broad, multidisciplinary and dynamic. They lie at the crossroads of frontier research in physics, biology, chemistry, and medicine. The Biological and Medical Physics, Biomedical Engineering Series is intended to be comprehensive, covering a broad range of topics important to the study of the physical, chemical and biological sciences. Its goal is to provide scientists and engineers with textbooks, monographs, and reference works to address the growing need for information.

Books in the series emphasize established and emergent areas of science including molecular, membrane, and mathematical biophysics; photosynthetic energy harvesting and conversion; information processing; physical principles of genetics; sensory communications; automata networks, neural networks, and cellular automata. Equally important will be coverage of applied aspects of biological and medical physics and biomedical engineering such as molecular electronic components and devices, biosensors, medicine, imaging, physical principles of renewable energy production, advanced prostheses, and environmental control and engineering.

T. Furukawa (Ed.)

Biological Imaging and Sensing

With 224 Figures

Professor Toshiyuki Furukawa, MD, D.Med.Sci.
President Emeritus
Osaka National Hospital
Hoenzaka 2-1-14, Chuo-ku, 540-0006 Osaka, Japan
e-mail: furukawa@onh.go.jp

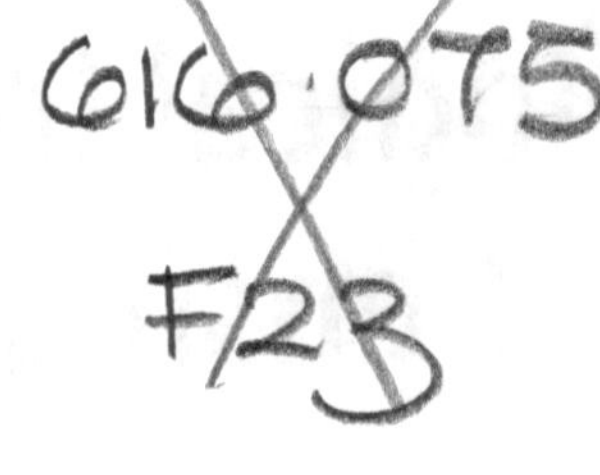

Library of Congress Cataloging in Publication Data:

Biological imaging and sensing / T. Furukawa, ed.
p. cm. - - (Biological and medical physics, biomedical engineering, ISSN 1618-7210)
Includes bibliographical references and index.
ISBN 3-540-43898-X (alk. paper)
1. Imaging systems in medicine. 2. Imaging systems in biology. I. Furukawa, T. (Toshiyuki), II. Series.
R857.O6 B556 2004
616.07'54- -dc21 2002030563

ISSN 1618-7210

ISBN 3-540-43898-X Springer-Verlag Berlin Heidelberg New York

Springer-Verlag is a part of Springer Science+Business Media.

springeronline.com

Printed in Germany

Final processing by PTP-Berlin Protago-TeX-Production GmbH, Berlin
Cover concept by eStudio Calamar Steinen using a background picture from The Protein Databank (1 Kzu). Courtesy of Dr. Antoine M. van Oijen, Department of Molecular Physics, Huygens Laboratory, Leiden University, The Netherlands. Reprinted with permission from Science 285 (1999) 400–402 ("Unraveling the Electronic Structure of Individual Photosynthetic Pigment-Protein Complexes", by A. M. van Oijen et al.) Copyright 1999, American Association for the Advancement of Science.
Cover production: *design & production* GmbH, Heidelberg

Printed on acid-free paper SPIN 10876380 57/3141/YU - 5 4 3 2 1 0

Preface

Recent biomedical research is directed at a new challenge to a microcosm of life and a deeper recognition of life by new technologies. Studies of the living body by new methods are a common idea, and new analytic methods have been developed to study objects that have not previously been visible. In Japan, there are several groups that are developing the required new methods. The latest physical approaches are used by these groups. This book is written by the Japanese leaders in this field for young biomedical researchers.

Medical informatics, biological engineering, and computer analysis for the next generation strategy of biomedical research. New developments such as nanomachines and intelligent materials are also included in this book. We have paid attention to the mathematical recognition that was lacking in biological and medical studies to develop a theory and technology for biofunction analysis. Five topics have been included in this book: photonics technology, ultrasonic techniques, solution of inverse problems, remote sensing and control, and biosensors. The chapters were organized by the following researchers: Satoshi Kawata, Professor at Osaka University, Department of Electric Engineering Postgraduate Course, for "photonic bio-information measurement and control"; Masatsugu Hori, Professor at Osaka University Medical School, Department of Internal Medicine and Therapeutics, for "development of innovative technology for ultrasonic diagnosis and treatment"; Hisashi Kado, Professor at Kanazawa University, for "development of magnetic source imaging for clinical applications"; Kunihiko Mabuchi, Professor at the University of Tokyo, International Industry–University Co-Operation Center, for "systemic research on sensing and information-processing techniques aimed at biological applications"; Eiichi Tamiya, Professor at Hokuriku Postgraduate Research Institute for Advanced Science and Technology, Department for "material science basic development to create the second-generation biosensor".

This book contains many interesting topics. One of the most unique articles concerns an implantable artificial heart pacemaker in which electrical energy can be charged with a near infrared laser. This product has been announced in several newspapers, but you will find a detailed description in this publication. By doing complex information processing of MEG using the personal computers in parallel, highly brain function of human being of thoughts is analyzable. It is remarkable that the three-dimensional ultrasound imaging of the human fetus was first invented in Japan. Now complete spherical ultrasound waves and a combination of composite aperture antennas have been developed. An extremely sharp eye watching the living body is going to be offered. These results emphasize the self-discipline of the Japanese biomedical engineer.

I believe that this book will provide cutting edge ideas for the development of new technologies in biomedical measurement and control, and facilitate the task of students and young scientists to maintain a keen interest in research work.

Osaka *T. Furukawa*
July 2001

Contents

1 Biological Imaging and Sensing from Basic Techniques to Clinical Application
S. Kawata, O. Nakamura, T. Kaneko, M. Hashimoto, K. Goto, N.I. Smith, T. Sugiura, I. Fujimasa, H. Matsumoto 1

1.1 Introduction: A General View of the Electromagnetic Waves That Pass through the Living Body 1
1.2 Imaging Cells through a Multi-Photon Process 2
1.2.1 Nonlinear Optics in Cells 2
1.2.2 The Imaging Property of Multi-Photon Microscopy 3
1.2.3 Instrumentation 5
1.2.4 Calcium Ion Dynamics Revealed by Multi-Photon Microscopy 6
1.3 Nonstaining Molecular Imaging (CARS) 9
1.3.1 The Fundamentals of Coherent Anti-Stokes Raman Scattering 10
1.3.2 3D Resolution by CARS Microscopy 11
1.3.3 High-Speed Image Acquisition 14
1.3.4 Molecular Imaging by CARS Microscopy 15
1.4 Transcutaneous Near-Infrared Light Power/Information Transmission for Implantable Medical Devices 19
1.4.1 Feasibility of Powering, Controlling, and Monitoring Implantable Medical Devices Using Near-Infrared Light 19
1.4.2 Transcutaneous Power Transmission by Near-Infrared Light 21
1.4.3 Transcutaneous Information Transmission by Near-Infrared Light 23
1.5 Cell and Nanosurgical Operation with Light 25
1.5.1 Introduction and Interactions between Ultra-Short Pulses and Biological Materials 25
1.5.2 Laser-Induced Disruption in Biomaterials 28
1.5.3 Cell Nanosurgery by Focused Light 28
1.6 The Manipulation of Living Bodies by Light 29
1.6.1 Photon Pressure 30
1.6.2 Three-Dimensional Laser Trapping 31

1.6.3 Force Measurement ... 32
1.6.4 Microscopy with a Laser-Trapped Particle ... 35
1.7 Physiological Function Analysis of the Human Body and Organ-Using Far-Infrared Imaging ... 40
1.7.1 Static Analysis of Abnormal Temperature Distribution on the Skin [57] ... 40
1.7.2 Dynamic Analysis of Abnormal Temperature Distribution on the Skin [70] ... 46
1.7.3 Dynamic Analysis of the Surface Temperature of Internal Organs ... 49
1.7.4 Unsolved Problems ... 51
1.8 A New Technology for Detecting Coronary Artery Disease ... 52
1.8.1 Coronary Artery Disease ... 52
1.8.2 The Detection of Subclinical Coronary Stenosis ... 52
1.8.3 A Noninvasive Physiological Approach to the Detection of Coronary Artery Disease ... 53
1.8.4 The Rheological Basis ... 53
1.8.5 The Impossibility of Conventional Standard Phonocardiography Technology ... 55
1.8.6 The Theoretical Basis for the New Technology ... 56
1.8.7 The Principles of the New Laser Phonocardiography Technology ... 57
1.8.8 The New Laser Phonocardiography Technology Design ... 58
1.8.9 The Details of the Prototype Device ... 59
1.8.10 Data Acquisition of the Vibratory Signal of the Anterior Chest Wall ... 62
1.8.11 Signal Processing ... 62
1.8.12 The Future of the New Laser Phonocardiography Technology ... 64
References ... 65

2 Imaging of Tissue/Organs with Ultrasound
M. Hori, T. Masuyama, K. Baba, O. Ohshiro, K. Ishihara, H. Kondo ... 69

2.1 Ultrasonic Biological Measurement (Ultrasonography) ... 69
2.1.1 The Principle of Ultrasonography ... 69
2.1.2 The Doppler Technique ... 71
2.1.3 Recent Advances in Ultrasound Imaging ... 72
2.1.4 The Ultrasonic Characterization of Myocardial Tissue ... 74
2.2 Three-Dimensional Ultrasound Imaging of the Fetus ... 76
2.2.1 Conventional Ultrasound Imaging of the Fetus ... 76
2.2.2 The Development of Three-Dimensional Ultrasound ... 79
2.2.3 Clinical Applications of Three-Dimensional Ultrasound in Obstetrics ... 81

2.2.4 The Advantages and Limitations of Three-Dimensional Ultrasound 82
2.2.5 The Future Development of Three-Dimensional Ultrasound 83
2.3 Imaging by a Spherical Ultrasound Wave 84
2.3.1 An Instantaneous Imaging Method 85
2.3.2 Laser-Induced Breakdown 86
2.3.3 Ultrasound Generated by Laser-Induced Breakdown 87
2.3.4 Imaging Using Laser-Induced Breakdown 89
2.3.5 Summary 93
2.4 Imaging Tissues Using an Ultrasound Wave and Light 93
2.4.1 An Ultrasound Wave and Light for Tissues 93
2.4.2 Ultrasound-Assisted Optical Imaging 94
2.4.3 The Experimental Setup 94
2.4.4 An Experiment for Weak-Scattering Samples 97
2.4.5 An Experiment in a Strongly Scattering Medium 100
2.4.6 Conclusions 101
2.5 An Ultrasonic Drug Delivery System Using Microcapsules 102
2.5.1 The Requirement for a Drug Delivery System 102
2.5.2 The Acoustic Characteristics of Microcapsules as Drug Carriers 103
2.5.3 A Noninvasive Measurement System for DDS 104
2.5.4 Collapse Monitoring of Microcapsules 105
2.5.5 Conclusion 107
2.6 The Biological Effects of Ultrasound 107
2.6.1 The Utilization of Ultrasound Energy for Therapeutics 108
2.6.2 The Biological Effects of Diagnostic Ultrasound 108
2.6.3 The Effect of Ultrasound on the Cell Membrane 109
2.6.4 Ultrasound for Gene Therapy 110
2.6.5 Direct Effects on Cell Components 111
2.6.6 The Stress-Induced Cellular Response 111
2.6.7 Potential Applications of Low-Power Ultrasound 113
References 114

3 The Imaging of a Magnetic Source
H. Kado, H. Ogata, Y. Haruta, M. Higuchi, M. Shimogawara, J. Kawai, Y. Adachi, C. Bertrand, G. Uehara 117

3.1 The Principle of Magnetic Field Measurement 117
3.1.1 The Magnetic Field 117
3.1.2 The Magnetic Dipole 118
3.1.3 Magnetic Flux 119
3.1.4 The Electromagnetic Coil 119
3.1.5 The Current Dipole 120
3.1.6 Time-Varying Magnetic Fields 120

3.1.7 The Search Coil Magnetometer 121
3.1.8 The Proton Magnetometer and Other Magnetometers 122
3.1.9 Magnetic Source Analysis 124
3.2 A High-Sensitivity Magnetic Field Sensor 125
3.2.1 The SQUID 125
3.2.2 A System for Biomagnetic Measurement 137
3.3 Magnetic Source Analysis 149
3.3.1 The Forward Problem 149
3.3.2 The Inverse Problem 154
3.3.3 Visualization 170
3.4 Biomagnetic Measurement 177
3.4.1 Magnetoencephalography 177
3.4.2 Other Biomagnetic Measurements 192
3.5 Other Applications of Magnetic Source Imaging 194
3.5.1 Field Observation 195
References 200

4 Bioanalyses Using Electrochemical and Electrophysiological Methods
E. Tamiya, K. Mabuchi, K. Yokoyama, Y. Murakami, M. Kobayashi, M. Suzuki, H. Suzuki, T. Suzuki, M. Kunimoto 205

4.1 Introduction 205
4.2 The Electrochemical DNA Chip Sensor 209
4.2.1 Introduction 209
4.2.2 The Multiplexed Electrochemical DNA Sensor 209
4.2.3 The Microfluidic PCR Chamber and the Electrochemical Detector 214
4.3 Enzyme-Based Electrochemical Biosensors 219
4.3.1 Introduction 219
4.3.2 Biosensors Based on Redox Active Polymers 220
4.3.3 Glucose Sensors on the Market 225
4.3.4 Conclusion 230
4.4 Cell and Tissue Monitoring and Their Applications: The Whole Cell Sensor 231
4.4.1 Overview of Cell-Based Sensor 231
4.4.2 The Surface PhotoVoltage (SPV) and Its Application 232
4.5 Neural Network, Neural, or Brain Analyses: Measurements of Neural Activity and Their Application for Analyses of the Neural Network System in the Living Body ... 239
4.5.1 Ultra-Microglutamate Sensors for Brain Analyses 239
4.6 The Application of Micromachining Techniques to Chemical Sensors, Biosensors, and Microanalysis Systems 242
4.6.1 Introduction 242
4.6.2 Basic Technologies 242

4.6.3 Microsensors for Dissolved Gases and Electrolytes 243
4.6.4 Microfabricated Biosensors 247
4.6.5 The Integration of Sensors and Micro-Electrochemical Systems 248
4.7 Microneurography: Measurements and Stimulation of a Single Peripheral Nerve Fiber 250
4.7.1 Introduction 250
4.7.2 The History of Microneurography 250
4.7.3 The Technique of Microneurography 251
4.7.4 The Advantages and Disadvantages of Microneurography . . 255
4.7.5 Summary 256
4.8 A Microelectrode for the Neural Interface 257
4.8.1 Introduction 257
4.8.2 The Nerve Electrode (Handmade) 258
4.8.3 A Microelectrode for the Neural Interface 260
4.8.4 Conclusion 263
4.9 Regulation: On-Line Measurements of Humoral and Neural Information from the Living Body and Their Application for the Control of Artificial Organs and Limbs 263
4.9.1 The Control of Artificial Hearts Using Humoral Information 263
4.9.2 Control of Artificial Hearts Using Autonomic Nervous Signals 270
4.9.3 The Control of Somatic Sensations and the Generation of Artificial Sensations by Direct Stimulation of the Neural System 275
4.9.4 Control of the Motor Function of Artificial Limbs (by Neural Signals) 286
References 287

Index 295

List of Contributors

Yoshiaki Adachi
Applied Electronics Laboratory,
Kanazawa Institute of Technology
3 Amaike, Kanazawa,
Ishikawa, 920-1331, Japan

Kazunori Baba
Center for Maternal, Fetal
and Neonatal Medicine,
Saitama Medical Center,
Saitama Medical School
1981 Kamoda, Kawagoe,
Saitoma 350-8550 Japan
baba-tokyo@umin.ac.jp

Cedric Bertrand
Applied Electronics Laboratory,
Kanazawa Institute of Technology
3 Amaike, Kanazawa, Ishikawa,
920-1331, Japan

Iwao Fujimasa
National Graduate Institute for
Policy Studies
2-2 Wakamatsu-cho Shinjuku-ku
Tokyo 162-8677, Japan

Kazuya Goto
Department of Applied Physics
Osaka University
Suita Osaka 565-0871, Japan
goto@ap.eng.osaka-u.ac.jp

Yasuhiro Haruta
Yokogawa Electric Corporation
2-9-32, Nakacho, Musashino,
Tokyo, 180-8750, Japan

Mamoru Hashimoto
Department of Systems
and Human Science
Graduate School
of Engineering Science
Osaka University
Toyonaka Osaka 560-8531, Japan

Masahiro Higuchi
Applied Electronics Laboratory,
Kanazawa Institute of Technology
3 Amaike, Kanazawa,
Ishikawa, 920-1331, Japan

Masatugu Hori
Department of Internal Medicine
and Therapeutics,
Osaka University
Graduate School of Medicine
2-2 Yamada-oka,
Suita 565-0871 Osaka, Japan
mhori@medone.med.osaka-u.ac.jp

Ken Ishihara
Ehime University Hospital
10-13, Doga-Hiyata Matsuyama,
790-8597 Japan

Hisashi Kado
Applied Electronics Laboratory,
Kanazawa Institute of Technology
3 Amaike, Kanazawa,
Ishikawa, 920-1331, Japan

Tomoyuki Kaneko
Department of Basic Science,
Graduate School
of Arts and Sciences,
The University of Tokyo
3-8-1 Komaba, Meguro-ku,
Tokyo 153-8902, Japan
ckaneko@mail.ecc.u-tokyo.ac.jp

Jun Kawai
Applied Electronics Laboratory,
Kanazawa Institute of Technology
3 Amaike, Kanazawa,
Ishikawa, 920-1331, Japan

Satoshi Kawata
Department of Applied Physics
Osaka University
Suita Osaka 565-0871, Japan

Masaaki Kobayashi
School of Materials Science
Japan Advanced Institute
of Science & Technology
1-1 Asahidai Tatsunokuchi
Ishikawa 923-1292, Japan

Hiroya Kondo
Department of Internal Medicine
and Therapeutics,
Osaka University
Graduate School of Medicine
2-2 Yamada-oka,
Suita 565-0871 Osaka, Japan
kondo@medone.med.osaka-u.ac.jp

Masanari Kunimoto
Department of Neurology,
International Medical Center
of Japan
1-21-1 Toyama, Shinjuku-ku,
Tokyo, 162-8655 Japan
mkunimot@imcj.hosp.go.jp

Kunihiko Mabuchi
Department of Imformation Physics
and Computing,
Graduate School of Information
Science and Technology,
The University of Tokyo
CCR Building
The University of Tokyo,
4-6-1 Komaba, Meguro-ku,
Tokyo 153-8904, Japan
mabuchi@ccr.u-tokyo.ac.jp

Thoru Masuyama
Department of Internal Medicine
and Therapeutics,
Osaka University
Graduate School of Medicine
2-2 Yamada-oka,
Suita 565-0871 Osaka, Japan
masuyama
@medone.med.osaka-u.ac.jp

Hiroshi Matsumoto
Center for Collaborative Research
The University of Tokyo
4-6-1, Komaba, Meguro-ku
Tokyo 153-890, Japan

Yuji Murakami
Pioneering Research Laboratories
Toray Industries, Inc.
3-2-1 Sonoyama Otsu
Shiga 520-0842, Japan

Osamu Nakamura
Graduate School
of Frontier Biosciences,
Osaka University
Suita Osaka 565-0871, Japan

Hisanao Ogata
Applied Electronics Laboratory,
Kanazawa Institute of Technology
3 Amaike, Kanazawa,
Ishikawa, 920-1331, Japan

Osamu Ohshiro
Research Center for Advanced
Science and Technology
Nara Institute
of Science and Technology
89165-5 Takayama
Ikoma 630-0192 Japan

Masahiro Shimogawara
Yokogawa Electric Corporation
2-9-32, Nakacho, Musashino,
Tokyo, 180-8750, Japan

Nicholas I. Smith
Department of Applied Physics
Osaka University
Suita Osaka 565-0871, Japan

Tadao Sugiura
Department of Bioinformatics
and Genomics
Graduate School
of Information Science
Nara Institute of Science
and Technology
Ikoma, Nara 630-0192, Japan
sugiura@is.aist-nara.ac.jp

Hiroaki Suzuki
Institute of Material Science,
University of Tsukuba
1-1-1 Tennoudai Tsukuba
Ibaragi 305-8573, Japan

Masayasu Suzuki
Department of Electric
and Electronic Engineering
Faculty of Engineering
Toyama University
3190 Gofuku Toyama
Toyama 930-8555, Japan

Takafumi Suzuki
Department of Information
Physics and Computing,
Graduate School of Information
Science and Technology,
The University of Tokyo
CCR Building
4-6-1 Komaba, Meguro-ku,
Tokyo 153-8904, Japan
suzuki@ccr.u-tokyo.ac.jp

Eiichi Tamiya
School of Materials
Science Japan
Advanced Institute
of Science & Technology
1-1 Asahidai Tatsunokuchi
Ishikawa 923-1292, Japan

Gen Uehara
Applied Electronics Laboratory,
Kanazawa Institute of Technology
3 Amaike, Kanazawa,
Ishikawa, 920-1331, Japan

Kenzi Yokoyama
Laboratory of Advanced
Bioelectronics
National Institute
of Advanced Industrial
Science & Technology
1-1-1 Higashi Tsukuba
Ibaragi 305-8562, Japan

1 Biological Imaging and Sensing from Basic Techniques to Clinical Application

S. Kawata, O. Nakamura, T. Kaneko, M. Hashimoto, K. Goto, N.I. Smith, T. Sugiura, I. Fujimasa, and H. Matsumoto

1.1 Introduction: A General View of the Electromagnetic Waves That Pass through the Living Body

In the last 20 years of the 20th century, amazing developments were accomplished in photonics technology, such as optical communications, optical data storage, short-pulse laser technology, and so on. In particular, the progress in the technologies related to semiconductor lasers and optical fibers is noteworthy. One of the keywords common to these technologies is "near infrared (NIR) light." The applications of these "NIR" technologies to biology and medicine have just begun at the beginning of the 21st century. "NIR" biomedical photonics is regarded as one of the most desirable future developments in science and technology.

NIR light is defined as light that has the wavelengths from 0.8 µm to 2.5 µm. Figure 1.1 shows the NIR band represented on the wavelength axis. The advantage of NIR light over visible and infrared light is its "transparency." This is because there are few atomic absorptions, and there is little molecular vibration, in the NIR band. Since NIR light is transparent to most materials, it is used for long-distance optical communications through optical

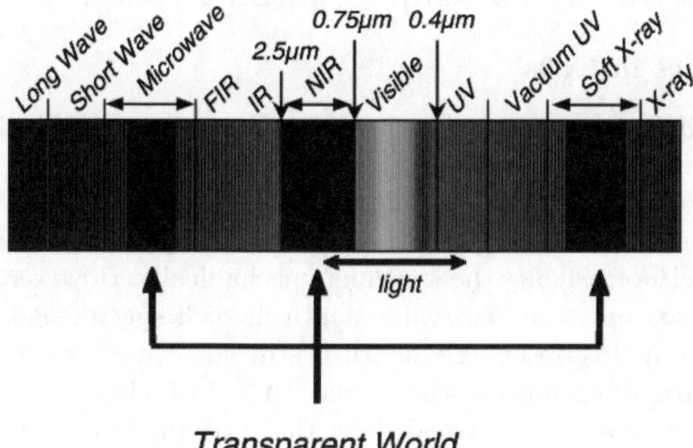

Fig. 1.1. The spectrum of light

fibers without much attenuation. NIR light is also used to remotely control televisions, air conditioners, and many other electric products due to its lower absorption by the surrounding materials. All these facts tell us that NIR light is tender and friendly to living bodies.

In this chapter, applications that utilize transparency of NIR light, such as multi-photon microscopy, coherent anti-Stokes Raman scattering microscopy, light-power transmission into a living body, optical communication through skin, laser nanosurgery, and laser trapping of cells, are presented. All of this work uses the "transparency", in other words less scattering and less absorption, properties of NIR light.

We have described four ways of applying NIR to living bodies. The first is to use the high transmittance of NIR light on living bodies. In Sect. 1.4, we report on transcutaneous optical power supply and optical communication methods using NIR light. A cardiac pacemaker implanted in a rat is driven by NIR light sent from outside the body. The second method is to sense the infrared vibration at a focused spot by using a set of NIR lasers. This is called CARS microscopy and is discussed in Sect. 1.3. The third way is to send and focus NIR short-pulse light into a deep portion of a living organ, and induce multi-photon absorption, whose energy corresponds to a photon in the visible range, by the resultant spatially and temporally condensed intense spot. This method is discussed in Sects. 1.2 and 1.5 as multi-photon microscopy and multi-photon surgery, respectively. The fourth way is to use the momentum of the NIR photons to move and control small particles such as cells, organellas, and latex beads attached to protein and DNA molecules. To move these kinds of particles, a force of the order of at least a piconewton is required. This means that the total irradiation power of the light is about 1 watt or more. Thanks to the transparency of NIR light, one can feed focused laser beams of some watts into biological specimens without thermally or photochemically destroying them.

1.2 Imaging Cells through a Multi-Photon Process

1.2.1 Nonlinear Optics in Cells

In the observation of a living specimen, it is important to reduce physical or physiological damage in the specimen during or even before the observation. For this reason, a confocal fluorescence microscope has been often used to observe those specimens, where the optically sectioned three-dimensional images can be obtained without slicing the specimen mechanically. However, even with a confocal microscope, the observable depth in such specimens is usually given as 50 μm, and this observable depth is still not enough to see the functions of whole parts of organs or other apparatuses in bodies.

Microscopic techniques have been advanced by the introduction of the multi-photon fluorescence microscope, which has been realized with the recent development of ultra-fast high-power laser systems [1,2]. Since the probability of two-photon excitation is proportional to the square of the excitation intensity, the fluorescence emission can be obtained only in the focal volume of an objective lens which illuminates the specimen. This localization of the fluorescence emission brings about three-dimensional resolution, the same as that in a confocal microscope, and reduces photobleaching without exciting out-of-focus planes. Furthermore, the near-infrared light used for two-photon excitation can penetrate more deeply into the specimen than visible light for single-photon excitation [3].

1.2.2 The Imaging Property of Multi-Photon Microscopy

Multi-point excitation is useful for increasing the amount of fluorescence at a moment in time without increasing the excitation power, because the fluorescence emission is less restricted the cell viability or nonlinear optical phenomena such as self-focusing, continuum generation arising in specimens with an increase in the excitation power [4]. The microlenses are arranged in a Nipkow-disk configuration that gives uniform illumination onto the specimen by its rotation [5]. The rotation of the microlens array brings about a real-time or faster scanning on the observing plane, and the fluorescence image can be captured by a CCD camera. Each focus is separated by around 6–10 µm so that the overlapping of light between the foci is restrained and does not reduce the spatial resolution. It is also possible to use galvanometer mirrors placed behind the microlens array for scanning the foci in a specimen [6]. The use of a pinhole-array disk in a multi-point multi-photon excitation microscope increases the lateral and axial resolution and enhances the contrast of images by reducing the fluorescence scattered within a specimen and that from reabsorption [7,8].

The pinhole-array disk eliminates undesirable fluorescence that is scattered or emitted by reabsorption in the specimen. Fluorescence excited in the specimen behaves just like a small light source and illuminates other parts of the specimen. The illuminated parts scatter or absorb the fluorescence and are imaged onto the CCD camera as a conventional scattering or fluorescence image. For this reason, fluorescence scattered or emitted by reabsorption blurs obtained images and makes the contrast of the images lower. This phenomenon appears more prominently in fluorescence images from deeper parts of a specimen, depressing the resolving power and the long-penetration advantages in multi-photon excitation with near-infrared light. We compared the imaging properties of the confocal and the nonconfocal systems by observing a specimen. Figure 1.2 shows images of the root of the convallaria obtained with our developed microscopes. An oil-immersion objective lens (Zeiss, NA 1.3 oil, ×40) was used for the observation. The total excitation power at the focal plane of the objective lens was 150 mW in both cases. Autofluorescence was obtained in the specimen by multi-photon excitation. The images in Fig. 1.2a were obtained with the confocal system and the images in Fig. 1.2b were obtained with the nonconfocal system. The z values in the images show

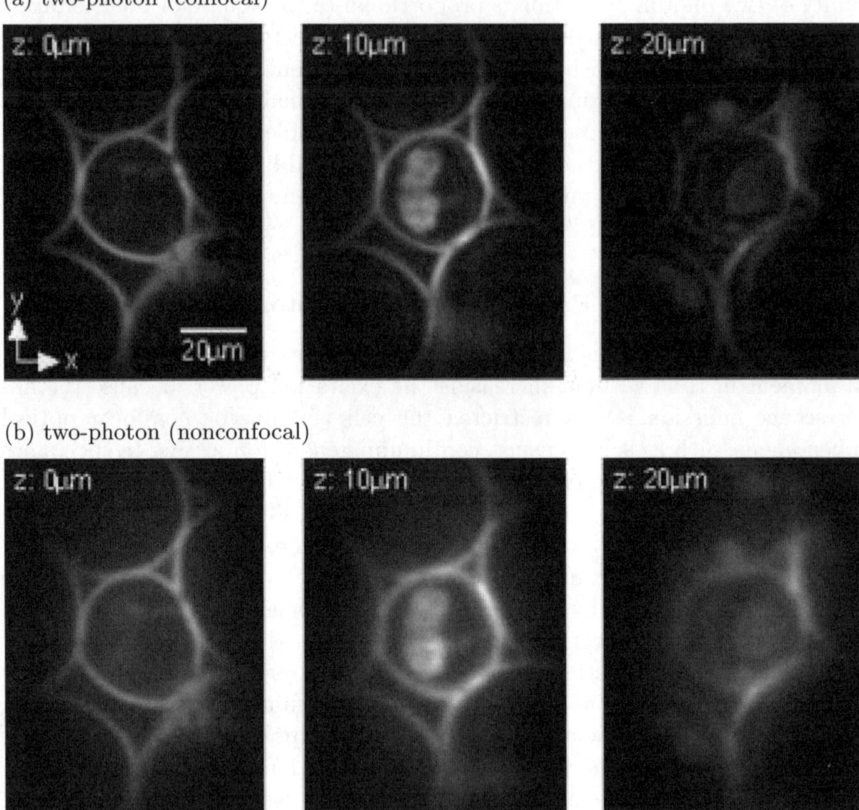

Fig. 1.2. Comparisons of the images obtained by **a** the confocal system and **b** the nonconfocal system

the depth of the observation planes in the specimen. By comparing these images, it is obvious that the fluorescence scattered within the specimen is effectively eliminated in the confocal images. The scattered fluorescence gives blurred images in the nonconfocal system and this becomes more prominent in the deeper part of the specimen. Figure 1.3a,b shows an x–z cross-section of images of the same specimen with and without the pinhole array, respectively. It can be seen that the deeper part of the image is more blurred in the nonconfocal image, but not in the confocal image. At the bottom of the specimen, the structure of the specimen cannot be recognized in the nonconfocal image. Although less fluorescence is detected with the confocal system, one can say that the penetration depth is longer with the pinhole array, because the structure in the deep part can be recognized with the pinhole array, but not without it. The fluorescence images at relatively shallow positions of the specimen are also blurred in the nonconfocal system. We think that this is

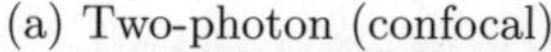

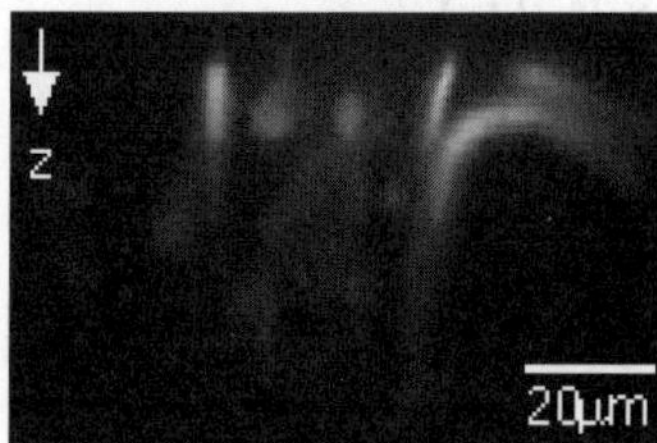

(b) Two-photon (nonconfocal)

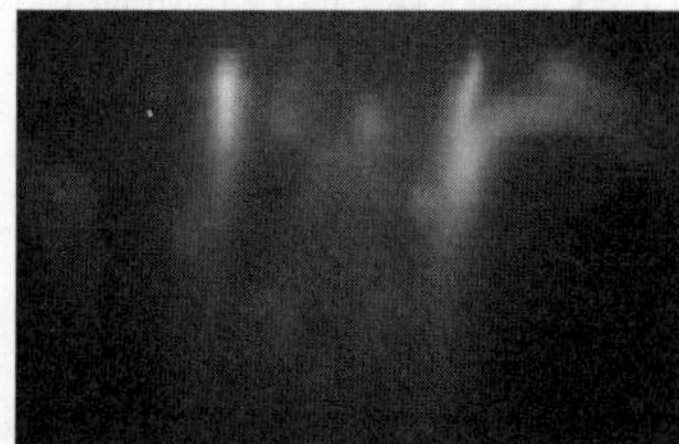

Fig. 1.3. x–y cross-section images obtained by **a** the confocal system and **b** the nonconfocal system

because of backward scattering of the fluorescence coming from the deeper parts of the specimen. This means that the use of confocal pinholes is also effective for observing not only deep parts of a specimen but also the shallow parts of it. Although the resolution of the CCD camera is not high enough to observe the enhancement of the lateral resolution in our current setup, the reduction of the scattered fluorescence helps one to recognize the details of the observed specimen.

The use of a pinhole array in a multi-point multi-photon excitation microscope increases the lateral and axial resolution and eliminates fluorescence scattered or emitted by reabsorption within a specimen. We demonstrated that the pinhole array eliminates the scattered fluorescence effectively and brings about higher contrast of images and a sharp depth-discrimination property.

1.2.3 Instrumentation

Figure 1.4 shows a schematic of a multi-point multi-photon excitation microscope. A collimated laser beam from a mode-locked Ti : Sapphire laser (Spectra-Physics, Tsunami, wavelength = 800 nm, pulse width = 80 fs, repetition rate = 82 MHz) is incident to a microlens-array disk and focused onto a pinhole-array disk. Each focused beam passes through each pinhole and is focused in a specimen by an objective lens so that the pinholes are imaged on a focal plane of the objective lens as multi-photon excitation points of fluorescence. Fluorescence from a specimen is introduced onto an intensified CCD camera (Hamamatsu, C-2400-35) by a dichroic mirror that is placed between the two disks and has around 95% reflectivity at a 400–580 nm wavelength of light.

Simultaneous rotation of the two disks scans a specimen to realize confocal fluorescence imaging of the specimen. The diameters of the microlenses and the pinholes are 250 µm and 50 µm, respectively. The microlenses and the pinholes are arranged in a helical order to achieve uniform illumination on a specimen and produce 12 images with one rotation of the disks. Both disks are precisely aligned so that each pinhole is placed on each focal point

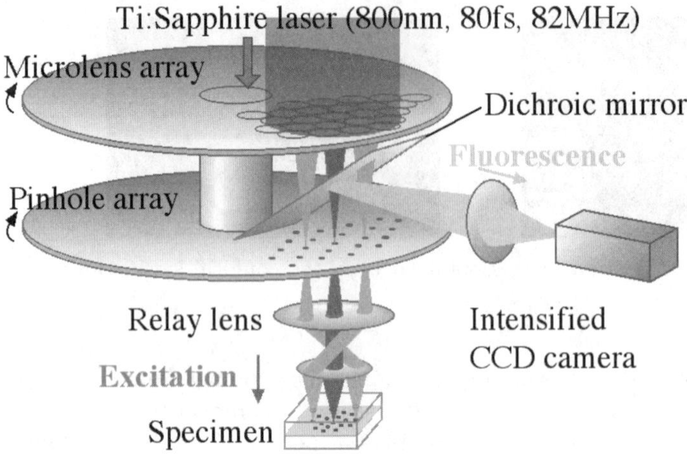

Fig. 1.4. An optical setup of a two-photon multi focus fluorescence microscope with a microlens and pinhole-array scanner

of the microlenses. In principle, simultaneous rotation of the two disks scans a specimen in 3 ms at 1800 rpm. However, in our current setup, the image acquisition time is limited to 33.3 ms because of the imaging speed of the CCD camera. Fluorescence imaging through the pinhole array is expected to yeild twice higher resolution compared with that of a nonconfocal system [3,9]. Although, even without a detection pinhole, multi-photon excitation microscopes have a three-dimensional resolution equivalent to that of confocal single-photon excitation microscopes, the lateral resolution is almost the same as that of conventional fluorescence microscopes because of the use of near-infrared light for excitation. The scattered fluorescence blurs the images in observing both shallow and deep parts of the specimen. From the comparisons between images obtained by the confocal and the nonconfocal system, the use of the pinhole array is useful irrespective of the observation depth and works more effectively for observing stronger scattering specimens. The elimination of the scattered fluorescence enables us to observe deeper parts of the specimen than without it. Except for the high spatial resolution, these advantages brought about by confocal detection do not appear in a typical multi-photon excitation microscope with single-focus scanning.

1.2.4 Calcium Ion Dynamics Revealed by Multi-Photon Microscopy

Modulation of the intracellular Ca^{2+} concentration ($[Ca^{2+}]_i$) constitutes a fundamental mechanism of signal transduction in excitable cells. A spontaneous increase in Ca^{2+} concentration can occur at a single focus or at multiple foci within a cell, and can lead to propagation of an elevation of

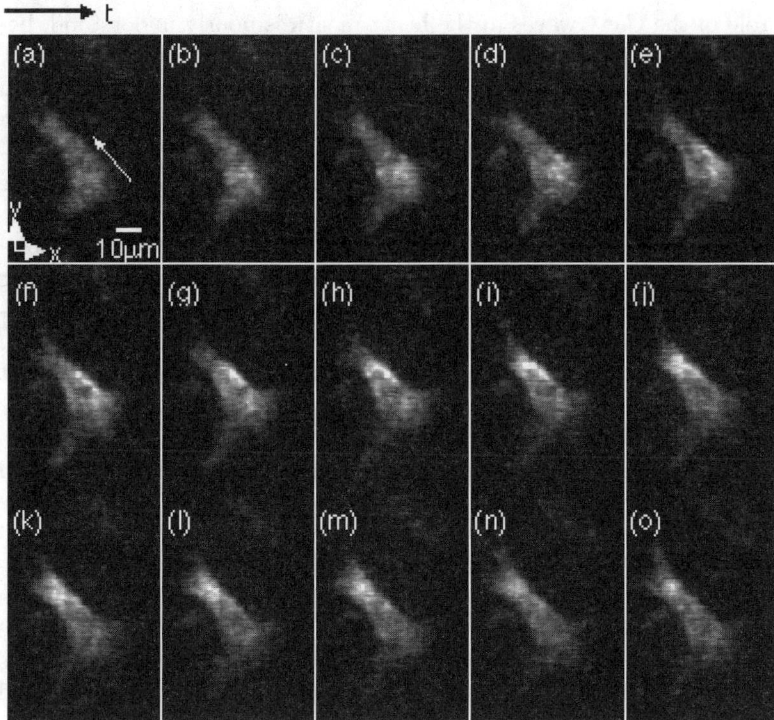

Fig. 1.5. Two-photon fluorescence images of a Ca^{2+} wave in a cultured rat cardiac myocyte

$[Ca^{2+}]_i$ throughout the cytosol in a wave-like pattern. Although these Ca^{2+} waves have been observed widely in both excitable and inexcitable cells, their role in the cardiac cell function remains unclear. For the past decade, after the advent of fluorescent Ca^{2+} indicators [10] and laser scanning confocal microscopy [11], the properties of Ca^{2+} waves in cardiac myocytes have been studied precisely.

We observed Ca^{2+} dynamics in a cultured rat cardiac myocyte with confocal two-photon microscopy [12,13]. Fluo-3/AM (Molecular Probes, 0.018 mM) was used for a fluorescent Ca^{2+} indicator and loaded into the cell for 30 min with a Tyrode solution containing 145 mM NaCl, 4 mM KCl, 1 mM $CaCl_2$, 1 mM $MgCl_2$, 10 mM glucose, and 10 mM HEPES (pH = 7.4 by NaOH). A water-immersion objective lens (Olympus, NA 0.8 water, ×40) was used for the observation. The total excitation power at the focal plane of the objective lens was 152 mW. Each image was obtained sequentially with a time resolution of about 33 ms. The brighter parts of the images show higher Ca^{2+} concentrations in the cell. In Fig. 1.5, a Ca^{2+} wave can be seen as the movement of higher Ca^{2+} concentration from the lower right edge of the specimen to the upper left.

The role of the Ca^{2+} waves in the heart in situ is poorly understood, because Ca^{2+} waves have been studied mostly in enzymatically isolated cells. We have developed a system for in situ imaging of $[Ca^{2+}]_i$ equipped with a multi-pinhole-type confocal scanning device, which enabled us to visualize real-time x–y images of Ca^{2+} waves [14,15]. Using this system on Langendorff-perfused rat hearts, with simultaneous recording of electrocardiograms, we found that Ca^{2+} waves were completely abolished by ventricular excitation, suggesting that the waves in the whole heart play little, if any, pathophysiological role. Nevertheless, it is possible that Ca^{2+} waves play some aggravating role in cardiac function if they occur frequently and propagate beyond individual cells on a large scale under certain Ca^{2+}-overloaded conditions. In this regard, quantitative analysis of Ca^{2+} waves in the working whole heart is essential in order to understand their functional significance.

Two-photon Ca^{2+} imaging was also performed in situ with muscle cells in a whole heart of a rat. The heart was excised under anesthesia with diethyl ether, and perfused on a Langendorff apparatus for 5 min with a Ca^{2+}-free Tyrode solution under 100% oxygenation. Thereafter, the heart was loaded with fluo-3/AM (100 µg) dissolved into the Ca^{2+}-free Tyrode (4 ml) containing 1% fetal calf serum and 0.06% pluronic F-127 on a recirculating system for 30 min. Following another 15-min perfusion with 0.5 mM-Ca^{2+} Tyrode to allow hydrolysis of acetoxymethyl esters within the cells, the heart was used for the experiments. The motion artifact on the image was prevented by 20 mM 2,3-butanedione monoxime (BDM), which was added to the perfusate. Unless otherwise specified, Ca^{2+} waves were analyzed under quiescence via blockade of atrioventricular conduction produced by mechanical ablation of the atrioventricular junction.

Figure 1.6 shows distributions of the Ca^{2+} concentration in the right ventricle of the heart. The total excitation power at the focal plane of the objective lens was 150 mW. Each image was obtained sequentially with a time resolution of about 33 ms. In Fig. 1.6, a Ca^{2+} wave propagating from the lower right to the upper left was observed.

In the experimental condition for observing $[Ca^{2+}]_i$, the excitation rate of fluo-3 in each focus can be estimated to be about 6% from its two-photon absorption cross-section, the numerical aperture (NA) of the objective lens used and the excitation power. This excitation rate is not high enough for this application, nor probably for many others. In addition, the field of view in our microscope cannot be considered to be wide enough. The penetration depth, the excitation rate and the field of view discussed above are currently limited by the maximum output power of the Ti : Sapphire laser used. The maximum output power of a commercially available Ti : Sapphire laser is about 2 W and this value is still not high enough in our calculation. We think that the development of a Ti : Sapphire laser with a higher output is necessary for further advancement of the multi-point multi-photon excitation microscope.

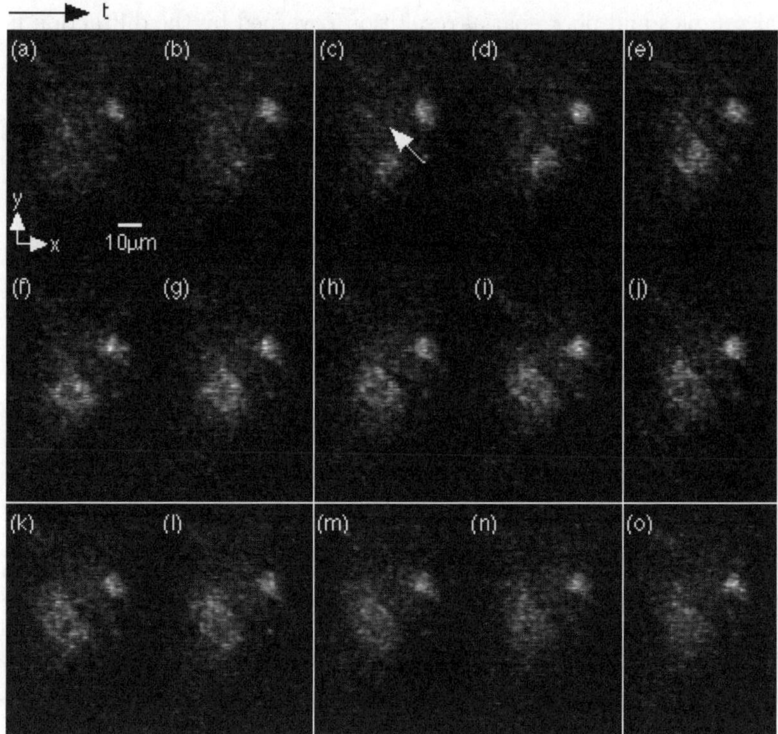

Fig. 1.6. Two-photon fluorescence images of a Ca^{2+} wave in the whole heart of the rat

1.3 Nonstaining Molecular Imaging (CARS)

Single-photon excitation confocal fluorescence microscopy is widely used because of the capability of 3D (three-dimensional) imaging of living cells [16]. Recently, multi-photon excitation fluorescence microscopy has been successfully developed owing to the progress of ultra-fast laser technology [17]. However, both microscopies fundamentally require a staining procedure with a suitable fluorophore. The ability for molecular discrimination depends on the dye. The staining procedure affects the biological sample, and hence the function of the cell is often lost. In addition, staining requires technical skill.

In analytical chemistry, vibrational spectroscopy observing molecular vibrations (Raman and infrared spectroscopy) is widely used for molecular identification, since the frequency of the molecular vibration is extremely sensitive to the molecular structure. Molecular mapping without staining, with micrometer or sub-micrometer resolution, is expected to be achieved through a combination of vibrational spectroscopy and microscopy. However, this combination is not so widely used for biological applications. Infrared microscopy suffers from the problems of strong absorption of water in the biological samples and poor spatial resolution governed by the diffraction limit. Although confocal Raman microscopy was developed with micrometer- or sub-micrometer-order 3D resolution [18], it requires a pinhole in the detector side and a high-resolution spectrometer. The weak intensity of the spontaneous Raman scattering is weakened by the pinhole and the spectrometer, and the Raman scattering is often disturbed by the autofluorescence of the biological samples.

In this section, another new nonstaining molecular imaging technique, coherent anti-Stokes Raman scattering (CARS) microscopy, will be described.

1.3.1 The Fundamentals of Coherent Anti-Stokes Raman Scattering

CARS microscopy is a combination of microscopy and CARS spectroscopy and is the most widely practiced technique of four-wave mixing spectroscopy.

Figure 1.7 shows the relation between photons and the molecular vibration in the CARS process [19]. In CARS, two laser beams whose angular frequencies are ω_1 and ω_2 are mixed into the substance. If the difference in angular frequency of these beams ($\omega_1 - \omega_2$) is consistent with that of the Raman active molecular vibration (Ω), the molecular vibration is coherently excited by the beating of the two beams. Then, the vibration mixed with the wave at ω_1 yields coherent radiation, which is CARS, at the anti-Stokes frequency $\omega_3 = 2\omega_1 - \omega_2$.

The advantages of CARS spectroscopy over conventional Raman (spontaneous Raman) spectroscopy are high intensity and separation from fluorescence. The high intensity of CARS shortens the exposure time, whereas conventional Raman spectroscopy often requires a long exposure time. The CARS radiation is separable from the fluorescence in CARS spectroscopy, since the wavelength of CARS is shorter than that of excitation beams.

CARS is a coherent process. The CARS radiation from different points interferes. It is known that this situation of mutual intensification each other is a phase-matching condition [19]. The phase-matching condition of CARS

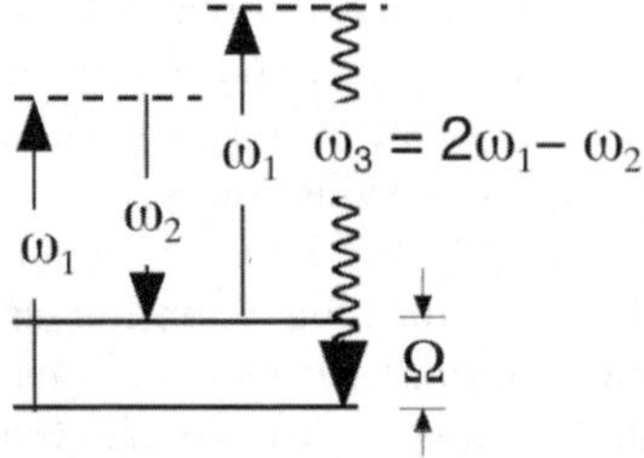

Fig. 1.7. A schematic showing a CARS process

Fig. 1.8. The phase-matching conditions of CARS

is given by

$$\boldsymbol{k}_3 = \boldsymbol{k}_1 + \boldsymbol{k}_1' - \boldsymbol{k}_2 \,, \tag{1.1}$$

where $\boldsymbol{k}_1$ and $\boldsymbol{k}_1'$ are the wave vectors of the ω_1 beam, $\boldsymbol{k}_2$ is that of the ω_2 beam, and $\boldsymbol{k}_3$ is that of the CARS radiation in the sample. In CARS spectroscopy, the configurations shown in Fig. 1.8 are used to satisfy the phase-matching condition to observe the isotropic sample.

1.3.2 3D Resolution by CARS Microscopy

The combination of CARS spectroscopy and microscopy was proposed by Duncan et al. [20]. Their system does not have axial resolution because the incident beams were not so tightly focused. We have proposed that three-dimensional imaging properties should be obtained to focus the excitation beams tightly [21]. Independent of us, A. Zumbusch et al. succeeded in obtaining the 3D resolved image and applied it to the observation of biological samples [22].

Figure 1.9 shows the optical layout of tightly focused CARS microscopy. The ω_1 and ω_2 beams are tightly focused into the sample by an objective with a high numerical aperture (NA), and then the yielded CARS radiation is collected by another objective though a filter. The tightly focused beam has a variety of wave vectors; the phase-matching condition is not so severe.

When the collimated beam is tightly focused by an objective, the electric field is given by gathering of plane waves that have various directions of progress. Since the wave vector indicates the direction of progress, the wave vector of the focused beam is distributed spherically, with radius $1/\lambda$, limited by the NA of the objective. Figure 1.10 shows this wave vector distribution of the focused beams, where n is the refractive index of the sample, μ is the

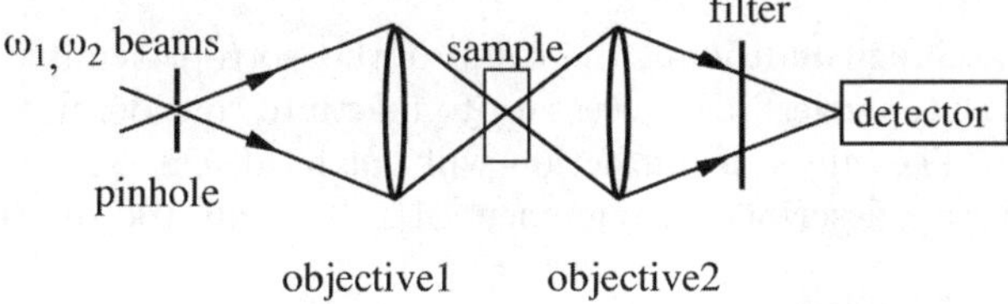

Fig. 1.9. The optical layout of a tightly focused CARS microscopy system

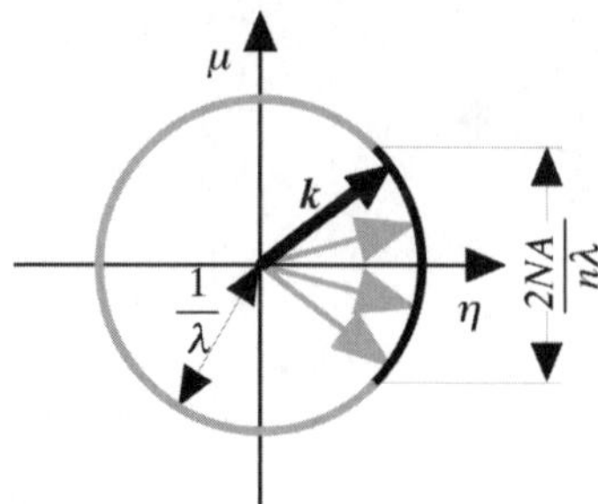

Fig. 1.10. The wave vector distribution of the focused beam

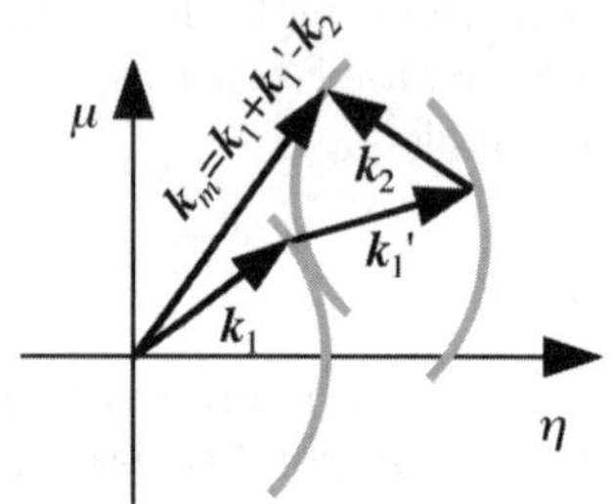

Fig. 1.11. The relations among $\boldsymbol{k}_m$, $\boldsymbol{k}_1$, $\boldsymbol{k}'_1$, and $\boldsymbol{k}_2$

axis of the lateral direction, and η is the axis parallel to the optical axis. In Fig. 1.10 one section is shown, because it is symmetrical about the η-axis.

If the vector $\boldsymbol{k}_m$ is defined by

$$\boldsymbol{k}_m \equiv \boldsymbol{k}_1 + \boldsymbol{k}'_1 - \boldsymbol{k}_2 \,, \tag{1.2}$$

it can take many kinds of combinations of $\boldsymbol{k}_1$, $\boldsymbol{k}'_1$, and $\boldsymbol{k}_2$, because each wave vector of $\boldsymbol{k}_1$, $\boldsymbol{k}'_1$, and $\boldsymbol{k}_2$ has a distribution like that shown in Fig. 1.10 (see Fig. 1.11).

The gray region in Fig. 1.12 shows the available $\boldsymbol{k}_m$. The observable wave vector of the yielded CARS is also expressed by Fig. 1.8 and drawn in Fig. 1.12 by an arc of a circle. The phase-matching condition is satisfied with the crossing of $\boldsymbol{k}_m$ and $\boldsymbol{k}_3$. It is found that the available region of $\boldsymbol{k}_m$ is so wide as to cross $\boldsymbol{k}_3$. Hence, the phase-matching condition is easily satisfied.

The above phase-matching condition is the result of the isotropic sample. To estimate the imaging properties, it is necessary to take into consideration the effect of diffraction. The effect of diffraction will not be discussed here (the detailed discussion was described in reference [21]), but only the result will be described.

Figure 1.13 shows the cutoff frequency of the transfer function of CARS microscopy with a confocal configuration. The cutoff frequency of the trans-

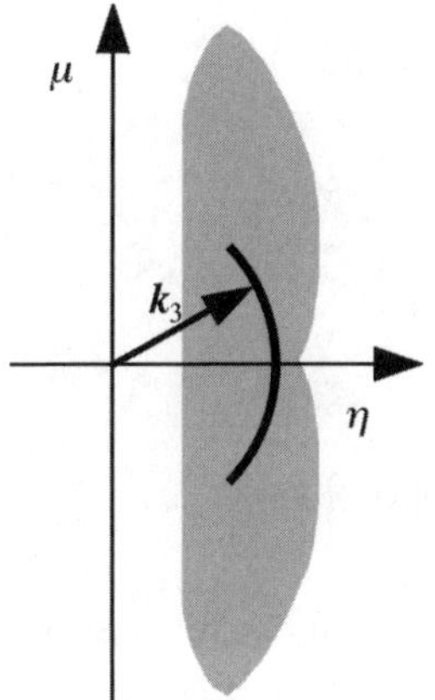

Fig. 1.12. The available area (gray region) of vector $\boldsymbol{k}_m$ ($\equiv \boldsymbol{k}_1 + \boldsymbol{k}_1' - \boldsymbol{k}_2$) and the observable wave vector of yielded CARS $\boldsymbol{k}_3$ on the circular arc. The phase-matching condition is satisfied by the crossing of these vectors

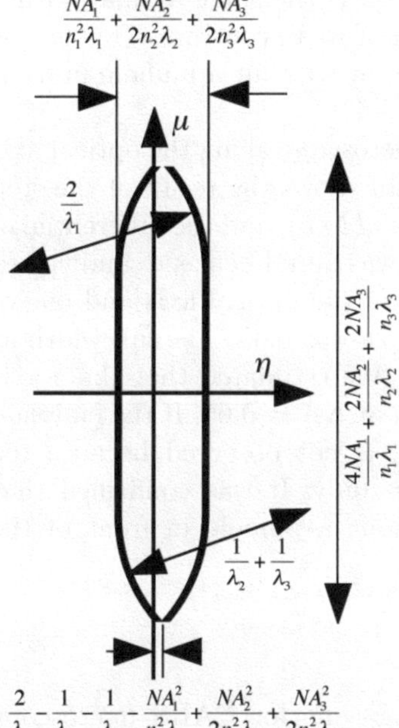

Fig. 1.13. The observable wave vector region of the sample in CARS microscopy

fer function shows the observable wave vector of the sample, taking into consideration that the sample is a set of plane-wave like distributions. It was found that CARS microscopy has 3D resolution, since the cutoff frequency in Fig. 1.13 extends from the origin of the axes in all directions, and is much the same as that of confocal fluorescence microscopy [16]. It is known that conventional fluorescence microscopy (not confocal fluorescence microscopy) or absorption microscopy have an insensitive region along the optical axis.

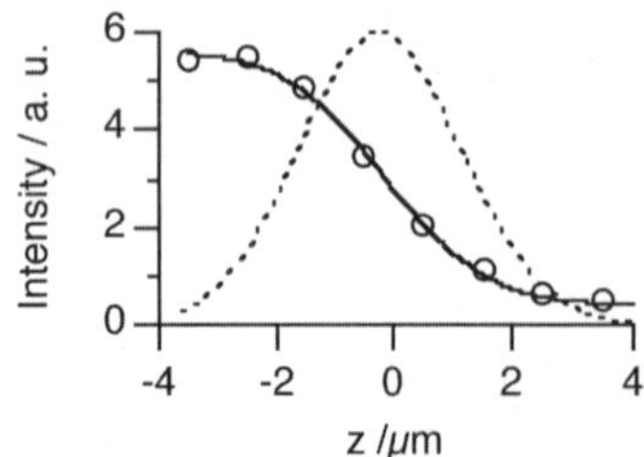

Fig. 1.14. The response of CARS microscopy along the optical axis

This fact means that the structure along the optical axis only is not observed by conventional microscopy.

It was also found that the cutoff frequency of the transfer function of CARS microscopy with a confocal configuration is the same as that with a nonconfocal configuration under the condition of weak contrast. This means that CARS microscopy has 3D resolution with or without a pinhole in front of the detector.

We observed the step response of CARS microscopy along the optical axis to confirm the 3D resolution power. Figure 1.14 shows the result of the step response (*circles*), the result of curve fitting (*solid line*), and the differential of the solid line (*broken line*). The sample, which was liquid benzene sandwiched between two glass plates, had a structure along the optical axis and one of the edges was shown. The NA of both objectives was 0.65. The full width at half-maximum of the broken line was 3.2 μm. We estimated that the z-axis resolution of the developed system was 3.2 μm at NA $= 0.65$. If the emission is a linear optical process, step-like response is not observed because the sample has a structure along the optical axis only. It was confirmed that CARS microscopy has depth resolution without a pinhole in front of the detector.

1.3.3 High-Speed Image Acquisition

W.W. Webb pointed out that it takes a long time for CARS microscopy obtain an image [23]. Zumbusch et al. took a few tens of minutes to obtain an image. Moreover, not only a CARS image but also a CARS spectrum is required to obtain a map of the molecular distribution. Hence shortening the observation time is a significant problem for CARS microscopy.

One method of high-speed imaging is the multi-focus system using a rotating microlens array [24]. We have proposed a multi-focus CARS microscopy system using a rotating microlens array [25]. Figure 1.15 shows the schematic layout of the multi-focus CARS microscopy system. The collimated ω_1 and ω_2 beams are collinearly overlapped and incident to the microlens array. The microlens array makes the collimated beam focus on many points through the objective, typically from a few hundred to a thousand. The yielded CARS

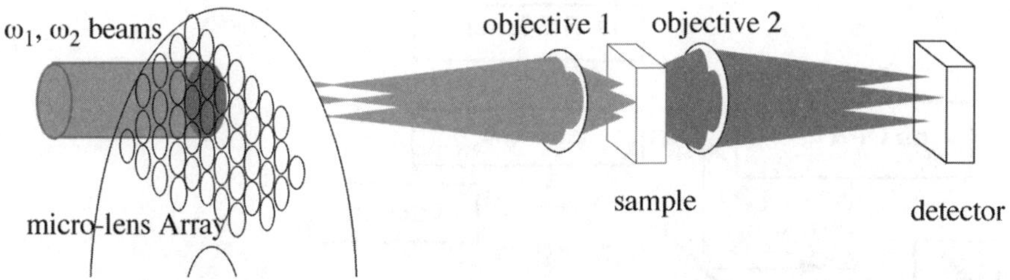

Fig. 1.15. The schematic layout of the multi-focus CARS microscopy system

radiation from these points is imaged by the other side of the objective onto the two-dimensional detector. These points are made to scan on the focal plane of objective 1 by rotating the micro-lens array.

A rotating microlens array may be suitable for CARS microscopy, to excite the sample and to detect the yielded CARS on a focal plane at a certain time. It is difficult to apply spectral imaging to the microlens array system, because a spectrometer is generally required to obtain a spectrum. On the other hand, CARS microscopy never requires a spectrometer. The CARS spectrum is given by scanning the wavelength of the ω_2 beam.

CARS radiation is sensitive to the intensity of excitation lasers, since CARS is a nonlinear process. It is difficult to obtain a CARS image when the pulse-to-pulse intensity of the excitation lasers is not very stable. Parallel excitation and detection make CARS imaging become insensitive to the intensity stability of lasers.

1.3.4 Molecular Imaging by CARS Microscopy

We produced a CARS microscopy system in the fingerprint region with a collinear configuration by using a tunable picosecond laser [26].

In the previous investigations, the observable Raman shift region was higher than 2000 cm^{-1} or around 3000 cm^{-1}. The molecules of biological samples consist of C, O, H, and N. It is difficult to distinguish the molecular species in the higher frequency region of the CH, OH, and NH stretching vibrations. The Raman bands in the fingerprint region, whose Raman shifts are typically from 500 to 1800 cm^{-1}, are much more sensitive to the functional groups of the molecule and the molecular conformations. Therefore, the extension of the observable Raman shift region into the fingerprint region is important.

Since CARS is a nonlinear process, the efficiency of CARS increases with the peak intensity of CARS excitation lasers. The ultra-fast laser with short pulse duration, high peak intensity, and low average power is suitable for CARS excitation. However, the spectrum of the femtosecond laser is too wide to identify the fine structure or the slight shift of the Raman bands. We used the picosecond laser to satisfy the high peak intensity and the narrow spectral bandwidth.

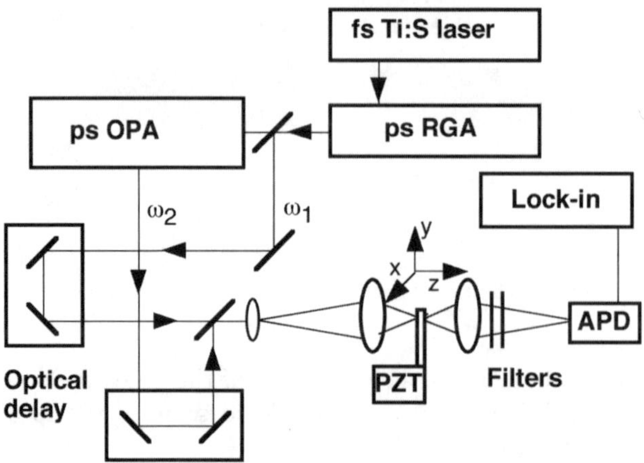

Fig. 1.16. The schematic layout of the CARS microscopy system

Figure 1.16 shows the developed CARS microscopy system for the fingerprint region. The system consists of a tunable picosecond laser system and a transmission-type scanning microscope with an xyz piezoelectric scanning stage.

The laser system is based on the femtosecond Ti : sapphire mode-locked (Ti : S) laser. The femtosecond pulses are used as a seed for a regeneration amplifier (RGA). The spectrum of the femtosecond pulses was limited by a mask placed in the pulse stretcher and the pulse duration was stretched to the picosecond region. The output pulses of the picosecond regeneration amplifier were divided in two. One is for the ω_1 beam and the other is for excitation of the optical parametric amplifier (OPA) for the ω_2 beam. The second harmonic of the idler wave from the optical parametric amplifier was used for the ω_2 beam. To obtain the CARS spectrum, the wavelength of the ω_2 beam must be scanning. The wavelength of the picosecond optical parametric amplifier is tunable by adjusting the angle of the OPA crystal and the angle of the grating. Stepping motors are attached to the crystal and the grating in order to control the wavelength of the ω_2 beam by a computer.

The ω_1 and ω_2 beams were collinearly superimposed and were temporally overlapped by optical delays. These beams were tightly focused by an objective. The yielded CARS was collected by another objective and detected by an avalanche detector (APD). The ω_1 and ω_2 beams were removed by filters, but CARS radiation was passing through them. The sample was moved by a piezoelectric stage (PZT) to obtain an image.

The dotted line in Fig. 1.17 shows the obtained CARS spectrum of polystyrene beads (diameter = 4.5 µm). The solid line is the Raman spectrum of the same samples observed by a conventional Raman microscope (JASCO NRS-2100).

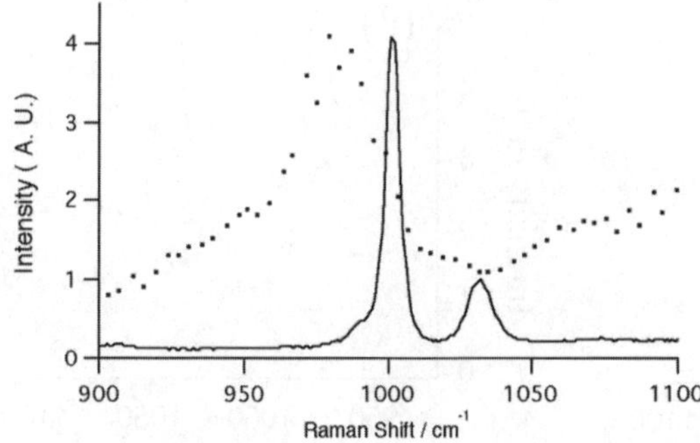

Fig. 1.17. The spectrum of polystyrene beads observed by conventional Raman microscopy (*solid line*) and CARS microscopy (*dotted line*)

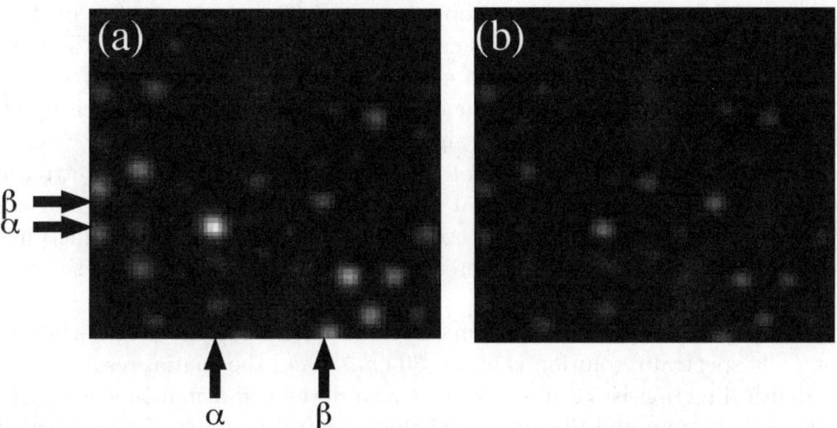

Fig. 1.18. The spectral images observed by the multi-focus CARS microscopy system at 990 cm^{-1} (**a**) and 1050 cm^{-1} (**b**)

A high-intensity band was observed around 1000 cm^{-1} in both spectra. This band is the phenyl breathing mode of polystyrene. From the bandwidth of this band it was found that the resolution of the developed system was about 30 cm^{-1}. In the CARS spectrum, an offset signal without a Raman shift dependency was observed and the phenyl breathing mode was about 15 cm^{-1} down shifted. The offset signal is the nonresonant background, and it is known that the intensity of the CARS spectrum is proportional to the square of the conventional Raman spectrum and the nonresonant background. We considered that the down shift is the result of interference with the nonresonant background.

We also constructed a multi-focus CARS microscopy system to shorten the observation time [25]. Figure 1.18 shows the image of a mixture of polystyrene (diameter = 4.5 μm) and glass beads (diameter = 3–5 μm) observed

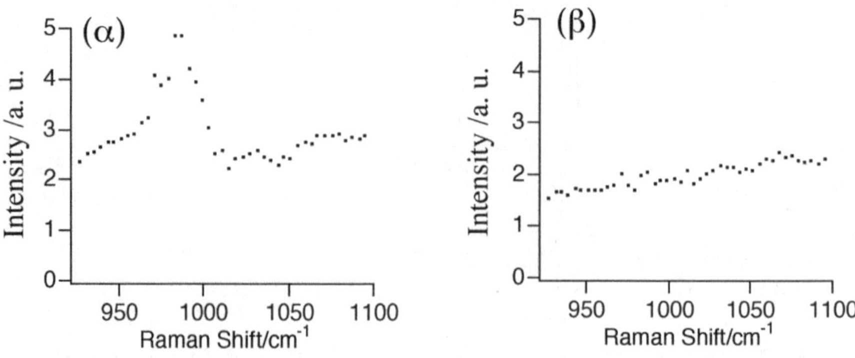

Fig. 1.19. The CARS spectra at α and β in Fig. 1.18

by the multi-focus CARS microscopy system at 990 cm^{-1} near the peak of the polystyrene band (a) and 1050 cm^{-1} away from that band (b). The laser system shown in Fig. 1.16 was also used, and the images were observed with a 2.5 cm^{-1} step. The exposure time of the each image was 20 s, which is 100 times shorter than for the stage scanning type.

Figure 1.19 shows the spectra at α and β in Fig. 1.18. These spectra are clearly different, and it is found that these are a polystyrene bead and a glass bead. The polystyrene has an intense band at 1000 cm^{-1}, but the glass has no high-intensity bands around there. It is possible to draw a clear distinction between polystyrene and glass.

We have developed a CARS microscopy system for the fingerprint region. The spectral resolution is about 30 cm^{-1}, and the spatial resolution in the depth direction is 3.2 µm. We have also developed a multi-focus CARS microscopy system and demonstrated that spectral imaging of a mixture of polystyrene and glass beads could identify these two substances.

In a CARS spectrum, the nonresonant background without any structure overlaps with the resonant CARS band, and many bands are closed in the fingerprint region. As the poor spectral resolution and thelow stability of laser intensity reduce the contrast of the CARS image nad spectrum, a laser of narrower spectral bandwidth and higher stability is required for biological applications.

We used near-infrared lasers to reduce the nonresonant background, because the intensity of the nonresonant background will be increased when two-photon absorption exists. However, if the wavelengths of the CARS excitation beams are close to the single-photon absorption of the molecule, the CARS emission is strongly enhanced by the electronic resonance. It is possible to obtain information about a specific molecule by electric resonance enhancement. The wavelength selection of excitation lasers is also important for CARS microscopy.

1.4 Transcutaneous Near-Infrared Light Power/ Information Transmission for Implantable Medical Devices

1.4.1 Feasibility of Powering, Controlling, and Monitoring Implantable Medical Devices Using Near-Infrared Light

Implantable medical devices assist or substitute for natural organs. Currently, cardiac pacemakers and cochlear implants are already in practical use. They are implanted to give signals to organs through electrical stimulation. In the near future, implantable microelectronic mechanical systems will become available, which will have more complete functions.

These devices need to be supplied with electric power and/or to communicate with instruments outside the body. In order to prevent infection, these two requirements should be met by wireless methods. One promising method is to use near-infrared light as a medium to transfer power [27,28] and information [29–31]. High-efficiency, noninvasive power/information transmission can be performed by using near-infrared light, because biological tissue exhibits a considerably high transmittance in the near-infrared region. Figure 1.20 shows a schematic diagram of implantable medical devices powered and controlled by near-infrared light. In Fig. 1.20, the implanted devices are equipped with photoelectric devices for receiving and emitting near-infrared light. For power supply, as shown in Fig. 1.20a, near-infrared light is sent from outside the body through the skin to a photodiode (PD) that powers an implanted device. Near-infrared communication has two functions, as shown in Fig. 1.20b: controlling and monitoring implanted devices. For controlling an implanted device, near-infrared light is aimed at a photodiode inside the body to send signals. For monitoring an implanted device, light transmitted from a light-emitting diode (LED) or a laser diode (LD) inside the body is detected by a receiver outside.

Figure 1.21 shows the spectral properties of human skin [32] and photoelectronic devices. As shown in Fig. 1.21, human skin shows a high transmittance in the near-infrared region. The sensitivity of Si PDs, the most widely used type of photodetectors, is high for wavelengths of about 900 nm. A light source can be selected from a wide variety of commercially available LEDs and LDs in the wavelength range of 700–1000 nm. The efficiency of transferring power to an implanted photodiode is determined by the product of the transmittance of the skin and the sensitivity of the photodiode at the wavelength of a given light source. It follows from Fig. 1.21 that high-efficiency transmission of light can be performed in the range of 800–1000 nm. In the following, techniques to transmit power and information are described.

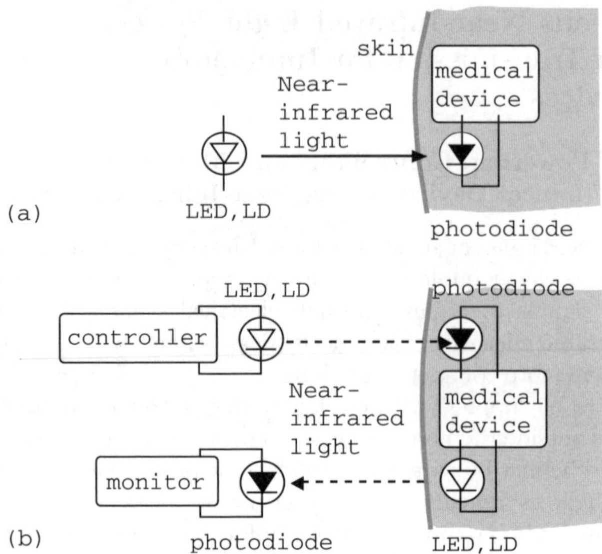

Fig. 1.20. A schematic diagram of near-infrared power/information transmission. Power transmission (**a**) and information transmission (**b**)

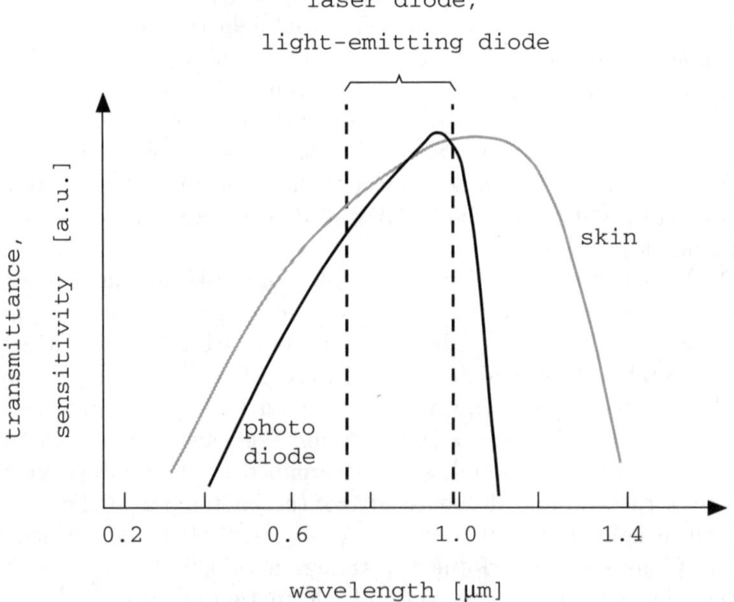

Fig. 1.21. The transmittance of human skin and the spectral sensitivity of Si photodiodes. The wavelength range of near-infrared laser diodes and light-emitting diodes is also shown

1.4.2 Transcutaneous Power Transmission by Near-Infrared Light

Once medical devices are implanted, they are expected to be used for a decade or more. For their continuous use, they should be powered indefinitely by a wireless method. To date, the following two methods have been considered to be practical. One method is to equip an implanted device by a primary battery. Cardiac pacemakers currently in use are powered with lithium primary batteries. The other method is to supply electromagnetic power through the skin, allowing implanted devices to be powered indefinitely from outside the body. In fact, power transmission by high-frequency electromagnetic induction is considered promising for driving artificial hearts [30]. However, these methods have the following disadvantages. Implanted devices using the former method require an operation every time the battery is replaced by a new one. The lifetime of batteries for cardiac pacemakers is in the range of 5–10 years. The latter method may cause electromagnetic interference with surrounding devices. In medical facilities, there are a lot of instruments that are susceptible to electromagnetic waves.

An alternative to these methods is to use near-infrared light as a medium to transfer power [27,28]. High-efficiency, noninvasive power transmission can be expected from this method, because biological tissue exhibits a considerably high transmittance to near-infrared light. To our knowledge, the first experiment on near-infrared power transmission was reported in 1999 [27]. In this experiment, a cardiac pacemaker was successfully powered by an implanted photovoltaic cell illuminated with near-infrared light through the skin of a guinea pig. The light source was an 810 nm near-infrared LD. The thickness of the skin was 2 mm. This technique is not only noninvasive to tissue but also almost free from electromagnetic interference with surrounding instruments.

A basic experiment was performed to investigate the efficiency of power transmission through tissue. Figure 1.22 shows the power density distribution of near-infrared light diffused by tissue. Samples of chicken were irradiated with a laser beam (140 mW, $\lambda = 810$ nm), the diameter of which was determined to be 10 mm by using a circular aperture. Then the diffused light was detected behind the samples with a PD having a detection area of 1 mm $\times$1 mm. The thicknesses of the samples were 5 and 10 mm. In Fig. 1.22, the lateral axis represents the distance between the beam axis and the position of the PD. As shown in Fig. 1.22, the beam spreads with increasing thickness of the tissue. The power transmittance can be defined as the integral of the power density over a sufficiently large area on the samples. From Fig. 1.22, the power transmittance is as high as 30% and 20% for thicknesses of 5 and 10 mm, respectively.

Figure 1.23 shows a rechargeable near-infrared power supply. It consists of an Si single-crystal solar cell array, a rechargeable battery, and a voltage regulator. The rechargeable battery is a polyacene capacitor (Kanebo). The voltage regulator is a step-up dc–dc converter (Ricoh). The output of the

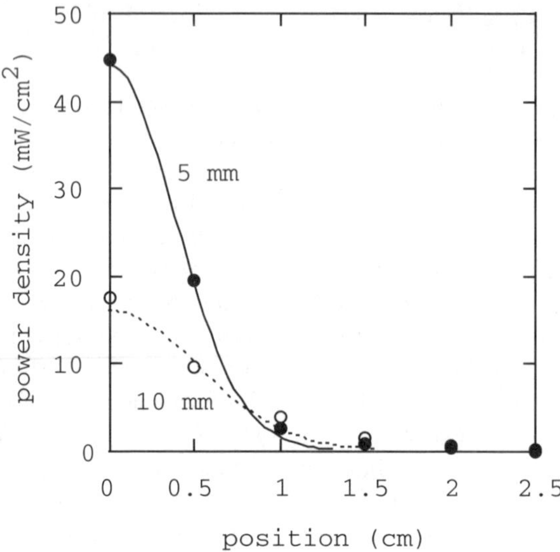

Fig. 1.22. Power transmission through chicken meat. Chicken samples 5 and 10 mm thick were used. The power and the diameter of the beam incident on each sample are 140 mW and 10 mm, respectively

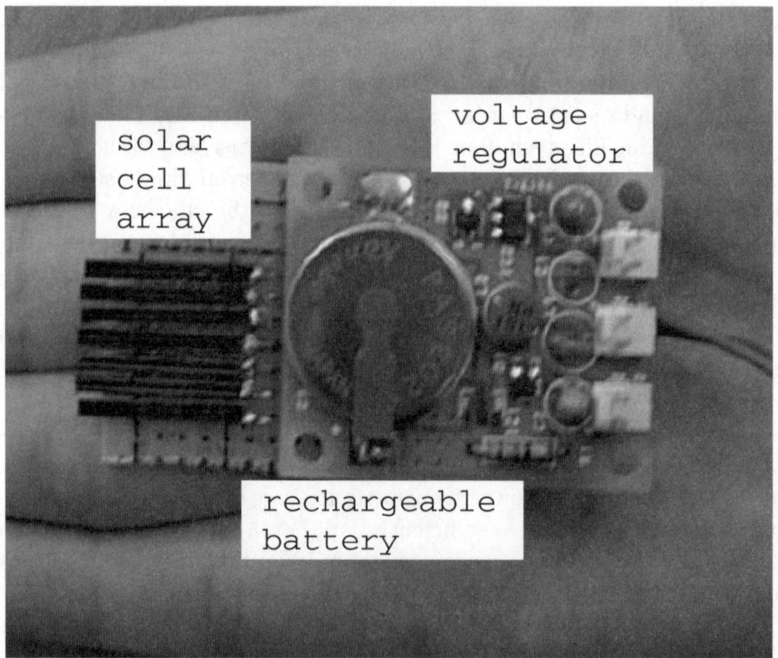

Fig. 1.23. A near-infrared power supply having a rechargeable battery

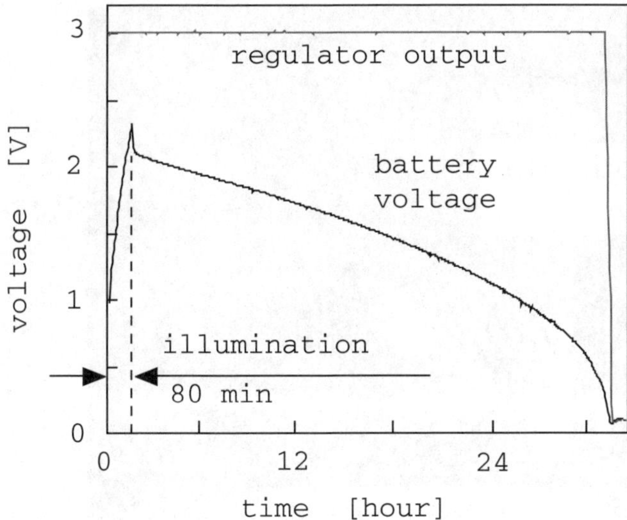

Fig. 1.24. The time-course of the battery voltage and the regulator output in the rechargeable power supply

voltage regulator is kept at 3.0 V if sufficient voltage is applied to its input. Figure 1.24 is a charge/discharge curve obtained for a 25 F polyacene capacitor. During the measurement, the regulator output was supplying a power of 0.2 mW. The power density of the light given to the solar cell array was 46 mW/cm^2. The period of illumination for the battery charge was 80 min. As shown in Fig. 1.24, the regulator output is continuously 3.0 V for more than 24 h after the discharge starts. Owing to the long life-time of the battery, this device can be recharged more than 100 000 times.

Recently, we have succeeded in developing an improved version of the rechargeable near-infrared power supply (Fig. 1.25). The solar cell array used in the previous version (Fig. 1.23) has been replaced by an Si PIN photodiode array. The power conversion rate of the photodiode array can be as high as 20%. With the new device, one can supply a charge of 0.5 mAh by illuminating the device with 25 mW/cm^2 for 15 min. A charge of 0.5 mAh can operate a commercial cardiac pacemaker, which generally consumes 0.02 mA, continuously for 24 h. The details of this new device will be reported elsewhere [28].

1.4.3 Transcutaneous Information Transmission by Near-Infrared Light

Near-infrared transcutaneous transmission [29–31] is a promising technique for communication between an implanted device and its controller or monitor placed outside the body. One of the greatest advantages of the use of near-infrared light as a medium is that the light emitted from transmitters does

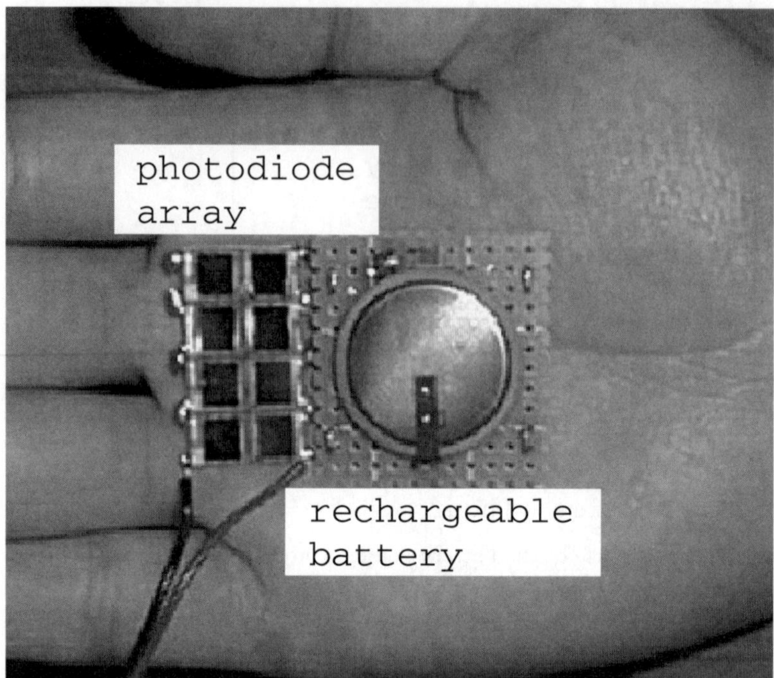

Fig. 1.25. A near-infrared power supply having a rechargeable battery charged by a photodiode array

not interfere with surrounding instruments; nor is the transmitter disturbed by the surroundings.

The principle of near-infrared transcutaneous transmission is simple, as shown in Fig. 1.20b. For a better signal-to-noise ratio, the light for transmission is usually modulated in the following way. First, the transmission light is intensity-modulated in order to produce a carrier wave. Light from LEDs and LDs can be easily intensity-modulated by changing the current injected to them. The carrier wave is then frequency- or phase-modulated according to a baseband signal. Also, pulse modulation such as pulse interval modulation is made possible by using a pulse wave as a carrier.

Recently, an implantable transmitter driven by near-infrared laser irradiation has been proposed. Figure 1.26 shows a schematic diagram of a near-infrared telemetry system using such a transmitter. The power necessary for operating the transmitter is sent to the body from outside it, by near-infrared light transcutaneous power transmission, as shown in Fig. 1.20a. In addition, a carrier wave is injected by illuminating the transmitter with intensity-modulated light. A PD in the transmitter receives the intensity-modulated light and converts it into an electrical carrier signal. The carrier signal is then phase-modulated by a modulator inside the transmitter. The

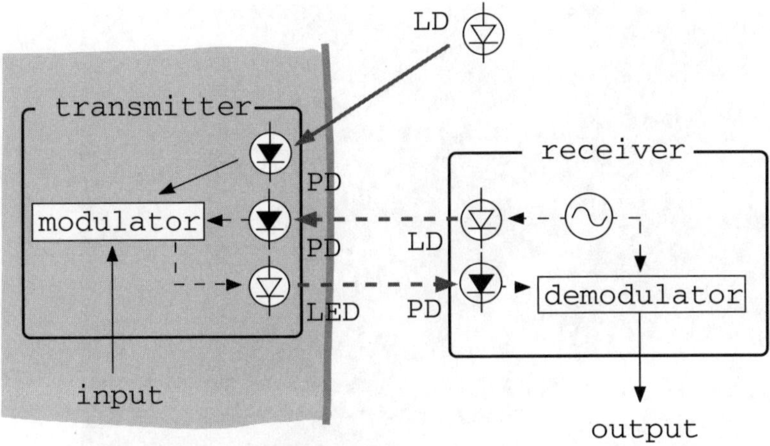

Fig. 1.26. A near-infrared transcutaneous telemetry system having a transmitter driven from outside the body

phase-modulated signal is transmitted with a near-infrared LED. For low-power operation of the transmitter, the intensity of the LED light is set as low as possible. For selective sensing of the LED light, it is separated from reflection of LD light by using optical filters. With a synchronous detection technique, the carrier-to-noise ratio can be improved even though the transmitted light is weak. A compact transmitter based on the above principle is shown in Fig. 1.27. Figure 1.28 is a result of transmission tests using the transmitter. It was covered with a chicken sample 3 mm thick, and transmission was performed through the sample. As shown in Fig. 1.28, the receiver output (phase) faithfully follows the input signal given to the transmitter (voltage). The details of the proposed transmitter will be reported elsewhere [31].

1.5 Cell and Nanosurgical Operation with Light

1.5.1 Introduction and Interactions between Ultra-Short Pulses and Biological Materials

Interactions between light and biological tissues have been studied with increasing interest since the invention of the laser. The ability to focus high concentrations of radiant power into a biological sample provides unique opportunities to examine and modify the biological samples. Lasers have become commonly used in fluorescence microscopy and, more recently, pulsed lasers have found widespread biomedical applications such as corrective eye surgery, and may prove to be suitable for dental drilling and macroscale surgery [33]. These examples are based on the phenomenon that occurs when light of a peak power density on the order of gigawatts/cm^2 to terawatts/cm^2 is

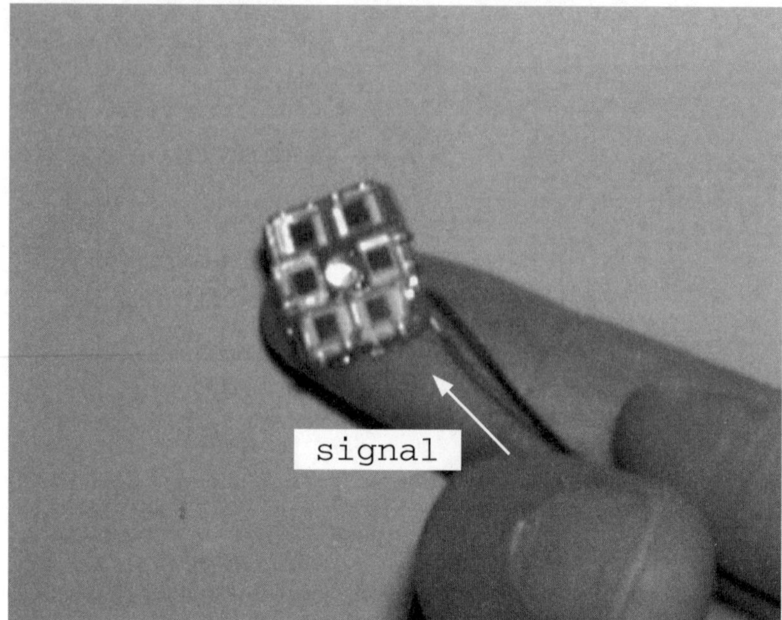

Fig. 1.27. An implantable near-infrared light transmitter driven by light. The thickness of the transmitter is about 4 mm

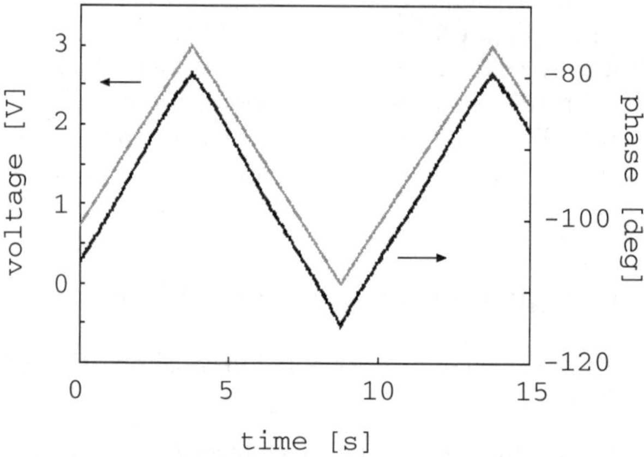

Fig. 1.28. A result of a transmission test using a triangular wave. The transmission was performed through chicken meat

absorbed locally by the sample. The energy transferred from the photons to the sample is sufficiently high to tear the sample apart, a process generally known as laser-induced ablation.

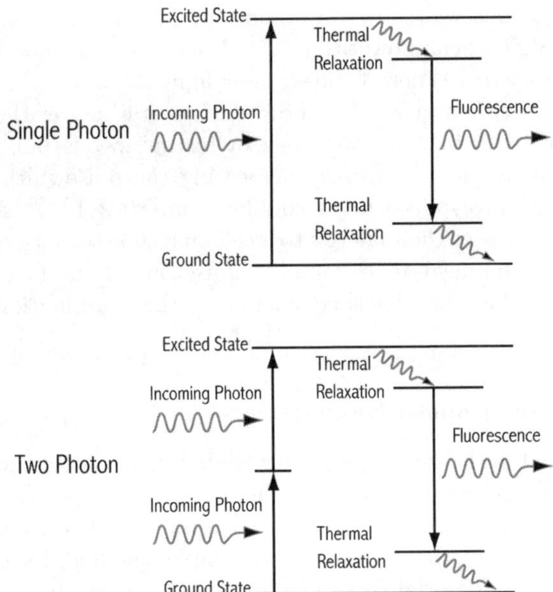

Fig. 1.29. One-photon versus two-photon absorption

Of particular interest to the biomedical field is that this damage can be restricted to a small focal zone, making possible a highly accurate method of cutting biological tissue. The amount by which damage can be restricted to the irradiated region depends on various parameters, including the illumination wavelength, pulsewidth, intensity, focusing conditions, and absorption properties of the sample.

The mechanism by which the sample absorbs light changes with increasing intensity. At a sufficiently high photon density, photons can be absorbed in simultaneous pairs or triplets (two- or three-photon absorption). Figure 1.29 illustrates the fundamental principle of two-photon absorption. The two photons must both be absorbed within 10^{-15} s and in the same region of the sample – a process that is low in probability [34]. This ensures that significant levels of two-photon absorption are only seen in regions of high photon density. The effect of this is to reduce the volume in which light is absorbed and therefore to decrease the volume in which laser-induced damage appears. Two-photon absorption can be applied in laser and biological material interactions by using near-infrared laser illumination. Many biological materials do not absorb strongly in the near-infrared, allowing transmission of laser power through the specimen except for the region in focus, where absorption of two or more photons occurs [35].

The peak power provided by pulsed laser illumination is sufficiently high to enable two-photon absorption. Pulsed illumination does, however, have other advantages over continuous wave (cw) illumination. At a given mean power, a pulsed laser has a higher peak power than a cw laser, the peak power rising as the pulsewidth decreases. Where focused laser light meets a living cell, the mean power of the laser is a measure of how much overall power the cell will be subjected to. We can increase the peak or instantaneous pulse power while retaining the same mean power by decreasing the pulsewidth. Femtosecond (fs) lasers may have pulsewidths smaller than 80×10^{-15} s. Focused femtosecond pulses transfer their energy to a cell on a time scale that is faster than the time taken for heat to thermally diffuse out of the focal volume [33], effectively confining the absorbed energy to the illumination zone.

1.5.2 Laser-Induced Disruption in Biomaterials

If sufficient energy is delivered to the sample, an irreversible transition occurs. Material inside the focal volume can be completely ionized to form plasma, and although some residual material remains, the focal volume is effectively destroyed. If the laser is focused at the surface of the sample, the high local temperature causes some of the material to be ejected at high speed. If, however, the focus is beneath the surface of the sample, material is not ejected, but still undergoes a transition, and the contents of the focal volume are destroyed.

These effects can be observed at peak power densities on the order of terawatts/cm^2 and exhibit fairly consistent threshold behavior in the dependence of damage on peak laser intensity. Laser ablation thresholds have been extensively studied in a variety of materials. However, biological materials are particularly difficult to characterize due to their wide variation in composition.

1.5.3 Cell Nanosurgery by Focused Light

Laser-induced disruption can be used as a tool to dissect or modify parts of a cell by focusing the laser in the region to be modified. Figure 1.30 shows incisions made in the surface of red blood cells using 800 nm 140 fs illumination at a mean power of 30 mW.

The damage threshold is higher in some areas of cells than in others. It also increases with depth beneath the cell surface: however, multi-photon absorption tends to minimize this effect. For characterization purposes, we measured the dependence of the damage threshold in dye-stained collagen on the depth beneath the sample surface [36]. Figure 1.31 shows the relation between the exposure time required to cause damage and the incident pulse energy (or peak power) for two different depths in a collagen sample. Although the damage threshold changes with depth, the functional relationship is the same, allowing well-controlled micro- or nanoscale manipulation within a cell.

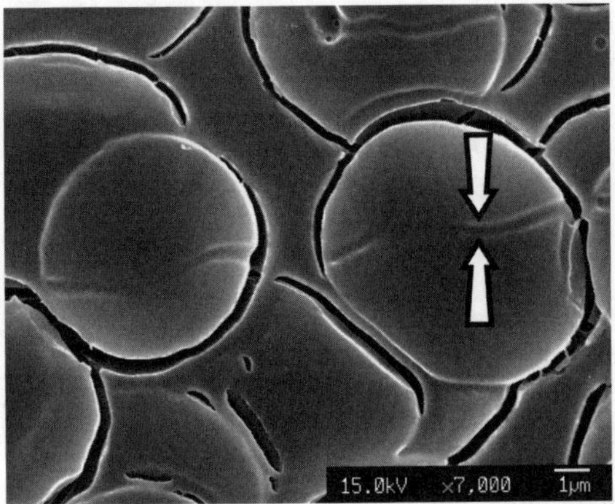

Fig. 1.30. Surface incisions created in rat erythrocytes (red blood cells) of submicron width by a NIR laser of 140 fs pulsewidth

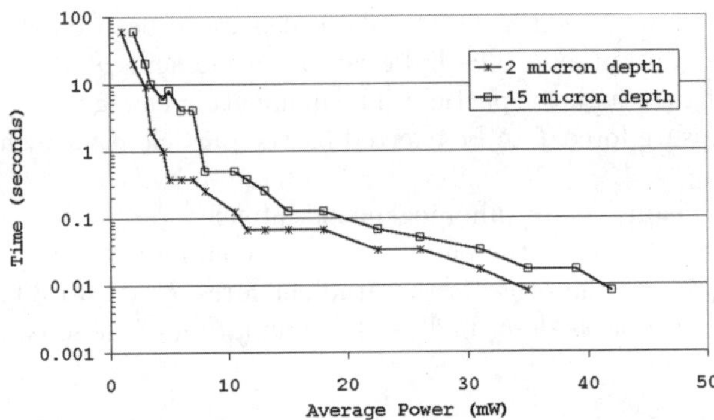

Fig. 1.31. The exposure time required to cause laser-induced disruption in Rhodamine G dye-collagen as a function of average laser power at 2 μm and 15 μm depths beneath sample surface

In the case of laser-induced disruption in living cells, the contents of the focal volume are destroyed but the cell as a whole can survive the disruption. This allows the dissection of cell components as well as investigation of a cell's resistance to and recovery from damage. The primary advantage of this type of submicron-scale surgery is that it allows the possibility of interacting with and modifying specific regions in a living cell without destroying the cell as a whole or even damaging the surface membrane of the cell.

1.6 The Manipulation of Living Bodies by Light

The manipulation of living bodies under a microscope is one of the essential techniques for experiments in biology and medicine. Recently, a manipulation scheme of such objects with laser light has started to be widely used [37,38]. This technique is called laser trapping, which enables us to capture a microscopic object by a radiation pressure force without mechanical contact. In this section, the mechanism of laser trapping and applications to biological and medical investigation are described.

1.6.1 Photon Pressure

Photon pressure is the force that is exerted on an object under manipulation by laser light [39]. The force is caused by momentum change of the photon when it is scattered by the object. The momentum of each photon is expressed as $h\boldsymbol{k}/2\pi$, where $\boldsymbol{k}$ is the wave vector of the photon and h is Planck's constant. The direction of the momentum is equivalent to the propagation of light. While the propagation direction of light changes through light scattering, the momentum of the photon changes, so that a force is exerted on the object according to the momentum conservation theorem. This is photon pressure.

Figure 1.32a shows a schematic diagram of photon pressure exertion, as a result of light refraction at the surface of a spherical particle both on the incidence to the particle and on exit. The light ray has a moment $\boldsymbol{p}$ and $\boldsymbol{p}'$ before and after travelling through the particle. The momentum change $\boldsymbol{p}'-\boldsymbol{p}$ causes the radiation pressure force $\boldsymbol{f}$ to be exerted on the particle, as shown in Fig. 1.32a.

If the incident laser beam has an inhomogeneous intensity profile like a Gaussian beam, the force exerted on the particle has two components; one is the scattering force $\boldsymbol{F}_{\text{scat}}$ and the other is the gradient force $\boldsymbol{F}_{\text{grad}}$ [37,39]. The net force $\boldsymbol{F}$ on the particle is described as the summation of these two forces, as follows:

$$\boldsymbol{F} = \boldsymbol{F}_{\text{scat}} + \boldsymbol{F}_{\text{grad}} \,. \tag{1.3}$$

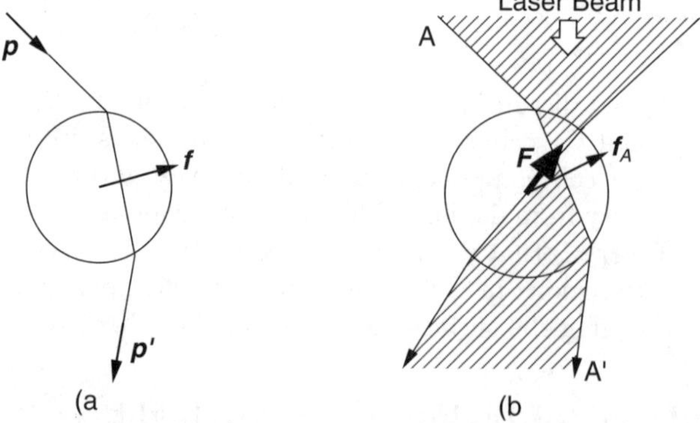

Fig. 1.32. A schematic of laser trapping. **a** The radiation force exerted on a particle by scattering. **b** The radiation force exerted by a converging laser beam acts to keep the particle at the spot

The scattering force $\boldsymbol{F}_{\text{scat}}$ pushes the particle along the propagation direction of light. In the case of a weak absorbing particle, such as a metallic particle, the force generated by light absorption is dominant in the scattering force [40]. The gradient force $\boldsymbol{F}_{\text{grad}}$ is exerted on a particle according to the gradient of the electromagnetic field around the particle. If the incident laser beam has a Gaussian beam shape, the particle is attracted to the center of the laser beam by the gradient force. By using this phenomenon, A. Ashkin, of the AT&T Bell Laboratory, developed an optical levitation technique to make a microscopic particle hover and translate from position to position, with a weakly focused laser beam coming from beneath the particle [39,41]. The particle is captured in the lateral direction of the laser beam owing to the gradient force. In the axial direction the particle is blown upwards by the scattering force and is stabilized at the point at which the scattering force is balanced by the gravity on the particle. With the use of optical levitation, dielectric particles sized from several micrometers up to 100 µm can be translated.

1.6.2 Three-Dimensional Laser Trapping

In 1986, A. Ashkin invented a method for trapping a small dielectric particle by the photon pressure exerted by a strongly focused laser beam under a microscope [37]. In this technique the particle can be trapped in three dimensions, and translated to an arbitrary location in a sample cuvette.

Figure 1.32b shows a schematic of radiation force generation on a particle with a strongly converging laser beam. In the figure, the particle is assumed to be larger than the spot size of the laser beam and to have a higher refractive index than that of the surroundings. A ray depicted as A–A′ in Fig. 1.32b generates radiation force $\boldsymbol{f}_A$ on the particle, as shown in the figure. Also, other rays generate radiation forces on the particle in the same manner. The net radiation force on the particle is given by the summation of such forces induced by all rays hitting the particle. The net radiation force becomes the force $\boldsymbol{F}$ shown in the figure. The force pulls the particle into the laser beam spot, even if the particle is located beneath the spot. The particle, which has a higher refractive index than that of the surroundings, can be trapped in three dimensions.

If the particle has a lower refractive index than that of the surroundings, it is hardly trapped because it is pushed away from the laser beam. Also, a metallic particle that has a size of the order of a micrometer is barely trapped in three dimensions, because of reflection on the surface of the metal. For lower refractive index particles and metallic particles, three-dimensional laser trapping was demonstrated by Sasaki et al. [42]. In their scheme the position of a laser beam spot was constantly moved to encircle the particle. The particle was captured three-dimensionally in the light field surrounding the particle, like being in a cage. In the case of a sufficiently small metallic particle compared to the wavelength of the laser beam, the particle can be

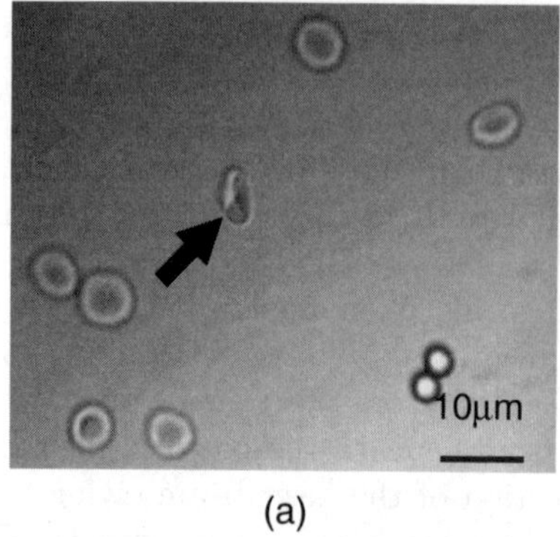

(a)

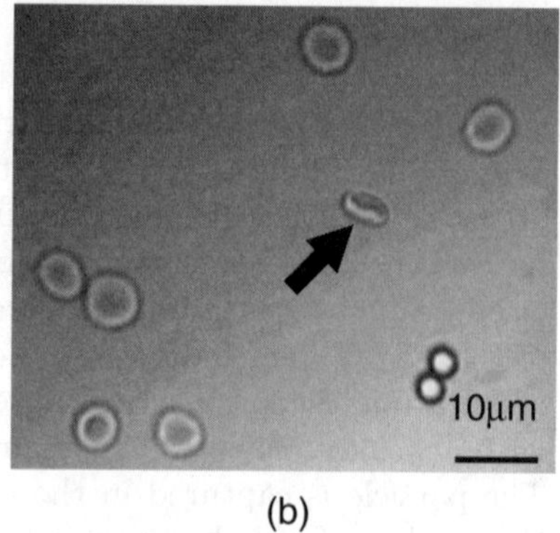

(b)

Fig. 1.33. Manipulation of a human red blood cell with a single laser beam. The arrows indicate the position of the laser beam spot. A trapped cell was first located near the center of the image (**a**). By shifting the spot position, the cell was moved to the right in the image with the spot movement (**b**). Other cells were not moved during the manipulation

trapped in three dimensions by radiation force [40,43]. For the trapping of such particles, the gradient force generated from the laser beam spot that pulls a particle must be stronger than the scattering force that expels the particle from the spot.

In the case of trapping in two dimensions, a micrometer-sized metallic particle on a substrate can be trapped by a strongly focused laser beam [44–46]. This is explained by optical absorption of the metallic particle. When the particle is located below the laser beam spot at the distance of the particle diameter, the net force on the particle by optical absorption is exerted to pull into the optical axis of the beam. In this case we can trap a metallic particle on a substrate in the lateral direction, but we cannot lift the particle away from the substrate in the axial direction.

Figure 1.33 shows the manipulation of a human red blood cell trapped by a Nd:YAG laser beam ($\lambda = 1064$ nm, 2 W). In the microscope images, first the red blood cell marked by an arrow is trapped around the center of the image (a), then the cell is translated to the right by moving the spot position of the laser beam (b). But other red blood cells in the image do not change their positions, because the laser beam is not irradiated onto these cells. Only the cell located at the laser beam spot can be manipulated with the light. This is an advantage of laser trapping. In this experiment the laser beam was focused with a microscope objective ($\times 100$, 1.3 numerical aperture (NA)) and the spot size of the beam was around 1 μm in diameter.

1.6.3 Force Measurement

In laser trapping the spring constant of the trapping force is small, in the μN/m order [47]. In comparison with a cantilever for an atomic force micros-

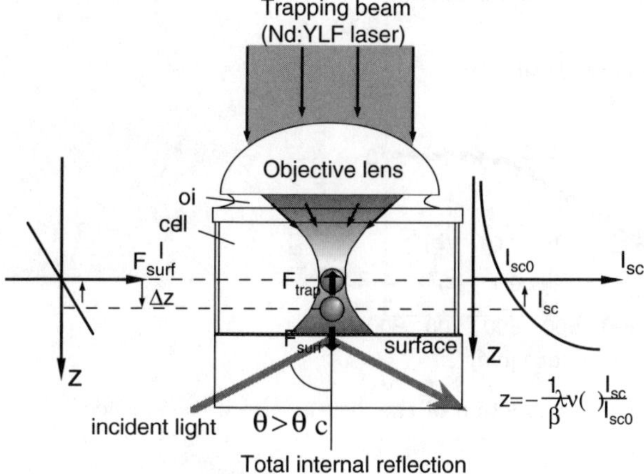

Fig. 1.34. A schematic of the surface force measurement

cope, the spring constant of the laser trapping force is more than two orders smaller. According to that property, laser trapping is utilized to measure a weak force, such as a surface force. For force measurement two schemes are used: one is by detecting the displacement Δx of the trapped particle to determine the force $\boldsymbol{F}$ with the equation $\boldsymbol{F} = -K\boldsymbol{\Delta x}$, where K is the spring constant of the laser trapping; and the other is by measuring the laser intensity variations to stabilize the particle position with laser intensity feedback. In the first scheme, the linearity of the trapping force to the particle displacement is assumed to determine the force, and this scheme can be applied for the measurement of a force that changes not only in strength but also in its direction. This scheme has been used for the measurement of a surface force [47] and a driving force on muscle protein [48]. The second scheme can sensitively detect small variations of force, which do not change direction. The second scheme has been applied to measure the tension on a string-like molecule; for example, the pulling force on DNA by RNA polymerase can be measured, during the transcription of DNA into RNA [49].

Surface force measurement on a small particle from a sample surface has been performed with laser trapping [47]. In general, the surface force attracts the particle from a distance of several hundred nanometers from the sample surface. Below a distance of several tens of nanometers, the particle is repelled from the sample surface. In the authors' experiments, a polystyrene particle was used as the probe particle to measure the surface force. Figure 1.34 shows the schematic of the surface force measurement by use of a laser trapped particle. First, the probe particle is trapped far from the sample surface to be measured and then moved down to the sample surface. At a certain distance, a surface force starts to work on the particle and the particle is slightly displaced. The displaced position of the particle is the balanced position

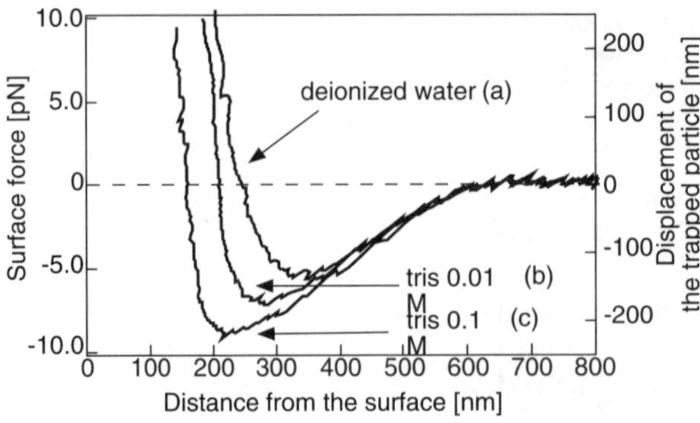

Fig. 1.35. The surface force as a function of the distance from the surface

between the three forces: the surface force from the surface, gravity, and the bouyant force on the particle. The trapping force exerted by the laser beam and the amount of displacement Δz of the particle are related by $F_z = -K_z \Delta z$, where K_z is the spring constant of the laser trapping in the z direction. Considering that the gravity and the bouyant force on the particle do not change during the measurement, the surface force is obtained from the displacement measurement.

To measure the particle displacement, evanescent illumination is used. In the evanescent illumination, a laser beam is incident to a prism attached to the sample substrate under a total internal reflection condition, and it generates an evanescent field on the sample surface. The evanescent field decays in an exponential manner with increase of the distance from the surface. The decay length, which is defined as the distance from the surface to the point at which the field intensity decays to $1/e^2$, is a fraction of the wavelength of the laser light. If a small particle is located in the evanescent field, the particle scatters the field and the scattered light can be detected far from the surface. The scattered light intensity also changes as a function of the distance between the surface and the particle in an exponential manner. The relation between the distance and the scattered light intensity is

$$I = A \exp\{-2\Delta z/\alpha\}\,, \tag{1.4}$$

where α is the decay length of the evanescent field and A is a constant. From this relation, the displacement of the particle can be obtained through scattered light intensity measurement.

Figure 1.35 shows a measured result of a surface force on a glass surface. The sample was a clean glass substrate and the probe particle was a polystyrene latex particle of 1 µm diameter, dispersed in water. The curves in the figure show the different results taken under three different solution conditions. Plot (a) is the case of deionized water as surroundings, plot (b) is that of 0.01 M tris-aminomethane solution, and plot (c) is that of 0.1 M tris-aminomethane solution. In all cases, the attractive force starts to work below a distance of 600 nm, and then the force becomes a maximum around the distance of 200–400 nm. After that, the force changes direction and the repulsive component increases. The maximum values of the attractive force are 5.50 pN, 7.57 pN, and 9.22 pN for (a), (b), and (c), respectively. From these results, the attractive force from a clean glass substrate is smaller in the solvent that has greater ion strength. In this experiment, the maximum and minimum detection limits were 21 pN and 32 fN, respectively. The spring constant of laser trapping in this experiment was 42 µN/m.

1.6.4 Microscopy with a Laser-Trapped Particle

By the use of a particle trapped by light as a probe, the local mechanical and optical properties of a sample can be investigated. The surface force measurement is an example of such an investigation. Furthermore, we may scan the particle over the sample to achieve local information at each position and make a map of the achieved information in two dimensions, to show the sample structure and the distribution of materials on the sample in more detail. This procedure is regarded as a scanning microscope with a laser-trapped particle probe. This idea was first demonstrated by Ghislain et al. as an atomic force microscope [50]. They successfully demonstrated force measurement by detecting the scattered light from a polystyrene particle suspended near a sample surface. Later, a near-field scanning optical microscope with a laser-trapped particle (laser trapping NSOM) was reported by Kawata et al. [51]. In their experiments, the near-field probe was also a polystyrene particle of 1 µ, irradiated by a laser beam for illumination in the total internal reflection condition. They measured the intensity of scattered light from the particle and fluorescence emitted from the sample, to perform scattering imaging and fluorescence imaging, respectively. Sasaki et al. [52] applied this probing method for a tiny chemical sensor which could detect pH in the nanometer region. They utilized specially designed fluorescence beads as a probe that changed the local fluorescence intensity with pH. They trapped a single fluorescence bead by a laser beam and detected the fluorescence emission from the bead. They showed that the fluorescence intensity changed with the changing of the pH of the surroundings.

In 1997, the authors developed a laser trapping NSOM with a three-dimensionally trapped metallic particle [40,53,54]. Metallic particles have many important features for near-field imaging, such as a high scattering efficiency and a large enhancement factor of the near field. In the authors' experiment, gold colloidal particles of 40 nm diameter were used as near-field probes. A gold particle has 350 times the scattering efficiency compared with a glass particle in water, when these particles are much smaller than the wavelength of light. Hence a gold particle gives a considerable amount of scattered light from the near-field region on the sample. Also, gold has important characteristics in that it is quite stable but is easily modified on its surface with molecular self-assembly. The surface modification of gold particles enables us to add specific functionalities on the particles, such as pH sensing by the change of fluoresence intensity. According to the characteristics, we may use the laser trapping NSOM for nanometric-chemical sensors which detect specific ions with ultra-high sensitivity, and fluorescence probing of specific biomolecules on a cell membrane by the use of conjugate materials fixed on the particle. In this section, the details of the laser trapping NSOM are described.

The Principle of the Laser Trapping NSOM. Figure 1.36 shows a schematic of the laser trapping NSOM with a metallic probe particle. An intense near-infrared laser beam is focused with a microscope objective to trap a metallic particle. A single metallic particle is suspended near the focal spot of the laser and approaches a sample surface from beneath. A gold colloidal particle, which has a diameter of several tens of nanometers, is used as a near-field probe.

Another laser beam which has a visible wavelength is also focused onto the particle simultaneously. This laser beam is for illumination on the sample. The laser beam for illumination is scattered by the probe particle, and then the scattered light diverges like radiation from a point light source as small as the diameter of the particle. When the particle is sufficiently close to the sample surface, the scattered light illuminates a part of the sample as small as the size of the probe particle and interacts with the material of this region. The scattered light is collected with the objective and is detected through a pinhole placed in the image plane of the objective. The detected light intensity of the scattered light contains local information about the sample, and therefore the microscope gives the structure and material distribution on the sample with a higher lateral resolution than the diffraction limit of light.

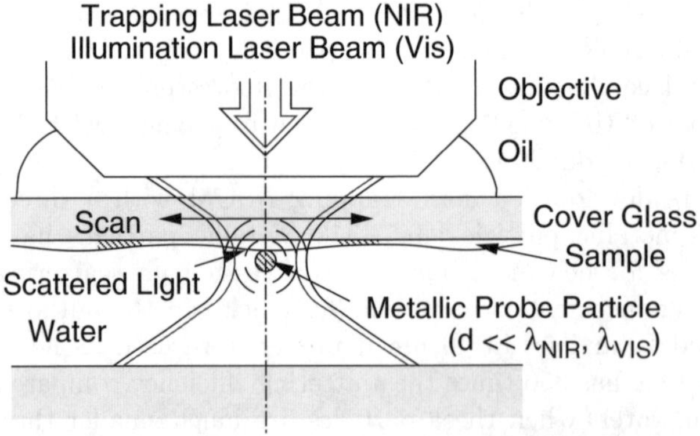

Fig. 1.36. Schematics of laser trapping NSOM with a metallic probe particle

This microscope has the following advantages, as follows. 1) Since the spring constant for holding a probe particle is so weak, a sample is rarely damaged mechanically. 2) There is no need to regulate the distance between the probe particle and the sample, because the probe particle touches a sample surface during scanning. 3) Observation with this microscope is done under water. It is quite useful for biological applications [55] such as fluorescence in situ hybridization (FISH) [56]. 4) According to the use of a metallic particle, light intensity scattered by the particle is expected to be much greater than that from a dielectric particle.

Experimental Setup. Figure 1.37 shows the experimental setup of the NSOM. The laser for optical trapping is a Nd : YLF laser (TFR, Spectra-Physics, 2.5 W, 1047 nm wavelength). This laser beam goes into microscope optics and is focused with an oil immersion objective (UPlan Apo, 100×, 1.35 NA, Olympus). DM1 in microscope optics is a dichroic mirror which reflects the light of the Nd : YLF laser and transmits visible light. Due to this mirror, almost none of the light from the Nd : YLF laser penetrates to the upper part of the microscope optics.

The laser for sample illumination is an air-cooled Ar ion laser (Uniphase, 20 mW, 488 nm wavelength). This laser is also incident on the microscope optics and is focused onto the probe particle. The spot position of the Ar ion laser corresponds to that of the Nd : YLF laser. Scattered light from the probe particle is collected with the objective, and an image is formed at the position of a pinhole and detected with a photomultiplier tube (PMT)

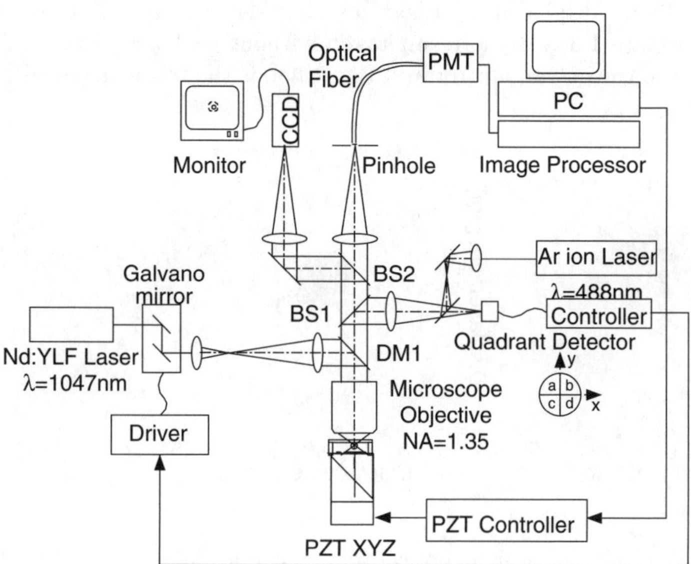

Fig. 1.37. The experimental setup of the laser trapping NSOM

behind the pinhole. The pinhole eliminates undesired light from other parts of the sample. Sample scanning is done by changing the applied voltage of a piezo tube. A CCD camera is used to observe the sample surface under a conventional microscope condition.

Feedback Stabilization of a Particle. According to thermal energy conditions, a probe particle moves continuously in a laser-trapping potential due to collisions with water molecules (Brownian motion). Due to this Brownian motion, the intensity of the light scattered from the particle changes over time. This will cause noise on measured signals when we use the trapped particle as a near-field probe. This noise affects the maximum resolving power of the laser-trapping NSOM. Consequently, it is necessary to reduce the fluctuation of the probe position in order to enhance the resolution of the NSOM. To reduce the fluctuation of the particle, a feedback stabilization technique is effective [53].

In the feedback stabilization technique, information about the particle position is given back to the spot position of the trapping laser beam. The position of the probe particle is monitored by the detection of the spot position. The spot position can be detected sensitively with a quadrant detector and two sets of deferential amplifiers to calculate the displacement signals of the spot in the x- and y-directions. To shift the spot position of the trapping laser beam, two galvanomirrors are used. The angles of the galvanomirrors are changed to minimize the displacement signals in both x- and y-directions. The displacement signals in the x- and y-directions are captured on a personal computer.

Figure 1.38 shows displacement diagrams that depict movements of a probe particle. Figure 1.38a is the result taken without feedback of the displacement signal to the galvanomirror and Fig. 1.38b is the result taken with

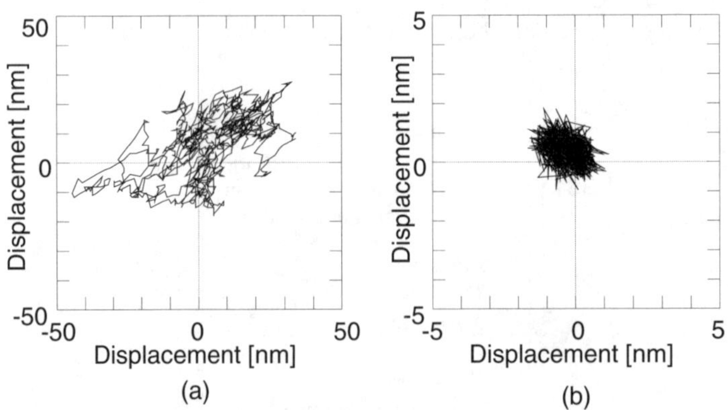

Fig. 1.38. Particle displacement in an optical trap (experimental data): **a** without feedback stabilization of a probe particle; **b** with feedback stabilization

feedback. The time interval of the displacement measurement was 10 ms. The probe particle was a polystyrene latex particle that had a diameter of 1 µm. The standard deviation of the particle position was about 13 nm in the case without feedback (a), although with feedback the standard deviation of the particle position was reduced to 0.45 nm. From this result, the feedback method is shown to be effective in stabilizing the position of the probe particle.

Observation of DNA Molecules with a Laser Trapping NSOM. The observation of DNA molecules was performed with the laser-trapping NSOM. DNA molecules were stained with YOYO-1 iodide dye (Molecular Probes, Inc.) in order to be visualized by fluorescence emission [53,54]. The dye has an absorption peak at 491 nm wavelength and an excitation peak at 509 nm wavelength and has a high affinity to bind DNA molecules. The efficiency of the fluorescent emission from the dye molecule increases significantly after binding to the DNA.

DNA molecules from calf thymus (Sigma Chemical Co.) were used. The sample was prepared as follows. First, the DNA was dissolved in water and stained with YOYO-1 iodide. Then the DNA solution was dropped onto a silane-coated cover glass. After 5 minutes the glass was rinsed with deionized water 10 times to wash away unadsorbed DNA molecules.

Figure 1.39 shows an observed result of DNA molecules stained with YOYO-1 iodide. A gold colloidal particle of 40 nm diameter was used as the near-field probe. In the middle of the picture an entangled DNA molecule can be seen. Also, in the left part many aggregated DNA molecules

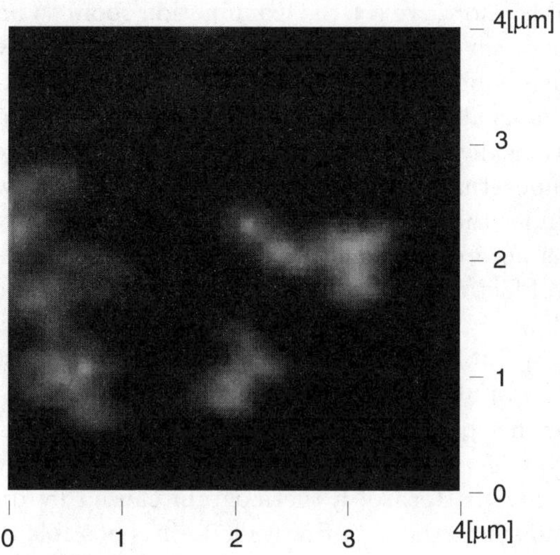

Fig. 1.39. DNA molecules observed with a laser-trapping NSOM

Table 1.1. Factors affecting skin surface temperature

	Thermatomes	Definitions
1	Angiological thermatomes	Abnormal temperature regions caused by organic vascular abnormalities
2	Functional angiological thermatomes	Abnormal temperature regions caused by vascular disfunctions
3	Neuro-dermatomal thermatomes	Abnormal temperature bands caused by somatosensory neuronal disorders
4	Myotomal thermatomes	Abnormal temperature regions suspected by abnormal muscular blood flow rate
5	Metabolic thermatomes	Abnormal hot and/or cold spots caused by excessive and/or lower heat production and blood flow

are observed. From these, the laser trapping NSOM can be seen to have the ability to image biomolecules fixed on a substrate.

1.7 Physiological Function Analysis of the Human Body and Organ-Using Far-Infrared Imaging

The surface temperature of the human body and organs is determined by the blood-flow rate, the structure of surface tissue, and the activities of the neurohumoral system, which regulate heat dissipation from the body surface. Because of this, we can detect distributions of many physiological functions from thermal images of the surface of the human body and organs obtained by far-infrared (FIR) imaging.

1.7.1 Static Analysis of Abnormal Temperature Distribution on the Skin [57]

In clinical practice, for thermal comfort, we set the examination room to an ambient temperature that is relatively warm. We refer to this as a thermally neutral condition, and the patient's body is able to maintain thermal equilibrium. The equilibrium condition is at 29–31°C when light clothes are worn. Under thermally neutral conditions, heat production and loss by the body are equal, and the skin-surface temperature is controlled only by the blood-flow rate of the cutaneous tissue. Under these conditions, it is possible to classify abnormal thermogram patterns according to their physiological origins. As shown in Fig. 1.40, many control factors are involved in determination of the skin-surface temperature [58,59].

The five parameters listed in Table 1.1 might be regarded as the most fundamental physiological causes of abnormality. In the table, the terminology used is of thermatomes for the specific temperature patterns. The word *thermatome* originated in *(neuro-)dermatome*, and was coined by P. LeRoy as defining an abnormal segmental pattern of a thermogram caused by disturbances of somatosensory and sympathetic pathways [60]. It is possible to expand the definition because the word *thermatome* is appropriate for describing a regional pathophysiological abnormality. Therefore, a *thermatome* is

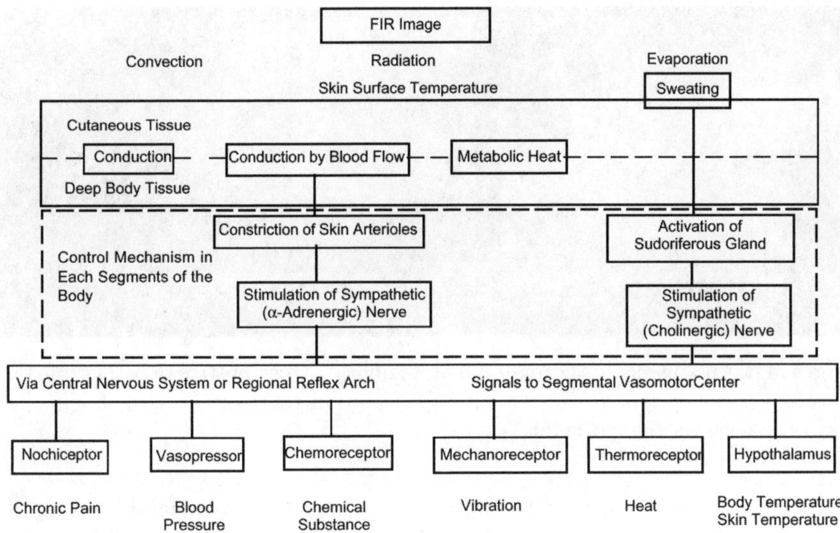

Fig. 1.40. Factors affecting skin surface temperature

Table 1.2. Factors affecting skin surface temperature

Main application area of thermographic examination	Diagnostic principles	Diagnosing techniques
Peripheral vascular diseases	To find abnormal angiologic thermatome	Sequential subtraction, thermal gradient and asymmetry detection
Local metabolic disorders	To find abnormal hot or cold areas and metabolic thermatome	Thermographic index
Chronic pain	To find abnormal dermatomal thermatome, myotomal thermatome, and angiologic thermatome	Asymmetry detection, and thermographic index
Autonomic nerve system disorders	To find abnormal dermatomal thermatome	Asymmetry detection, thermographic index and stress test
Inflammatory diseases	To find abnormal hot area and metabolic thermatome	Asymmetry detection, and thermographic index
Tumors	To find hot or cold areas, metabolic thermatome, and abnormal vascular pattern	Sequential subtraction, thermal gradient and asymmetry detection
Body temperature abnormalities	To analyze abnormal body temperature and discrepancy between skin and body temperature	Thermo-physiological test

a region within an abnormal thermal image and is associated with a thermo-physiological disorder [61]. When we classify the applications of clinical thermography, we can indicate each pathophysiological diagnostic principle using such thermatomal expressions (Table 1.2).

Fundamental Thermo-Physiological Image Expressions

Angiologic Thermatomes. Skin-temperature distribution is determined primarily by the skin blood flow. Arterial obstruction usually causes a severe drop of peripheral skin temperature in cool and neutral thermal environments. It is proposed that such images should be called *angiologic-thermatomes* because they reflect the local blood-flow distribution. However,

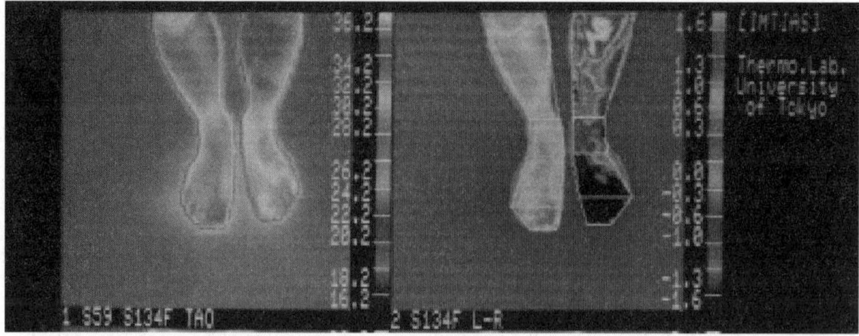

Fig. 1.41. An angiologic thermatome of a femoral artery obstruction. *Left:* a standardized thermogram. *Right:* asymmetry detection thermogram ([right leg] – [left leg], displayed on the left side)

it must be emphasized that the name is not based on the original concept of the angiosome, which is derived from the region of the body used for source arteries in skin flap transplantation [62]. Figure 1.41 shows a low-temperature area on the left foot caused by obstruction of the left femoral artery.

Functional Angiologic Thermatomes. The skin blood flow is controlled by the adrenergic sympathetic nervous systems. In particular, the vasomotor sympathetic nerves in the distal regions of the extremities are composed only of contractile nerve endings, which control body temperature by vascular contraction or dilatation. Images with lower temperatures are sometimes not a result of mechanical obstruction of peripheral arteries, but are due to functional obstructions, such as the case of Raynaud's syndrome. The term *functional angiologic thermatome* indicates that the lower- or higher-temperature areas are a reflection of functionally reduced or increased blood flow, respectively. In Fig. 1.42, the left half of the body is at a higher temperature than the right. The patient suffered from cervical syringomyelia and shows left-side Horner's syndrome. The pattern indicates a lack of sympathetic control in the left half of the body, and resembles a thermogram in stellate ganglion block.

Dermatomal Thermatomes. The sensory nerves are distributed at a particular vertebral level, with the segment referred to as a *sensory dermatome.* As the vasomotor centers of the vertebrate are present in each somatic segment, skin-temperature distribution within the segments is usually called a thermatome. Here, they are referred to as *dermatomal thermatomes*; i.e., thermally expressed sensory dermatomes. The terms are usually applied to the lower-temperature regions produced as a result of radiculopathy. The sensory segment of skin exhibiting chronic pain coincides with the lower-temperature area of the extremities (especially the legs and feet). Figure 1.43

shows a case of L5/S1 lumbar-disk herniation. The area of pain coincides with the area of lower temperature indicated by the L5/S1 sensory dermatome. It has been agreed that the area of chronic radiculopathic pain coincides with the lower-temperature regions and that the *dermatomal thermatome* is a good screening tool for the detection of radiculopathic complaints.

Myotomal Thermatome. Recent research has shown that some low-temperature areas can indicate muscular lesions. In particular, areas of low-temperature skin on the surfaces of the body trunk and the medial side of the extremities may indicate a reduced blood flow or the lower temperature of a muscle under the skin. Low-temperature zones produced as a result of whiplash injury or cervical radiculopathy indicate the muscle lesion directly. The *myotomal thermatome* is defined as a thermogram expression of muscle lesions.

Metabolic Thermatome. Thermographic observation can detect abnormal temperature distributions. Many thermographers consider that hot or cold spots indicate metabolic abnormalities, and the detection and measurement of hot spots is still an important factor in the differential diagnosis of tumors and for making prognoses for inflammatory diseases. The term *metabolic thermatomes* refers to a metabolically dominant thermal image. However, the metabolic image is usually combined with the skin blood-flow image, which may lead to ambiguity. A thermal stress test is therefore required in order to distinguish between those two phenomena.

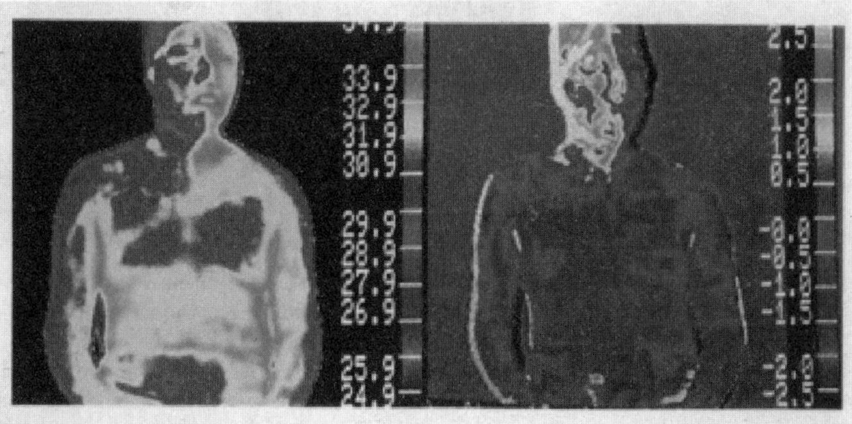

Fig. 1.42. A functional angiologic thermatome of a left-side Horner's syndrome by cervical syringomyelia (the *right-side* image is a time-sequential subtraction image between the before and after images of the left-side stellate ganglion block)

Algorithms to Convert Static Physiological Functions. When we describe the physiological function of a unit element of tissue (a boxel), we must define at least two more dimensions to an anatomical image, such as the environmental and intrinsic parameters of the boxel [63]. We should fix or measure these two extra parameters to diagnose the status of a physiological function.

Asymmetry Detection Algorithm. Many abnormal temperature distributions have been observed on one side of the body. Breast cancer, Horner's symptom and lumbago usually exist on one side of the body. After we developed a computer thermography system (CTS) [64], we developed a method for deleting the environmental and intrinsic parameters from a thermal image. In this method, the thermal image of one side of the body was subtracted from the corresponding part of the other side [65]. This can be done because the temperature distribution of the skin shows a symmetric pattern along the sagittal plane [66]. Thus, we can observe asymmetric abnormalities with-

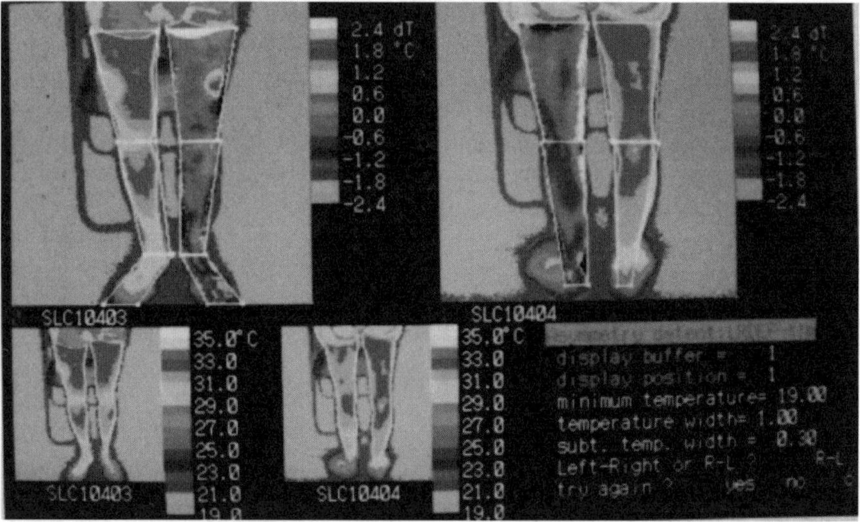

Fig. 1.43. The patient suffers from pain in the left back lower extremity and foot. We can detect a low-temperature part from two standardized thermograms (*lower left* and *lower right*). However, if we subtract the temperature of the right leg from that coincident part of the left leg, and print on the left side, we can see more clearly the temperature differences between the two sides. The patient was suffering from sclerotic vertebral canal stenosis at the L5/S1 level and told us that his painful parts were coincident with the parts where the temperature was lower by 0.6°C or more than the right side. If the site shows that an abnormal temperature exists on one side of the body, we can eliminate personal differences and environmental effects from the thermograms using this method

out interference from individual, environmental and anatomical variations (Fig. 1.43).

Heat-Conduction Formulas at the Skin Surface. Under a thermally neutral condition (around 29°C in the nude, or around 25°C in half-body exposure), we can write static heat-conduction formulas for the skin surface [67,68]. The parameters of skin-temperature control are composed of three heat-flux components from the body – radiation (Qr), basic evaporation (Qe), and heat convection (Qf) – and three heat-production components – conduction from deep body (Qc), metabolic heat production (Qm), and blood-flow convected heat (Qb). In the thermally neutral condition, equilibrium is established at the skin surface:

$$Qr + Qe + Qf = Qc + Qm + Qb\,. \tag{1.5}$$

The parameters are expressed by the following equations:

$$Qr = Kr\varepsilon(Ts^4 - Tw^4) = 0.805 \times 10^{-10}(Ts^4 - Tw^4)\,, \tag{1.6}$$

$$Qf = 0.58 \times 10^{-2}(Ts - Ta)^{1.25}/D^{0.25}\,, \tag{1.7}$$

$$Qe = 1.85 \times 10^{-2}\,, \tag{1.8}$$

$$Qc = Kc(Tc - Ts)/3d = 0.93 \times 10^{-2}(Tc - Ts)/d\,, \tag{1.9}$$

$$Qm = Mo \times S \times 2^{(Ts-Tm)/10} = 1.44 \times 10^{-2} \times S \times 2(Ts - Tm)/10\,, \tag{1.10}$$

$$Qb = \alpha\rho cVs(Tb - Ts) \times S\,, \tag{1.11}$$

where $Kr = 4.88 \times 10^{-3}$ (kcal/m^{2}hK); Ts (K) is the skin temperature; Tw (K) is the wall temperature; $\varepsilon = 0.98$ (emissivity of skin); D (cm) is the diameter of a model; Ta (K) is the room temperature; $Kc = 0.168$ (kcal/mhK) (thermal conductivity of skin); d (cm) is the depth of the core temperature point from the skin surface; $Tm = 309$ (K) (standard tissue temperature); S (cm) is the thickness of the skin; $\alpha = 0.8$ (counter current heat exchange ratio in warm conditions); $\rho c = 0.92$ (cal/mlK) (heat capacity of blood); $Tb = 310$ (K) (blood temperature in the core); and Vs (ml/100 g tissue min) is the skin blood-flow rate.

These equations become a basic mathematical model for skin temperature analysis. We can obtain the following and calculate the blood-flow rate of cutaneous and subcutaneous tissue (Vs):

$$\begin{aligned} Vs = {} & [Kr\varepsilon(Ts^4 - Tw^4) + 0.58 \times 10^{-2} \times D^{-0.25}(Ts - Ta)^{1.25} \\ & +1.85 \times 10^{-2} - Kc(Tc - Ts)/3d - Mo \times S \times 2^{(Ts-Tm)/10}]/ \\ & \alpha\rho cS(Tb - Ts)\,. \end{aligned} \tag{1.12}$$

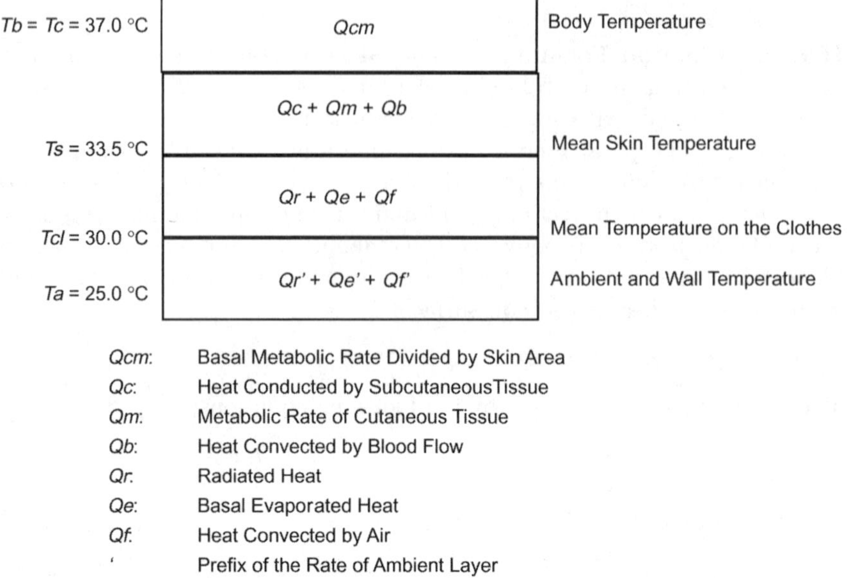

Fig. 1.44. In the thermally-neutral condition, heat production and convected heat are equal within the four layers, and the total system is in equilibrium. Metabolic heat is produced in the first layer (Qcm) (the body core). This heat is convected to the second layer of cutaneous and subcutaneous tissue [$Qc + Qm + Qb$ in (1.5)], it radiates or is convected to the third layer between the clothing and the skin surface [$Qr + Qe + Qf$ in (1.5)], and it dissipates into the environment ($Qrf + Qef + Qff$). Therefore, in the neutral condition, equations $Qcm = Qc + Qm + Qb = Qr + Qe + Qf = Qf + Qef + Qff$ are valid

If we can measure deep body temperature and other anatomical parameters, we can convert a thermogram into a blood-flow rate image from (1.12) (Fig. 1.45) [69]. Theoretically, functional images of other parameters can be obtained by those formulas.

1.7.2 Dynamic Analysis of Abnormal Temperature Distribution on the Skin [70]

The physiological procedures described above are performed at thermal equilibrium. Under thermal nonequilibrium conditions, the dynamic thermal control process can be observed on the skin. In clinical trials, one solution for detecting hidden pathophysiological function is to introduce some physical, chemical, or neurohumoral stress to a patient and then take time-sequential thermal images.

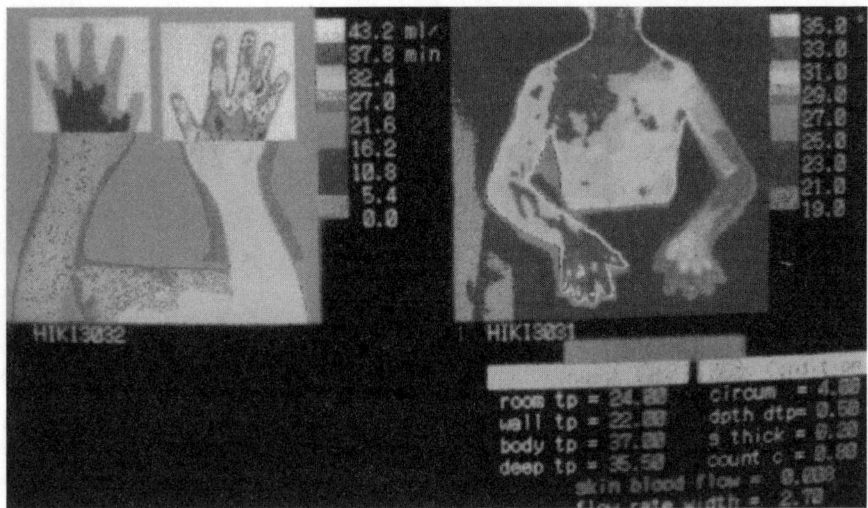

Fig. 1.45. The blood-flow rate distribution of cutaneous tissue of hands. The blood-flow rate (ml/min/100 g cutaneous tissue) of both hands is calculated by (1.12). The patient was injured on his left hand by pulling it violently in a motorbike accident. The *right* image shows a thermogram and the *left* image shows the blood-flow rate calculated by (1.12) using the parameters written in the right corner of the figure. Deep body temperature was measured by a deep body thermometer, using the Fox's method. The cutaneous tissue blood flow of the left fingers decreased to under 5.4 ml/min/100 g tissue

Time-Sequential Image Processing. If the thermal production of a region does not balance the thermal dissipation, we must add a heat accumulation factor in (1.12). If the observation time is short and the body temperature and the environmental temperatures (wall temperature and ambient temperature) are kept constant over this period, we can estimate the quantitative changes of some physiological parameters under a transient status. The most dominant parameter is blood flow.

When we administer a vasodilator as a load (Fig. 1.46) [71], or induce reactive hyperemia using vascular occlusion [72], the environmental temperature change has negligible effects on skin-temperature control. In such cases, we should add thermal accumulation item to the equilibrium equation (1.5) of the thermally neutral condition. If we observe skin-temperature change (ΔTs) in a short period (Δt), the following equation results:

$$Cs \times \Delta Ts = \Delta Qc + \Delta Qm + \Delta Qb - (\Delta Qr + \Delta Qe + \Delta Qf)\,, \qquad (1.13)$$

where Cs is the heat capacity of skin. In this equation, we can neglect metabolism (ΔQm) and evaporation (ΔQe). We analyzed the effect of the regional skin-temperature change on those parameters. If the change is more than 1°C and we calculate the contribution rate of each of the parameters

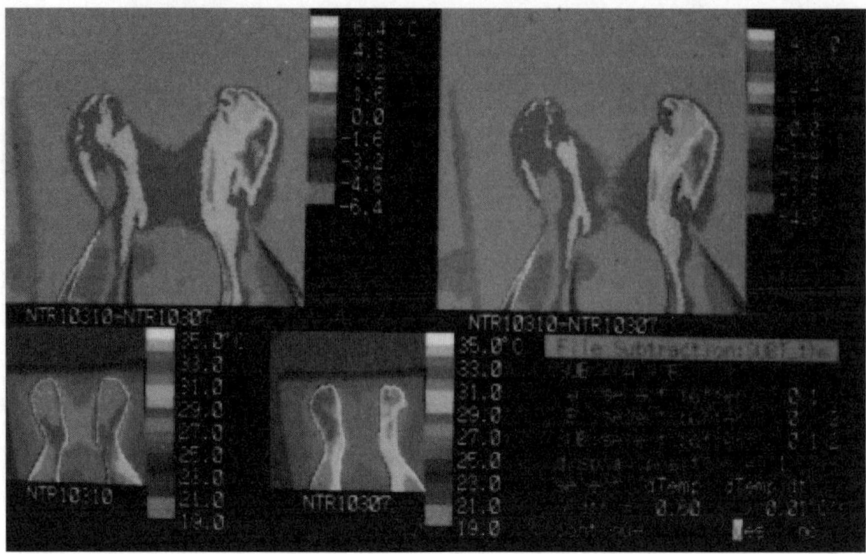

Fig. 1.46. Time-sequentially subtracted thermograms before and after analgesic drug injection. The patient suffered from left-foot pain caused by lumbar disk herniation. The lower right thermogram shows baseline data. She complained of right foot pain where the thermogram shows a hypothermic region. Three minutes after injection of an analgesic (Neurotropin), her pain was diminished and the skin temperature of her foot was elevated (*lower left*). When we subtract the baseline image from the image after injection, we obtain the left upper image. The subtracted image is called the difference thermogram. The image clearly indicates to us regions where the temperature was elevated. From (1.15), we understand that the temperature difference of the image was proportional to the cutaneous blood flow. The patient's vascular bed responded to the analgesic and dilated the arterial bed. We can easily see that the blood-flow rate of the region decreases by functional constriction of arteries and not by organic obstruction. The upper right image shows the mean temperature elevation rate distribution in this three minute perior. If the interval between the images is shorter and the sensitivity of the machine higher, the image might show us sympathetic vascular control signals

in (1.6)–(1.11), quantities of heat production and dissipation (i.e. of ΔQr, ΔQe, and ΔQf) are calculated to be 1/100 of ΔQb [73]. Therefore, the rate of skin-temperature change in a short time period is calculated by the following equation:

$$Cs \times \Delta Ts = \Delta Qb + \Delta Qc = A\rho c[\Delta Vs(Tb - Ts) + VsTs] + Kc\Delta Ts\,. \tag{1.14}$$

Therefore,

$$C_1 \Delta Ts = C_2 \Delta Vs + C_3 \,, \tag{1.15}$$

or, after differentiating, we obtain the following equation:

$$\frac{\partial Vs}{\partial t} = C \times \frac{\partial Ts}{\partial t} \,. \tag{1.16}$$

If we subtract two thermograms after imposing a stress, and consider the meaning of (1.15), the subtracted image shows us the amount of regional blood-flow rate change between the two images.

Strong Thermal Stress Application to the Body. There have been many reports relating to vascular diseases. The cold-water immersion test is the most prevalent test for identifying peripheral arterial obstruction. However, when we apply the water-immersion test to a patient, the skin-surface temperature is forcedly changed within a very short period (Fig. 1.47). In such a case, we must make use of two Newton cooling equations. One is from the environment to cutaneous tissue and the other is from cutaneous tissue to subcutaneous tissue. However, we cannot estimate each parameter separately in clinical use. Therefore, we must make a complicated sensitivity-analysis table to analyze the parameters jointly [74].

1.7.3 Dynamic Analysis of the Surface Temperature of Internal Organs

Recently, following the development of new FIR sensors, which are able to work without liquid nitrogen cooling, FIR devices have been applied to measure the surface temperature of internal organs or tissues in an operating theater. Blood flow, metabolic heat, and environmental temperature determine the surface temperature of an internal organ. When we expose it to an environmental temperature, the surface temperature decreases in accordance with Newton's cooling law, because its surface lacks thermal insulation tissue. If there is an arterial network on the surface, we can easily detect the image and evaluate the status of the blood-flow rate. A coronary arterial network exists on the cardiac surface. After exposing the heart in the open air, it shows up clearly. When a clip on a branch of a coronary artery is released after an anastomosing coronary bypass-graft operation, we can observe the pattern of warm blood flow running in the peripheral coronary arteries (Fig. 1.48). The coronary blood-flow rate can be observed remotely and noninvasively. If a thermal coronary angiography system is installed in the cardiac operating room, not only the coronary bypass graft operation but also other operations of open heart surgery can be carried out safely.

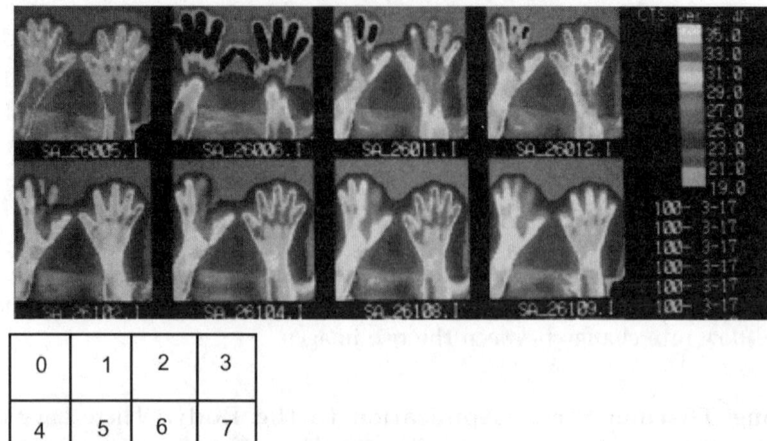

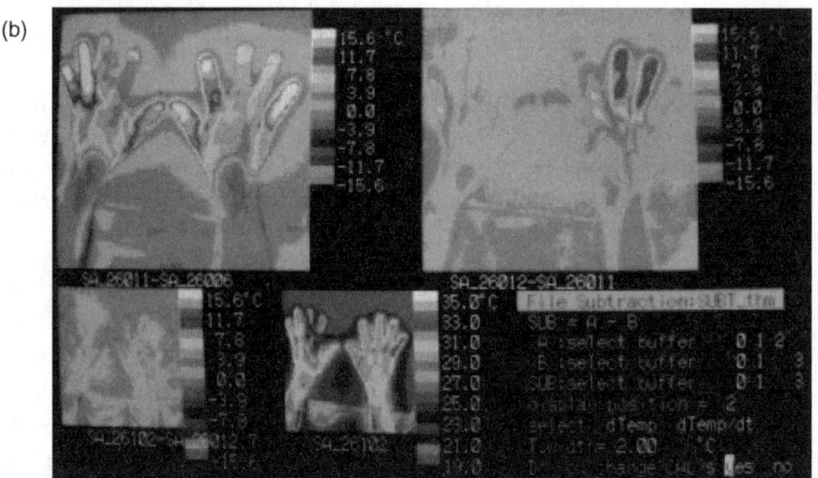

Fig. 1.47. Time-sequentially subtracted thermograms after cold-water immersion stress to both hands of a Raynaud's syndrome patient. **a** The upper row of images shows, from left to right, the baseline thermogram (0), directly after immersion in 4°C water for 10 s (1), 5 min after immersion (2), 10 min after immersion (3), 20 min after immersion (4), 30 min after immersion (5), and 1 hour after immersion (6). Before applying the cold-water immersing stress, we cannot detect the abnormal temperature pattern. However, after applying the load, we can analyze each finger condition precisely. If we apply image subtraction to the thermogram, we can find the difference in the blood-flow recovery for each finger. **b** Time-sequentially subtracted thermograms: [(a)2 − (a)1] (*upper left*), [(a)3 − (a)2] (*upper right*), [(a)4 − (a)3] (*lower left*), and the same image of (a)5 in the *lower right* image. However, in the cold-water immersing test, not only the skin surface temperature but also the deep body temperature usually change. Thus, many parameters in (1.6)–(1.11) might also change. Considering the equations, we should use a milder thermal load than in this trial

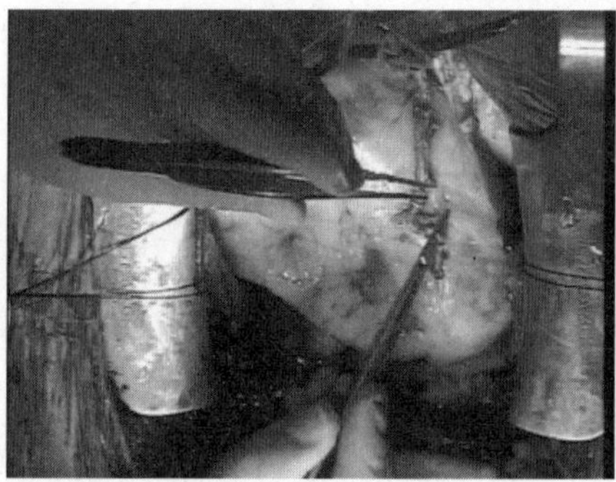

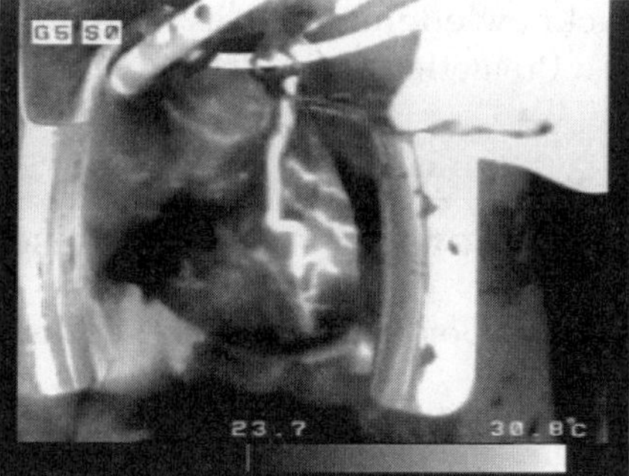

Fig. 1.48. The blood-flow pattern in a coronary arterial graft and the peripheral coronary arterial network. An ordinary video image is shown on the left and a thermal image on the right

Thermal Coronary Angiography (TCA). Thermal coronary angiography (TCA) is a cardiac imaging system to apply a far-infrared sensing technology. The main purpose is to display the coronary arterial network during a grafting operation on the coronary artery using a noninvasive method. Three cases of coronary bypass-graft operation were applied with a median section of sternum for the feasibility study. The TCA system was composed of a far-infrared camera (FIR camera), Thermal Vision LAIRD S270 (Nikon), and a video camera, Handyscope HVS-10 (Aishin Cosmo Co. Ltd.). The FIR camera has an image sensor with 410,000 picture elements of a PtSi Shottky-barrier charge-couple device cooled with a Stirling engine, and has a camera head of $140(W) \times 390(D) \times 175(H)$ mm in size and 9.2 kg in weight. The FIR camera and the video camera were mounted on an aluminum board and were assembled into an optical head system [75]. The fused images were displayed on television monitors in an operating theater through a camera control system. Images and control signals between the optical head and the camera control system were connected by an electrically shielded line of 20 m in length.

1.7.4 Unsolved Problems

There are many applications of FIR imaging. The analysis algorithms for each system are still being developed. One of the most important problems is to develop an analysis algorithm for indirect temperature reactions. We have observed that skin-temperature change occurs at a location that is distant from the stimulation site. The autonomic nervous system and neuro-humoral system control cutaneous blood-flow distribution. If we can find the control algorithms and install them in the image-processing system, we will be able to detect autonomic nervous system disorders noninvasively. These demands already exist in the clinical setting, so clinicians should directly communicate their requirements to thermographers.

Acknowledgment. This research was supported by the Japan Society for the Promotion of Science "Research for the Future" program.

1.8 A New Technology for Detecting Coronary Artery Disease

1.8.1 Coronary Artery Disease

Coronary artery disease is life threatening and happens due to the gradual progression of coronary artery stenosis. For the prevention of coronary artery disease, the detection and suppression of coronary artery stenosis is of most importance. In general, diagnosis of coronary artery disease is invasively done by coronary arteriography, following the onset of symptoms of coronary artery disease. A reliable noninvasive method is required for the early detection of coronary artery disease and for repeated monitoring of the flow state of the coronary artery. It is reported that arterial narrowing as small as 25% can produce post-stenotic turbulence [76–78]. The related auditory response can be used to detect lesions of the coronary artery [78]. However, occlusions larger than 95% may not produce noises, since in such instances blood flow is drastically reduced. In spite of an ongoing effort to develop an instrument to noninvasively detect coronary disease, a noninvasively detecting device that is commercially available for clinical use has not yet been developed. Because coronary artery disease is one of the major causes of mortality in the developed world, the prompt dissemination of research results is critical to advancing patient care. In particular, better methods for identifying asymptomatic persons at high risk of developing future clinical coronary artery disease are needed, so that we can target preventive monitoring methods, such as noninvasive examination, at those most likely to benefit from such examinations in a cost-effective fashion.

We have developed a new technology of a noninvasive laser phonocardiograph, which is made up of a noncontact ultra-supersensitivity and small mass laser sensor with low noise and a high gain amplifier and signal processing approach using a maximum entropy method [79–81].

1.8.2 The Detection of Subclinical Coronary Stenosis

For the prevention of coronary artery disease, the detection and suppression of subclinical coronary artery stenosis induced by subclinical atherosclerosis is of most importance. Methods for detecting subclinical atherosclerosis can be grouped into physiological and anatomical methods. Physiological methods require some functional impairment of blood flow or another arterial vascular function. Because acute coronary events are frequently caused by lesions that represent less than 75% stenosis of the vessel, the goal in identifying persons at high risk of coronary events is not to identify those with flow-limiting coronary stenoses (as does stress testing) but, rather, to identify persons who are at higher statistical risk of having or developing vulnerable coronary lesions. Therefore, the most promising methods are likely to be those that sensitively and specifically detect and quantify atherosclerosis rather than stenosis. Although physiological tests clearly have a role in symptomatic persons and could have a role in asymptomatic persons, they are not fundamentally tests designed to detect and quantify subclinical atherosclerotic stenosis.

1.8.3 A Noninvasive Physiological Approach to the Detection of Coronary Artery Disease

The use of noninvasive physiological methods to examine vascular beds for the presence of suclinical stenosis prior to the development of symptoms is an attractive way of identifying high-risk individuals who may be candidates for more aggressive therapy. Anatomical methods designed to directly detect and quantify subclinical stenosis noninvasively, that have been tested in large populations, and clinically available technique, used for intima-medial thickness are conceptually attractive methods of stratifying selected asymptomatic persons according to coronary heart disease risk, as a guide to more informed decisions regarding individualized preventive measures, such as drug therapy for atherosclerosis reduction. A noninvasive diagnostic system for coronary artery disease using a specially designed accelerometer has been reported by only one research group in the United States of America [82]. In this system, the vibration sensor is not so supersensitive and has not been put into practical use.

1.8.4 The Rheological Basis

The theoretical background of the acoustic diagnosis of coronary artery stenosis with our invention is described in the following.

The rheology of the coronary and other arteries as blood shows a feature in which the viscosity increases with the shear rate increasing: it is handled as a non-Newtonian incompressible fluid. In the case of fluid with such a viscosity flowing in a circular pipe the diameter and flow rate of which completely change, the flow pattern is divided into two different flows, namely laminar flow and turbulent flow.

Given that the maximum diameter of the right or left coronary artery in the trunk is about 5–6 mm, it is considered that the Reynolds number in the trunk of the right or left coronary artery is about 200–300 in the rest condition and about 1500 or less in intense motion: therefore turbulent flow is not generated in the coronary artery through all the cardiac cycle. However, in the case that a local change of a diameter of the blood vessel is generated by deformation of the vessel wall.

Local Turbulence of a Blood-Flow "Jet." The factor causing local turbulence of blood flow is described in the following. When the blood vessel in the heart ventricle wall is strongly compressed by contraction of the cardiac muscle, the coronary blood flow decreases. Therefore, the coronary blood flow mainly flows during cardiac muscle diastole, which is different from other arteries. Also, the left ventricle wall is thicker than that of the right ventricle: therefore, the left coronary artery blood flow is accelerated at diastole, and it reaches the maximum blood flow from almost zero. When stenosis occurs, the fluid is accelerated at diastole, and it is immediately decelerated afterwards.

Transmission of the Jet Vibration to the Chest Wall. There have been several cases in the past in which the stenotic murmur of the coronary artery could be detected in a clinical demonstration [83–86]. The murmur generally cannot be detected, as it is very minute and is attenuated in inverse proportion to involution of a frequency from a source of vibration. It is considered as a cause that the murmur has been buried in other biological vibrations with the attenuation. The vibration of coronary artery stenosis, which is very minute, intermingles with the displacement vibration of the surface are amplifier processing, signal processing are carried out by using the high-resoluble

In what follows, the measuring principle is described. The vibration signal, which is a mechanical and physical signal, arises in the body.

On the vibration signal, there are heart sounds that derives from switching of the heart valve (the sound generated as the atrioventricular valve between the atrium and the heart ventricle closes), heart sounds generated as an arterial valve between a blood vessel and the heart ventricle closes, extra heart sounds, valve abnormal sounds, and a murmur that derives from coronary artery disease. The vibration signal travels to the body surface, causing biotissue to vibrate. The mechanical vibration transmitted to the chest wall becomes a displacement signal. At the body surface, the heart sound and cardiac murmur that are detected by a stethoscope as vibration of the surface are caught and are recorded by a phonocardiograph, which produces a displacement signal.

The turbulent flow murmur vibration based on stenosis, diastole coronary artery and other sources is buried in the overall noise because the amplitude is very weak, compared with the usual heart sound and murmur, and the frequency band of the vibration shifts into the high-frequency area. There is also the displacement signal on the chest wall. Generally, in the record using the phonocardiograph, the stenotic vibration is buried in the basic line, and it cannot be detected and distinguished as a stenotic murmur because the vibration is very minute.

1.8.5 The Impossibility of Conventional Standard Phonocardiography Technology

It is regulated in the conventional industrial standard that a phonocardiograph can detect a heart sound signal with frequency between 20 and 600 Hz, and the heart sound signal is that vibration signal deriving from the heart and blood vessels and transmitted to the body surface, which contains any cardiac murmur.

The phonocardiograph comprises a heart sound microphone, an equalizer (in the case of using a direct conductive microphone, such as an accelerometer microphone or a velocity microphone), a heart sound recorder, an electrocardiographic signal power supply, and it records the heart sound signal and the electrocardiographic signal simultaneously.

The measurement of the phonocardiograph in accordance with the international industrial standard is for detecting actuation abnormalities of valves and the existence of an intracardiac shunt.

However, in regard to the very minute signal driving from the coronary artery stenoses during diastole and having a wide frequency band between 200 and 1200 Hz, it is impossible to match the acoustic impedance between the heart sound microphone and body surface if the weight of the microphone is not 5 g or less and ideally 1 g [78]. Therefore, it is impossible for the phonocardiograph, in accordance with the international industrial standard, to detect the very minute acoustic vibration signal between 200 and 1200 Hz, as mentioned above.

1. In conventional practice, this very minute vibration signal in the frequency band between 200 and 1200 Hz is outside the international industrial standard, and the signal is outside the frequency range of the measured object of the phonocardiograph. As the vibration intensity of such a very minute vibration signal is very weak, it needs to be amplified to over 100 dB in order to make the signal the measured object. In the case of using an acoustic vibration sensor, such as a microphone and an accelerometer, as the detecting sensor of the phonocardiograph for the above amplification, the matching of acoustic impedance between the sensor and the body surface cannot be undertaken, since the weight of the sensor becomes over 200 g.
2. Although the sensor technology by the displacement gauge principle is the conventional practice, satisfactory detection cannot be carried out because the sensitivity is insufficient and the vibration signal of the object is buried in the noise.
3. The time resolution is also insufficient in conventional practice, using a vibration sensor such as a microphone with vibrating plate resonance or an accelerometer with charge generation based on enclosure resonance.
4. There is no system that can perceive and detect very minute displacement vibration signals such as the coronary artery stenotic, diastolic and other murmurs. To detect the very minute signal, transmission of the stenotic

vibration is needed, which is difficult to achieve in a conventional phonocardiograph in view of the objective vibration frequency band, the degree of amplilfication, the signal-to-noise ratio and other factors. The essential problem is that the microphone, which captures the vibration signal, has an important defect. This defect is that the vibration transmitting characteristics of the chest structure are lowered and changed by the weight of the microphone; in other words, a mismatch of the acoustic impedance is generated, because the microphone contacts the body surface (using the contacting technique of a vibration perception detector).

1.8.6 The Theoretical Basis for the New Technology

It has been known in our previous research that the resolution of the vibration detector is not sufficient even if such a supersensitive accelerometer is used. A noncontacting technique has to be chosen as a structure for the vibration perception detector in order to detect the vibration at the body surface. In order to detect the very minute displacement amplitude accurately, a laser beam of which the monochromaticality, directionality, and convergence are excellent is the only choise at the present time. The transitivity of that laser beam must be low, and the reflectivity high, while the light source focus should be as small as possible.

The displacement signal measured by the laser displacement gauge is integrated by the chest tissue, while the acceleration signal due to stenosis and diastole are transmitted to the chest wall. The frequency domain can distinguish the detected displacement signal on the chest wall, since the frequency of the displacement signal of the vibration does not change: only the phase changes. Thus, coronary artery disease can be diagnosed through distinguishing the displacement signal containing the turbulent flow vibration deriving from coronary artery stenosis in the frequency domain. The amplitude strength can be distinguished as a power.

The new technology is a high-resolution phonocardiography technique that allows detection and analysis of low-amplitude phocardiographic signals that may not be detected on the body surface by routine measurements.

Our aim is to provide a state-of-the-art review of the noninvasive detection of coronary stenotic murmurs. Theoretical and technical principles are discussed in detail. It is a potentially powerful technique, but one that is still in need of further technical refinement before it can be introduced into routine clinical use. While the light microscope reveals important cell structures, such as the cell membrane, cytoplasm, and nucleus, it gives us little, if any, information on mitochondria, the Golgi apparatus, ribosomes, and so on, and conventional phonocardiograms only provide heart sounds and cardiac murmurs that correspond to blood flows in cardiac cavities in that order. Only with higher-resolution techniques, the electron microscope for anatomical studies and high-resolution phonocardiography in our case, can we detect other crucial events in coronary blood-flow states, which are not represented by waveforms in conventional standard phonocardiography.

1.8.7 The Principles of the New Laser Phonocardiography Technology

It is a primary objective of this invention to provide an acoustic and noninvasive diagnostic system for coronary artery disease and other conditions, which can detect the vibration signal of a murmur deriving from a shape abnormality such as stenosis of a blood vessel like the coronary artery. Detection of the vibration signal of the murmur deriving from stenosis of the coronary artery in its early stages provides the possibility of diagnosing abnormal condition, preventing and treating heart disease and other conditions. According to the present invention, in one aspect thereof, there is provided an (acoustically noninvasive) diagnostic system for coronary artery disease and other conditions comprising a detector for the vibration signal from a subject using pulsed laser beam, which is placed apart from the subject, and a detector of vibration signal detected by the detector of vibration signal of environmental noise, and vibration signal detected by the detector of vibration signal of subject and the detector of vibration signal of environmental noise is filtered for canceling internal noise and external noise, and filtered vibration signal is amplified and recorded. That is to say, it comprises a complex multiple vibration sensor system with a detector of the vibration signal from the subject using the pulsed laser beam and a detector for the vibration signal from environmental noise, and a processing facility that cancels the internal noise of the measuring instrument and the external noise, amplifies only the very minute vibration signal deriving from stenosis of the coronary artery on the body surface, and records the vibration signal as data.

According to the present invention, in another aspect thereof, a diagnostic system for coronary artery disease and other conditions is provided, which is characterized in that the said detector of the vibration signal of the subject has one or several laser source heads and vibration detective sensors with a laser displacement gauge and a three-axis accelerometer. In our invention, perception and detection of the stenotic vibration signal on the body surface is conducted using a noncontacting technique, and a pulsed laser beam is used in the vibration detecting sensor for the input vibration signal, which is harmless to the skin. The laser source head is fixed to a support or is slid along the support, and is incorporated in the sensor system.

According to the present invention, in still further aspect thereof, there is provided a diagnostic method for coronary artery disease and others comprises the steps of detecting vibration signal by a detector of vibration signal of subject using pulsed laser beam which is placed apart from the subject and a detector of vibration signal of environmental noise, filtering said vibration signal to cancel internal noise and external noise, amplifying the filtered vibration signal and recording said filtered vibration signal. That is to say, the diagnostic method for coronary artery disease and other conditions perceives and detects the very minute vibration signal of the murmur deriving from stenosis of the coronary artery on the body surface in a noncontacting technique by using the above diagnostic system for coronary artery disease and other conditions using a pulsed laser beam that is harmless to skin tissue.

The foregoing objects, other objects as well as the specific construction and function of the present invention will become apparent and understandable from the following detailed explanation thereof, when read in conjunction with the accompanying drawing.

1.8.8 The New Laser Phonocardiography Technology Design

The present invention, with a complex, multiple sensor system, has a vibration displacement gauge that utilizes the Doppler phenomenon and regular reflection and diffuse reflection of the laser diode visible pulsed laser beam on the body surface. Based on the data gathered by basic research done up to now, in order to set the measuring range, the working distance, the resolution for entering the frequency and the amplitude of the displacement element of the very minute stenotic murmur of the coronary artery containing each displacement element of respiratory movement, the heartbeat fluctuation component, and the heart murmur fluctuation component in the displacement of the chest wall in the dynamic range, the wavelength of the pulsed laser beam should be 760 nm, the pulse duration 10 nm, and the sampling frequency 50 kHz. The laser safety standard is class 2 ("class 2" in the following), for example.

A supersensitive and high-resolution three-axis accelerometer is attached to the laser source head in order to the detect vibration signal from the internal noise and the external noise deriving from the environment of the laser source head, which is included in a complex sensor. Since measurement at the bedside is disturbed by noise, the accuracy of the measurement is reduced or the measurement itself cannot be carried out. Therefore, the supersensitive three-axis accelerometer is incorporated in the system as a vibration sensor, in order to monitor the vibration signal from the internal noise and the external noise deriving from the environment of the laser source head directly and visually as a noise signal to correct, and cancel, the above vibration signal from the vibration signal of the body surface as a vibration signal to be measured by using time averaging, an ensemble mean, the differential composition of the bridge circuit, and other techniqes.

The detector of the vibration signal of the measured object with the united complex sensor is fixed to a support, and is allowed to slide along this support. The measuring equation is set in the complex-multiple sensor system or the diagnostic system in order to remove environmental noise: accordingly, the specific fixed apparatus or device which has been considered up to now is not required. A supersensitive microphone is incorporated in the system in order to monitor the environmental noise deriving from the measuring environment.

Therefore, the system can monitor the vibration signal of the environmental noise by visualizing it as a referred vibration signal of the environmental noise, and cancel the vibration signal of the environmental noise from the vibration signal of the body surface as a vibration signal to be measured.

A number of high damping gradient filters (about 1000 dB/octave) are put on an amplifier with low noise and a high amplification factor. This may form a system that measured displacement of such a system from the laser source is extracted as an acceleration signal by combining differentiating circuits, and is displayed and recorded. As mentioned above, the noise is removed on the hardware. Each objective amplified signal is converted into the frequency domain by a signal processing method such as linear prediction processing and is made to be a power spectrum. Thereafter, the noise signal is canceled from the objective measured signal and only the frequency of the objective signal is extracted, displayed, and recorded.

The coronary artery blood flow causing the objective measured signal is the flow during the heartbeat diastole. The frequency band is designed to be of 100 Hz width (recently of 10 Hz width), as it is an unsteady quantity with fluctuation. The domain from (II) heart sound being stop consonant of arterial valve in the heart sound to the time after 100 ms of the (II) heart sounds in the maximum is recorded for ten heartbeats. Therefore, the measuring time is about 10 s, and this time is much shorter than the measuring time of the electrocardiogram, the 24 h electrocardiogram (conventionally called the Holter ECG), the motion loaded electrocardiogram, or the scintigram in the conventional noninvasive examination of coronary artery disease.

The software supplements the low signal-to-noise ratio in the hardware in the measuring system, and the noise element entering from the commercial power supply is removed by using a filter transformer. There, the measuring system is the sensor fusion measuring system as a whole.

1.8.9 The Details of the Prototype Device

The measuring principle of displacement of vibration utilizing a pulsed laser beam is shown in Fig. 1.49 (currently, combined with a pulse Doppler method).

A class 2 laser diode is used as a laser beam, being harmless to humans, and the near infrared area of 670 nm wavelength is utilized. The displacement is outputed as a displacement magnitude signal (shown as AB and A′B′ in Fig. 1.49), which is based on the principle of triangulation. The minute vibration of the light source is generated in the environment even if the laser beam is used as the signal detecting medium, and this becomes internal noise deriving from the measuring system. Therefore, measurement with monitoring is also carried out to cancel this internal noise. Furthermore, as the minute vibration of the body of the subject also becomes a kind of internal

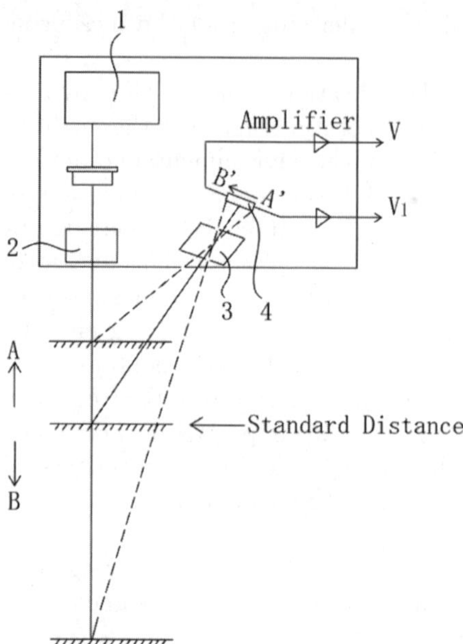

Fig. 1.49. A diagram showing the measuring principle of displacement of vibration utilizing pulsed laser beam. 1: laser diode driving circuit, 2: a projection of the laser beam, 3: reception lens, and 4: light position detecting element

noise deriving from the measuring system, measurement with monitoring is also carried out in order to cancel this internal noise.

The displacement of the body surface due to respiration and the displacement of the body surface due to heartbeats and other movements become mixed in the detected displacement magnitude signal as internal noise of the measured object. In order to narrow down the signal frequency, which is necessary for the distinction of the signal of the vibration deriving from the coronary artery disease, the displacement signal is passed through a low-pass filter and a high-pass filter which have sufficient attenuating gradients.

As described above, the acceleration signal deriving from coronary artery stenosis during diastole becomes the displacement signal by being passed through the body tissue. Therefore, it is necessary to differentiate the displacement signal of the body surface, which is detected by using the laser beam as a signal-detecting medium on the principle of triangulation, in order to convert it into an acceleration signal. And the frequency domain as the frequency of that does not change can distinguish the displacement signal by the differential and integral operation.

The signal from the measured object can be taken out of the time domain as necessary. As the signal of the measured object deriving from coronary artery disease is based on the beating period of the heart, an electrocardiogram is used as the scale of the time domain.

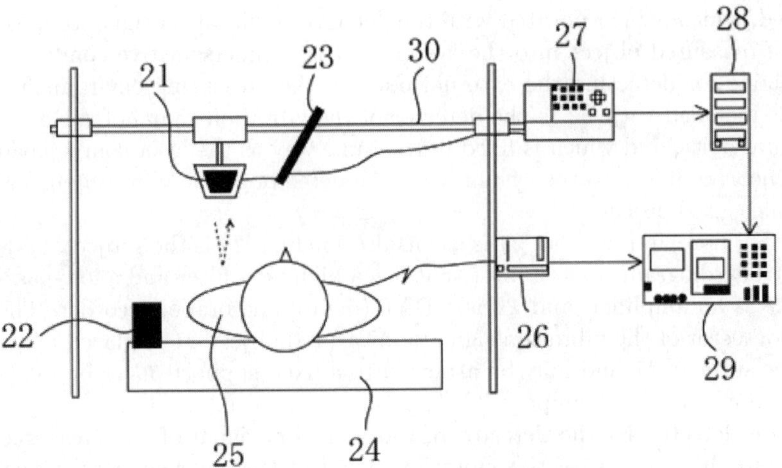

Fig. 1.50. A diagram showing the layout of the diagnostic system for coronary artery disease and other conditions in the present invention

In the electrical processing of the measurement signal, in order to reduce systematic errors, a power supply that sufficiently takes into consideration the internal noise deriving from the measuring apparatus and the noise entering the measuring system and the ultra-low noise and supersensitive amplifiers and others are used, and the connection method and others are done.

The stenotic murmur signal in the measured signal of the outputting object is formed as a whole, and contains the remaining noise signal deriving from the measuring system, which cannot be removed, and the noise signal deriving from the measured object. In order to acquire the signal of the measured object, which is buried in the noise by exceeding the signal-to-noise ratio of the system, it is separated from the other noise by using a signal-processing technique, which follows our earlier research. It has known from our earlier research that the frequency of the noise vibration of the stenotic murmur and its power value are determined by the internal diameter of the stenotic site, the flow velocity, the viscosity of the blood, and other factors. Therefore, conversion into the frequency domain compares the signal of the object. Then, it is displayed as the power value of the generating frequency after conversion into the frequency domain by using the maximum entropy method (MEM) as a means of linear prediction. In this form, it is measured by using the complex multiple sensor system shown in Fig. 1.50 as a basic measuring system. In Fig. 1.50, *21* is the detector of the vibration signal of the measured object, which has one or a number of laser source heads and a vibration detecting sensor of the measured object (complex sensor) with a laser displacement gauge and a three-axis accelerometer, *22* is a vibration-detecting sensor of environmental noise for a bed using a three-axial accelerometer in a detector of vibration signal of environmental noise for a bed, which is incorporated with the detector of vibration signal of measured of measured object into the system, *23* is a supersensitive condenser microphone for detecting the external noise of the measuring environment, which is included with *22* in the detector of the vibration signal of the environmental noise, and which is fixed in the same way as the laser source head and is incorporated into the system with the detector of the vibration signal of the measured object.

Furthermore, in Fig. 1.50, *24* is a consultation bed, *25* is the subject, *26* is an electrocardiogram, *27* is a filter that has a high-pass filter and a low-pass-filter, *28* is an amplifier, and *29* is a DAT (digital audiotape) recorder. The united detector of the vibration signal of measured object *21* is placed apart from the subject *25*, and may be attached to a fixed support *30* or be made to slide along the fixed support *30*. The murmurs and internal and external noises are detected by the detector of the vibration signal of the measured object and the detector of the vibration signal of the environmental noise, and the vibration signal containing the murmur and the internal and external noises is sent to the filter *27*. The filtered vibration signal is amplified by the amplifier *28* and is recorded on the DAT recorder *30*.

1.8.10 Data Acquisition of the Vibratory Signal of the Anterior Chest Wall

It is possible to measure in a lying position (supine) or in a sitting position and others because the measurement is done using a noncontacting technique. It is measured by fixing a projecting laser beam to the left sternal border of the left fourth intercostal space, which corresponds to the front of the heart, or by sliding a number of laser source heads or one laser source head over the left back in a position without lung respiration. The reason for measuring without respiration is to avoid the effect of vesicular sound from the anatomical respiration. In our earlier research, it has been found that the frequency band of the stenotic vibration deriving from the coronary artery stenosis ranges from almost 200 Hz to 1000 Hz. The signal is filtered by the high-pass filter at 200 Hz and by the low-pass filter at 1000 Hz, and it is possible to set the system from 100 Hz to 1000 Hz. According to the superscription, it is possible to acquire the above frequency band with an excellent signal-to-noise ratio. The gain of the amplifier *28* is approximately set from 200 times to 1000 times. The measurement is done for ten heartbeats of a subject; that is, for about 10 s. This measured signal is recorded with the electrocardiogram signal in the DAT 29.

1.8.11 Signal Processing

The acquired signal is converted from analog to digital at 2 kHz as a sampling frequency, and is incorporated into the computer. The maximum entropy method (MEM) is used for analyzing in the frequency domain. The objective

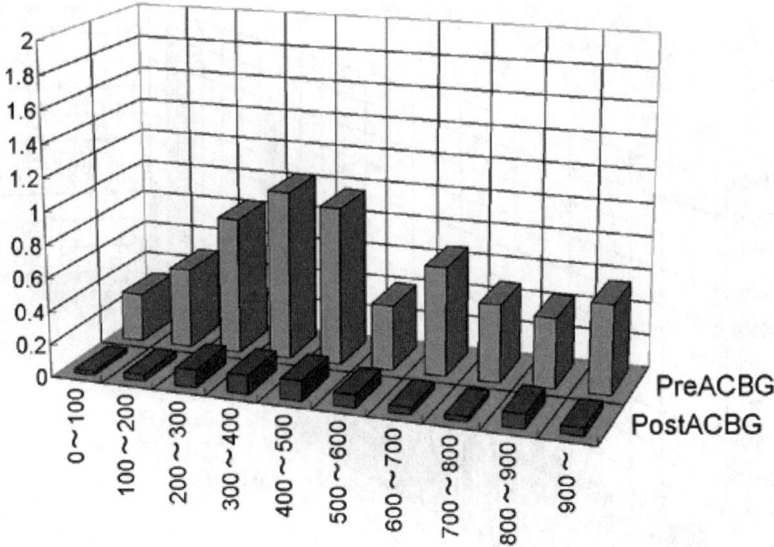

Fig. 1.51. A histogram showing several instances of mean values of the power spectrum strength of ten heartbeats

position of the frequency (spectrum) analysis is set at auxocardia from the R-R interval on the electrocardiogram. Although the spectral power by the MEM is shown as a relative value, it needs to be compared not by a relative value but by an absolute value. Then, the power spectral density is integrated on each 100 Hz frequency band. The reason for the 100 Hz interval is that it seems that the coronary rate of the blood flow in heart diastole is changeable and unsteady.

The histogram shows several instances of the mean value of the power spectrum strength of ten heartbeats, as shown in Fig. 1.51. F, A, B, and others in Fig. 1.51 are the testee, and after B is a testee after a coronary artery bypass operation. In Fig. 1.51, the abscissa shows the frequency and the ordinate shows the power spectrum. The rows of the amplitude strength correspond to the number of stenotic sites. By comparing the histogram with the findings of the coronary angiography that is done independently, it is found that a vibration with a strong power spectrum is the vibration of the stenotic murmur.

A histogram showing a preoperative and a postoperative instance of the mean value of the power spectrum strength of ten heartbeats of the coronary artery bypass operative patient is shown in Fig. 1.52. B shows the preoperative stage and after B shows the postoperative stage in Fig. 1.52. It is known that the noise at the stenotic site disappears due to the superfluous vibration, since the stenosis is canceled by formation of the bypass and the blood flow increases.

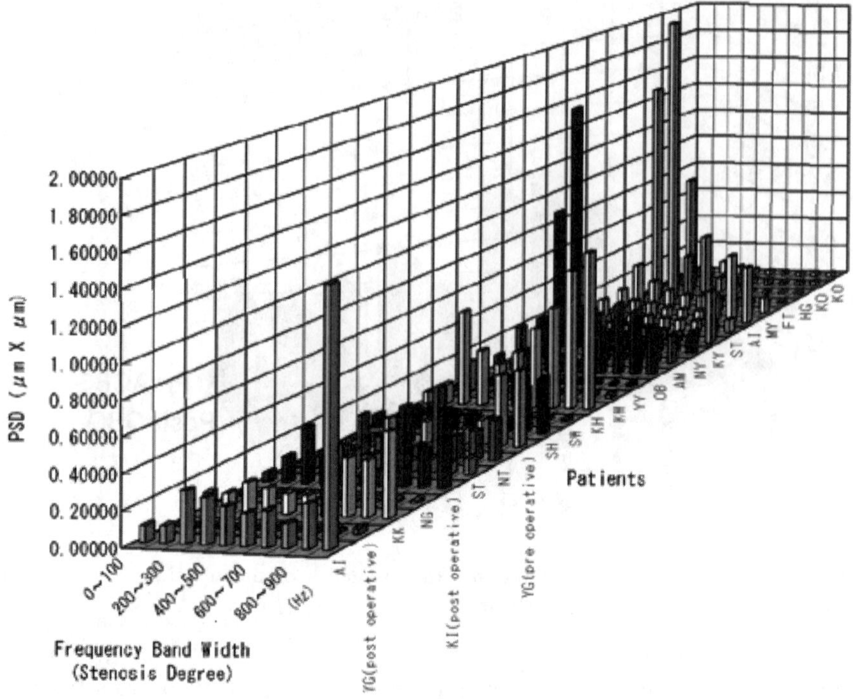

Fig. 1.52. A histogram showing a preoperative and postoperative instance of the mean value of the power spectrum strength of ten heartbeats of a coronary artery bypass patient

The dynamic flow imaging of coronary blood flows can be analyzed by the new technology using calculations of time serial signals, power spectral densities, and trajectories. Detailed data showed that the turbulence flows due to stenosis and calcification of the coronary artery, restenosis, or remnants of stenosis after PTCA and little improvement of coronary artery flow following CABG had produced chaos in coronary blood flow. The diastolic noises detected on the anterior chest wall were analyzed to determine whether or not multi-periodic or chaotic signals had been contained. In the normal coronary artery, the flow dynamics showed quasi-periodic cycles. Turbulent flows produced chaos in the coronary blood flow in the atherosclerotic coronary artery. The trajectory is a useful tool as an indicator of the coronary flow dynamics.

1.8.12 The Future of the New Laser Phonocardiography Technology

The present invention is aimed at providing a diagnostic system for coronary artery disease and other conditions, and a diagnostic method using the above diagnostic system, which can detect vibration signals of murmurs deriving from stenosis of the coronary artery in their early stages, that it is possible to diagnose abnormal conditions, and prevent and treat heart disease. What has been shown is a diagnostic system for coronary artery disease comprising a detector of the vibration signal of environmental noise, and vibration signal detected by the detector of vibration signal of subject and the detector of the detector of vibration signal of environmental noise is filtered to cancel internal noise and external noise, and the filtered vibration signal is amplified and recorded. The above detector of the vibration signal of the subject has one or a number of laser source heads and a vibration detecting sensor with a laser displacement gauge and a three-axis accelerometer, and the detector of the vibration signal of the environmental noise has a three-axis accelerometer and a supersensitive microphone.

The new technique is a potentially powerful technique, but it is still in need of further technical refinement before it is introduced for routine clinical use.

A visual presentation of comparativ signal spectra can be provided on the CRT in the form of a side-by-side window or split-screen presentation of the earlier spectra and the later spectra, or an overlay presentation.

References

1. W. Denk, J.H. Strickler, W.W. Webb: Science **248**, 73 (1990)
2. W. Denk, D.W. Piston, W.W. Webb: in: *Handbook of Biological Confocal Microscopy*, ed. by J.B. Pawley (Plenum Press, New York, 1995)
3. C.J.R. Sheppard, M. Gu: Optik **86**, 104 (1990)
4. J.C. Diels, W. Rudolph: in: *Ultrashort Laser Pulse Phenomena* (Academic Press, San Diego, 1996)
5. G.Q. Xiao, G.S. Kino: Proc. SPIE Int. Soc. Opt. Eng. **809**, 107 (1987)
6. A.H. Buist et al.: J. Microsc. **192**, 217 (1998)
7. K. Fujita et al.: Proc. SPIE **3740**, 390 (1999)
8. K. Fujita et al.: J. Microsc. **194**, 528 (1999)
9. O. Nakamura: Optik **93**, 39 (1993)
10. T. Takamatsu, W.G. Wier: FASEB J. **4**, 1519 (1990)
11. T. Takamatsu et al.: Cell Struct. Funct. **16**, 341 (1991)
12. K. Fujita et al.: Opt. Comm. **174**, 7 (2000)
13. K. Fujita et al.: Proc. SPIE **3921**, 305 (2000)
14. T. Hama et al.: Cell Signal. **10**, 331 (1998)
15. T. Kaneko et al.: Circ. Res. **86**, 1093 (2000)
16. E.H.K. Stelzer and R.W. Wijnaendts van Resandt, Chapter 7, "Optical cell splicing with the confocal fluorescence microscope: Microtomoscopy", in: *Confocal Microscopy*, ed. by T. Wilson (Academic Press, London, 1990)
17. W. Denk, J.H. Strickler, W.W. Webb: Science **248**, 73 (1990); W. Denk, K.R. Delaney, A. Gelperin, D. Kleinfeld, B.W. Strowbridge, D.W. Tank, R. Yuste: J. Neurosci. Meth. **54**, 151 (1994)
18. G.J. Puppels, F.F.M. de Mul, C. Otto, J. Greve, M. Robert-Nicoud, D.J. Arndt-Jovin, T.M. Jovin: Nature **347**, 301 (1990)

19. Y.R. Shen: *The Principles of Nonlinear Optics* (John Wiley and Sons, New York, 1984); M.D. Levenson: *Introduction to Nonlinear Laser Spectroscopy* (Academic Press, Orlando, 1988)
20. M.D. Duncan, J. Reintjes, T.J. Manuccia: Opt. Lett. **7**, 350 (1982); M.D. Duncan, J. Reintjes, T.J. Manuccia: Opt. Eng. **24**, 352 (1984); M.D. Duncan: Opt. Comm. **50**, 307 (1985)
21. M. Hashimoto, T. Araki: J. Opt. Soc. Am. A, **18**, 771 (2001); M. Hashimoto, T. Araki: SPIE Vol. **3749**, 496 (1999)
22. A. Zumbusch, G.R. Holtom, X.S. Xie: Phys. Rev. Lett. **82**, 4142 (1999)
23. M. Antina: Science **284**, 1445 (1999)
24. H.J. Tiziani, M.H. Uhde: Appl. Opt. **33**, 567 (1994); A. Ichihara, T. Tanaami, K. Isozaki, Y. Sugiyama, Y. Kosugi, K. Kimuriya, M. Abe, I. Umemura: Bioimages **4**, 52 (1996); J. Bewersdorf, R. Pick, S.W. Hell: Opt. Lett. **23**, 655 (1998); K. Fujita, O. Nakamura, T. Kaneko, M. Oyamada, T. Takamatsu, and S. Kawata: Opt. Comm. **174**, 7 (2000)
25. M. Hashimoto, K. Inoue, T. Araki, K. Fujita, O. Nakamura, S. Kawata. "Multifocus CARS microscopy", in: Focus on Microscopy 2001, Amsterdam, the Netherlands, April 1–4, 2001
26. M. Hashimoto, T. Araki, S. Kawata: Opt. Lett., **25**, 1768 (2000); M. Hashimoto, T. Araki, S. Kawata: "Molecular vibrational imaging in the fingerprint region by CARS microscopy", in: *Focus on Microscopy 2000*, Shirahama, Japan, April 9–13, 2000
27. K. Murakawa et al.: IEEE Eng. Med. Biol. **18**, 70 (1999)
28. K. Goto et al.: OEEE Trams- Bop,ed- Emg- **48**, 830 (2001)
29. N. Kudo, K. Shimizu, G. Matsumoto: Fron. Med. Biol. Eng. **1**, 19 (1988)
30. Y. Mitamura, E. Okamoto, T. Mikami: ASAIO Trans. **36**, M278 (1990)
31. K. Goto et al.: Rev. Sci. Instrum. **72**, 3079 (2001)
32. R.R. Anderson, J.A. Parrish: J. Invest. Dermatol. **77**, 13 (1981)
33. M.D. Feit et al.: Appl. Surf. Sci. **127–129**, 869 (1998)
34. O. Nakamura: Microsc. Res. Technol. **47**, 165 (1999)
35. K. Konig: J. Microsc. **200**(2), 83 (2000)
36. N.I. Smith et al.: Appl. Phys. Lett. **78**, 999 (2001)
37. A. Ashkin, J.M. Dziedzic, J.E. Bjorkholm, S. Chu: Opt. Lett. **11**, 288 (1986)
38. A. Ashkin, J. Dziedzic, T. Yamane: Nature **330**, 769 (1987)
39. A. Ashkin: Science **210**, 1081 (1980)
40. T. Sugiura, T. Okada, Y. Inouye, O. Nakamura, S. Kawata: Opt. Lett. **22**, 1663 (1997)
41. A. Ashkin: Phys. Rev. Lett. **24**, 156 (1970)
42. K. Sasaki, M. Koshioka, H. Misawa, N. Kitamura, H. Masuhara: Appl. Phys. Lett. **60**, 807 (1992)
43. K. Svoboda, S.M. Block: Opt. Lett. **19**, 930 (1994)
44. S. Sato, Y. Harada, Y. Waseda: Opt. Lett. **19**, 1807 (1994)
45. H. Furukawa, I. Yamaguchi: Opt. Lett. **23**, 216 (1998)
46. P.C. Ke, M. Gu: Appl. Opt. **38**, 160 (1999)
47. T. Ohta, T. Sugiura, S. Kawata: Appl. Phys. Lett. **80**, 3448 (2002)
48. K. Kitamura, M. Tokunaga, I.A. Hikikoshi, T. Yanagida: Nature **397**, 129 (1999)
49. Y. Hong, M.D. Wang, K. Svoboda, R. Landick, S.M. Block, J. Gelles: Science **270**, 1653 (1995)

50. L.P. Ghislain, W.W. Webb: Opt. Lett. **18**, 1678 (1993)
51. S. Kawata, Y. Inouye, T. Sugiura: Jpn. J. Appl. Phys. **33**, L1725 (1994)
52. K. Sasaki, Z. Shi, R. Kopelman, H. Masuhara: Chem. Lett. **1996**, 141 (1996)
53. T. Sugiura, T. Okada: Proc. SPIE **3260**, 4 (1998)
54. T. Sugiura, S. Kawata, T. Okada: J. Microsc. **194**, 291 (1999)
55. G.A. Valaskovic, M. Holton, G.H. Morrison: Ultramicroscopy **57**, 212 (1995)
56. M.R.P. Moers, W.H.J. Kalle, A.G.T. Ruitter, J.C.A.G. Wiegant, A.K. Raap, J. Greve, B.G. Grooth, N.F. Van Hulst: J. Microsc. **182** 40 (1996)
57. I. Fujimasa: IEEE Eng. Med. Biol. **17**, 34 (1998)
58. I. Fujimasa: "Physiological function imaging", in: *Thermography*, ed. by I. Fujimasa (Syujyun-sha, Tokyo, 1988)
59. I. Fujimasa: "Pathophysiological understanding of thermograms. Physiological function imaging", in: *Thermography*, ed. by I. Fujimasa (Syujyun-sha, Tokyo, 1988)
60. P.L. LeRoy et al.: "Thermography as a diagnostic aid in the management of chronic pain", in: *Evaluation and Treatment of Chronic Pain* (Urban & Schwarzenberg, Baltimore, 1985)
61. Y. Miki, I. Fujimasa: Biomed. Thermol. **9**, 281 (1989)
62. G.I. Taylor, J.H. Palmer: Brit. J. Plastic Surgery **40**, 113 (1987)
63. I. Fujimasa et al.: Biomed. Thermol. **8**, 34 (1988)
64. I. Fujimasa et al.: Thermol. **1**, 221 (1996)
65. K. Mabuchi et al.: Biomed. Thermol. **10**, 286 (1990)
66. S. Uematsu: Thermol. **1**, 4 (1985)
67. M. Iwatani: Jap. J. Med. Electron. Biol. Eng. **20**, 249–255 (1982)
68. I. Fujimasa: Jap. J. Med. Electron. Biol. Eng. **23**, 503(1985)
69. I. Fujimasa et al.: BME **2**, 179 (1988)
70. I. Fujimasa, T. Chinzei, I. Saito: IEEE Eng. Med. Biol. **19**, 71 (2000)
71. I. Fujimasa and K. Motomura: *Proceedeings of the 8th Asian and Oceanian Cogress of Neurology Satellite Symposium*, pp. 102–105 (Excerpta Medica, Tokyo, 1992)
72. Y. Hyakuna: Biomed. Thermogr. **5**, 7 (1985)
73. H. Miyake: Biomed. Thermogr. **6**, 182 (1986)
74. W.B. Hobbins: Biomed. Thermol. **15**, 229 (1995)
75. I. Fujimasa et al.: in: *Proceedings of the 20th Annual International Conference of IEEE EMBS* **20**, 950 (1998)
76. J.J. Fredberg: Bull. Math. Biol. **36**, 143 (1974)
77. A.M.A.S. Khalifa, D.P. Giddens: J. Biomech. **14**, 279 (1981)
78. J.-Z. Wang, B. Tie, W. Welkowitz, J.L. Semmlow, J.B. Kostis: IEEE BME **37**, 1087 (1990)
79. H. Matsumoto, H. Ozaki, T. Kisi, K. Iwamoto, M. Tiba, K. Mabuti, K. Yagyuu: Resp. Circ. **47**, 289 (1999)
80. H. Matsumoto, K. Mabuti, K. Yagyuu, Y. Matsumoto: Resp. Circ. **47**, 1265 (1999)
81. H. Matsumoto, K. Yagyuu, K. Mabuti, Y. Matsumoto: Circulation **100**, I–648 (1999)
82. J. Semmlow, W. Welkowitz, J. Kostis, J.W. Mackenzie: IEEE BME **30**, 136 (1983)
83. W. Dock, S. Zoneraich: Amer. J. Med. **42**, 617 (1967)
84. P.G. Lund-Larsen: Acta Med. Scand. **182**, 433 (1967)
85. R.E. Fearson, L.S. Cohen et al.: Amer. Heart J. **76**, 252 (1968)
86. J.R. Burg, K.A. Weaver, T. Russel II, D.G. Kassebaum: Chest **63**, 440 (1973)

2 Imaging of Tissue/Organs with Ultrasound

M. Hori, T. Masuyama, K. Baba, O. Ohshiro, K. Ishihara, and H. Kondo

2.1 Ultrasonic Biological Measurement (Ultrasonography)

2.1.1 The Principle of Ultrasonography

By definition, ultrasound is sound with a frequency greater than 20 000 cycles per second; that is, the sound is above the audible range. The principal advantages of high-frequency sound or ultrasound as a medical diagnostic tool are: 1) ultrasound can be directed in a beam, 2) it obeys the laws of reflection and refraction, and 3) it is reflected by objects of small size. The principal disadvantage of ultrasound is that it propagates poorly through a gaseous medium. When discussing any type of sound, one must understand what a cycle, wavelength, velocity, and frequencies are. A sound wave is a series of compressions and rarefactions, and these changes are frequently depicted as a sine wave, with the peak of the hill representing the pressure maximum and the nadir of the valley the pressure minimum. The combination of one compression and one rarefaction represents one cycle, and the distance between the peak compression of one cycle to the next is the wavelength. The velocity represents the speed at which sound waves travel through a particular medium. The frequency is the number of cycles in a given time. Thus, the velocity is equal to the frequency times the wavelength. The velocity at which sound travels through a medium depends on the density and elastic properties of the medium. In other words, sound travels faster through a dense medium than through a less dense substance. Velocity also depends on temperature. The velocity of sound is fairly constant for human soft tissue, approximately 1540 m/s (meters per second).

When sound travels through a medium, it shows different forms of behavior depending on the acoustic impedance of the medium. By definition, the acoustic impedance is the density of the medium times the velocity with which sound travels through that medium. As a sound wave travels through a homogeneous medium, it essentially continues in a straight line. When the beam reaches an interface between two media with different acoustic impedances, it undergoes reflection and refraction. The amount of sound that is reflected depends on the degree of difference between the two media; i.e., the greater the acoustic mismatch, the greater the amount of sound reflected.

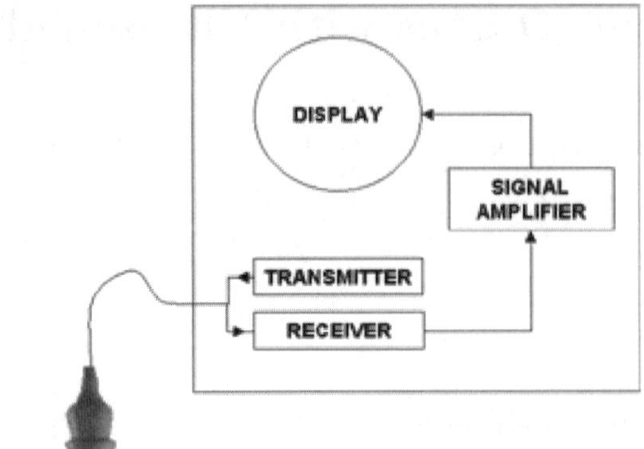

Fig. 2.1. The block diagram of the components of an ultrasonic echograph

For example, more sound is reflected from an interface between gaseous and solid media than from a liquid–solid interface. Whether the ultrasound is reflected by an interface also depends on the relative sizes of the mismatched media and the wavelength. The total thickness presented to the ultrasonic beam must be at least one fourth the wavelength of the sound. Thus, ultrasound with a higher frequency or a shorter wavelength can reflect sound from smaller objects. The loss of ultrasound as it traverses a medium is known as attenuation, which is a combination of absorption and scattering.

The instrument used to create an image using ultrasound is known as an echograph. Figure 2.1 shows a block diagram of an echograph. The transducer both sends and receives the ultrasound, and the transmitter regulates the sending of the ultrasound by way of a timer that controls the duration and frequency of the ultrasonic pulse. The transducer converts the returning echoes to electrical impulses, which in turn go to the receiver and the signal amplifier. The returning echoes or impulses are processed so that they can be displayed. The impulse is converted from a spike to a dot and, within limits, the taller the echo, the brighter the dot. This presentation is known as B-mode, the B standing for brightness. This type of display was the backbone of ultrasonography for many years. If the interface from which the echo is derived is constantly moving, then the echo position changes constantly with reference to the transducer. The echo signal moves back and forth on the face of the display. If the intensity-modulated dots sweep from left to right, M-mode tracing, where M stands for motion, is obtainable. If the spatial orientation of the transducer is tracked electronically, one could obtain a spatially oriented M-mode examination (Fig. 2.2). In this figure, when the ultrasonic transducer is close to the top of the beaker, it traverses the circular object at the point at which the two echoes are relatively close together (A). When the transducer is moved downward, the two B-mode echoes are farther apart and in the center of the display (B), and then the two echoes are closer

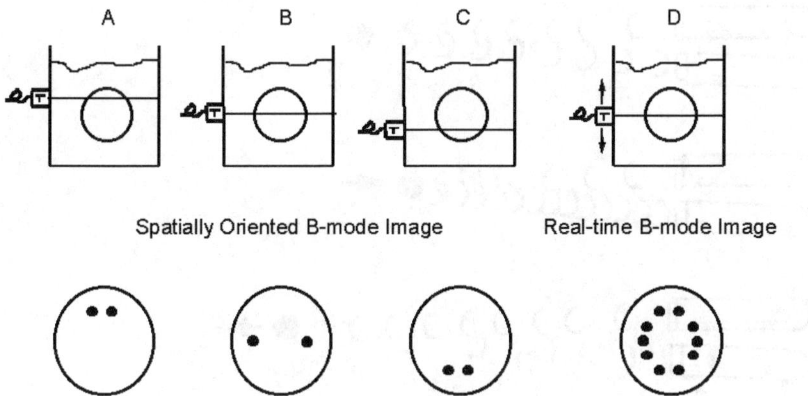

Fig. 2.2. A spatially oriented M-mode examination for two-dimensional imaging

together (*C*). In combination, these dots reveal the shape and size of the object being examined (*D*).

2.1.2 The Doppler Technique

The Doppler technique is playing an increasingly important role in ultrasound examination, particularly of the heart. There have been significant advances in Doppler instrumentation and in our understanding of how this information can be clinically useful. The Doppler examination is based on the Doppler effect first described by Christian Johann Doppler in 1842. If a source of sound is stationary, then the wavelength and frequency of the sound emanating from that source are constant. If, however, the source of the sound is moving toward one's ear, then the wavelength is decreasing and the frequency is increasing. If the source of sound moves away from the ear, then the wavelength is increasing and the frequency decreasing. Figure 2.3 demonstrates how one can use reflected sound to determine the motion of a target that reflects the ultrasound. Briefly speaking, the Doppler shift, or frequency, represents the difference between the received and the transmitted frequencies. The relation between the reflected frequency (f_r) and transmitted frequency (f_t) differs depending on the direction of the movement of the target. The faster the movement, the larger the difference is between f_r and f_t.

There are at least three kinds of Doppler ultrasound: continuous-wave Doppler, pulsed-wave Doppler, and color Doppler techniques. Figure 2.3 illustrates the principle of continuous-wave Doppler ultrasound. One uses two transducers, a transmitter and a receiver, or a transducer with both transmitter and receiver elements. Because the ultrasonic beam is constant, all the moving targets within the beam produce Doppler signals. There is no way knowing where the individual target might be with relation to the transducer. In addition, one cannot determine how many moving targets there are

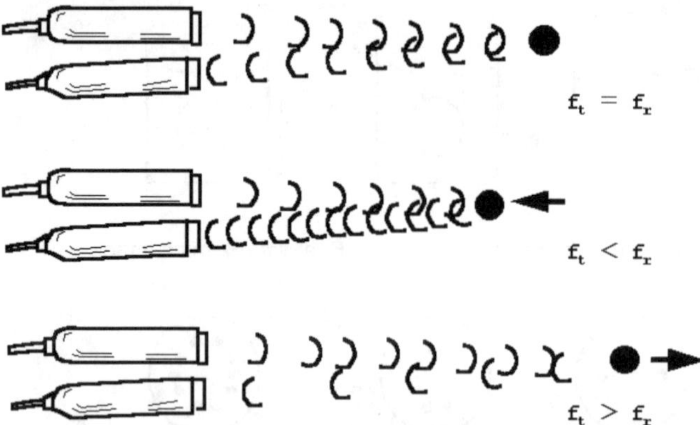

Fig. 2.3. Drawings demonstrating the Doppler effect using reflected sound from a target

within the beam. Thus, to overcome these problems, the pulsed-wave Doppler technique has been invented. By sending a short burst of ultrasound, the frequency of that burst is distorted if the target from which it is reflected is moving. Similarly to the continuous-wave Doppler technique, when the target is moving, the frequency of the received burst of ultrasound alters, reflecting the direction of the movement and the velocity. One may determine velocity just as with the continuous wave approach. The major disadvantage of pulsed-wave Doppler is that the velocity that one can measure is limited. The pulsed system inherently has a pulse repetition frequency (PRF). The PRF determines how high a Doppler frequency the pulse system can detect (the Nyquist limit). A combination of Doppler and two-dimensional echo was developed for flow mapping (color Doppler). This approach utilizes multiple Doppler gates, which register the Doppler signals rapidly. Because the flow patterns are dramatically displayed using the flow mapping technique, this approach is currently the most frequently used Doppler technique among the three techniques in ultrasonography laboratory.

2.1.3 Recent Advances in Ultrasound Imaging

Ultrasound imaging has seen dramatic improvements in image quality over recent years with the utilization of harmonic frequencies in the imaging of both tissue and contrast agents. Although clinical validation of this new technology is ongoing, it is worthwhile to consider some of the physical principles underlying its use. When sound traverses through the body, it is composed of a group of frequencies that define its spectral content. If the ultrasonic wave consists of a single frequency (the fundamental frequency), then it forms a sine wave passing through the tissue. Harmonic frequencies are those that occur at a multiple of the fundamental frequency, with the second harmonic referring to twice the fundamental frequency. Ultrasonic energy

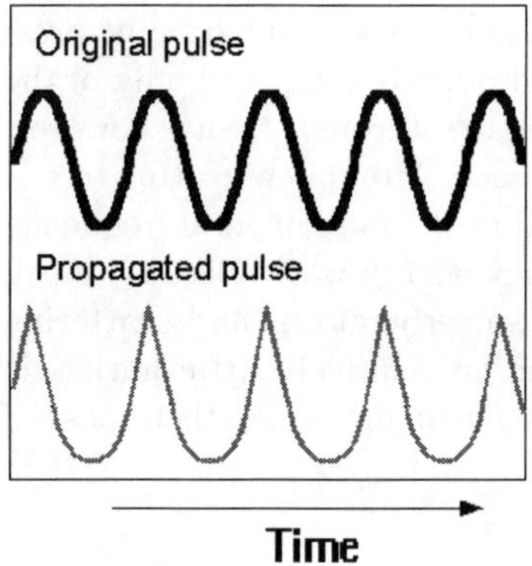

Fig. 2.4. Nonlinear propagation of ultrasound through water. Features at peaks are harmonic components and travel faster than rounded nadirs

was originally considered to propagate by a linear process, meaning that new frequencies cannot be created. Attenuation may differentially affect some frequencies more than others. Therefore, the spectral content may change with propagation, but frequencies that were absent in the original wave cannot subsequently appear. However, it has become apparent recently that there are nonlinear acoustic effects active in ultrasonography. The nonlinear effects of the interaction of ultrasound with matter (both on propagation and reflection) may create frequencies that were not present in the incident beam. New harmonics are considered to be generated because water or tissue is not a completely incompressible medium. Thus at the peaks of pressure, water is slightly denser than it is at the nadirs of pressure. This change in density also affects how sound travels through water. Sound travels just a little faster at the peaks of waves than in the nadirs. This leads to a very slight change in the shape of the wave as it propagates. At each instant in wave propagation, an infinitesimal amount of harmonics is generated. Such a shape change is shown in Fig. 2.4, which reveals that sharp features in the waveforms require harmonic frequencies beyond the fundamental. As the fundamental frequency propagates through a compressible medium, nonlinear interactions – i.e., the generation of harmonic frequencies – occur. The range of frequencies that accumulate along the distance of propagation change the wave shape to include sharp features. These sharp features at peaks are harmonic components and travel faster than rounded nadirs. Anyone who has ever been to a beach has witnessed a shape change with wave propagation. As the sinusoidal swells of the ocean move toward the shore, the swells develop into sharp, breaking waves. The sharp peaks of the waves represent the higher harmonic frequencies, whereas the swells approaching the shoreline represent the fundamental frequencies.

Two aspects of harmonic generation are critical to their utility in improving image quality: 1) the growth with distance of propagation; and 2) the nonlinear relation between the source pressure and the second harmonic pressure. At the transducer surface, the ultrasound pulse is composed only of the fundamental frequencies. As soon as it propagates through tissue, however, energy builds up at the second harmonic frequency. After a few centimeters of distance, enough energy has been converted from the fundamental frequency to yield a significant second harmonic frequency energy wave. Because much of the artifact in an echo image is related to reverberations and scattering at or near the chest wall, these artifacts contain relatively little harmonic frequency energy. If imaging is confined to the harmonic range, then most of the near-field artifacts are eliminated.

2.1.4 The Ultrasonic Characterization of Myocardial Tissue

Abundant indirect information is obtainable by analysis of ultrasonographically (echocardiographically) assessed abnormal regions of the myocardium; however, direct, noninvasive identification of the tissue composition of cardiac structures is still one of the important but unmet goals of cardiac diagnosis. Ultrasonic cardiac tissue characterization may be defined as the identification and characterization of abnormalities in the physical or physiological state of the myocardium based on analyzing interactions between ultrasound and tissue [1]. The rationale for this field of study is that sufficient information is available in the ultrasound signal passing through or returning from myocardial tissue to identify the tissue as normal or abnormal, and to indicate the nature of the abnormality. Progress in quantitative myocardial tissue characterization to date has been possible because of significant advances in electrical engineering and physics, applied to improving and modifying ultrasonic measurement and imaging systems. Quite a few groups have approached the problem of ultrasonic myocardial characterization with different instrumentation techniques and analysis systems, and one important direction of research has relied on the analysis of unprocessed radio-frequency signals returning from the myocardium. Initially, the degree of ultrasonic attenuation was quantified in transmission studies, but there are more reflection studies now in which the extent of ultrasonic backscatter is quantified. This is largely attributed to the recent advent of a real-time backscatter imaging system.

Biological determinants of tissue acoustic properties have not been fully understood. As mentioned earlier, when sound waves traverse through a material, they are reflected or scattered where local regions of different acoustic impedance are encountered. Thus, tissue elements responsible for scattering represent local regions of acoustic impedance mismatch. Early work in this field identified collagen as a primary determinant of both scattering and attenuation of myocardial tissue [2]. The geometric attributes of myocardial scatterers have been investigated by a number of groups, who have postulated that myocardial scatterers are comparable in size to cardiac myocytes [3,4]. Myocardial acoustic properties are also influenced by the orientation of ventricular muscle fibers. Mottley and Miller have demonstrated that the

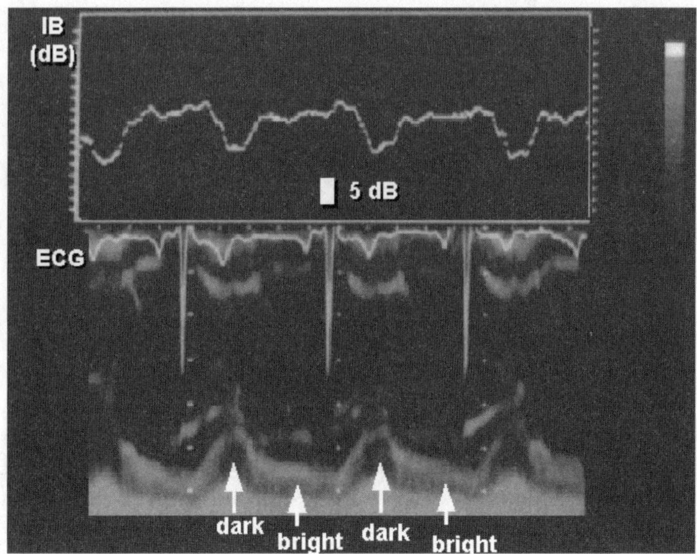

Fig. 2.5. Cardiac cycle-dependent variation in ultrasonic integrated backscatter from the myocardium of the left ventricular posterior wall in a human subject

magnitude of both ultrasonic attenuation and backscatter in excised heart tissue depends critically on the angle of insonification [5]. Tissue water content and hematocrit both influence myocardial scattering and attenuation; however, nutritive blood flow per se may not be necessary for the production of cardiac cycle-dependent variation of myocardial backscatter [6]. The dynamic aspect of ultrasonic scattering was first described by Madaras and associates [7]. They observed that backscatter intensity varies throughout the cardiac cycle, with maximal levels at end-diastole and minimal levels at end-systole (Fig. 2.5). This variation is observed in both experimental animals and human subjects. The magnitude of cyclic variation of backscatter is related to intrinsic myocardial contractile performance, but is independent of perfusion with blood [8,9]. Several hypotheses have been advanced to explain the cyclic alteration of myocardial acoustic properties: these include cyclic alterations of myocardial elastic characteristics and alteration of myocardial scatterer geometry. However, no simple explanation yet appears sufficient to account for the entire range of evidence from experimental animals and human subjects.

2.2 Three-Dimensional Ultrasound Imaging of the Fetus

2.2.1 Conventional Ultrasound Imaging of the Fetus

Ultrasonography, or two-dimensional (2D) ultrasound, has been widely used in obstetrics, since fetal assessment by using fetal images is indispensable for modern practice, and ultrasound is the only imaging modality proved to be safe for a fetus. More than one million babies are born in Japan every year and almost all of them undergo ultrasound examination before birth. Two approaches, transabdominal and transvaginal, are available in ultrasound examination for a fetus. The transabdominal approach is easy to perform, and is used for fetuses mainly after 12 weeks of gestation. The transvaginal approach is not easy to perform but resolution of the image near the vagina is high because a high frequency is applied in a transvaginal probe. The transvaginal approach is used for fetuses before 12 weeks of gestation.

A gestational sac (GS) is seen in the uterus by transvaginal ultrasound in the middle of 4 weeks of gestation (Fig. 2.6). This finding usually excludes the possibility of ectopic pregnancy. A fetus can be depicted at the end of 5 weeks of gestation as a tiny circle. Multiple pregnancy is diagnosed at this stage by counting the number of GS's or fetal images. Morphological abnormalities of a fetus cannot be depicted, because it is too small at this stage, but its heart can be seen beating. Fetal biometry is strongly recommended in the first trimester to confirm or determine the gestational age (Fig. 2.7). But an

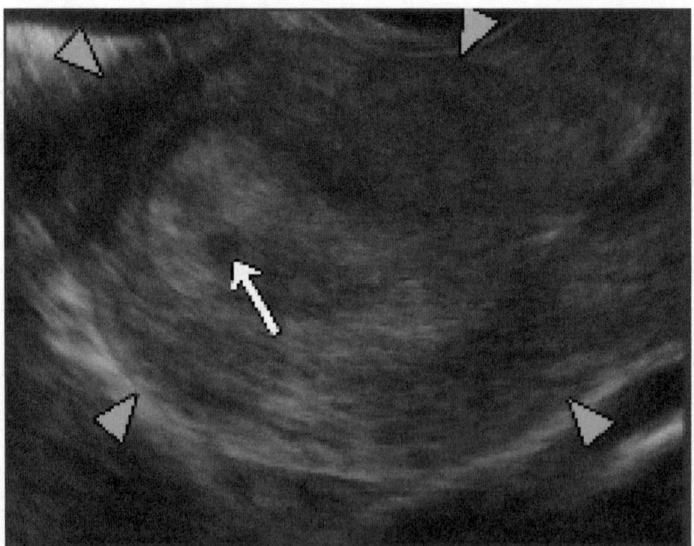

Fig. 2.6. A gestational sac (*arrow*) in the uterus (*arrowheads*) in the middle of the fourth week of gestation

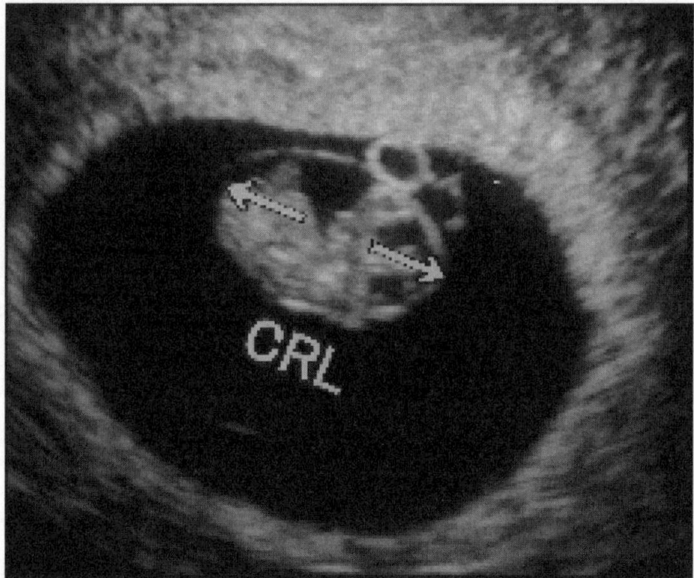

Fig. 2.7. Gestational age is confirmed or determined by fetal biometry in the first trimester. *CRL:* Crown–rump length

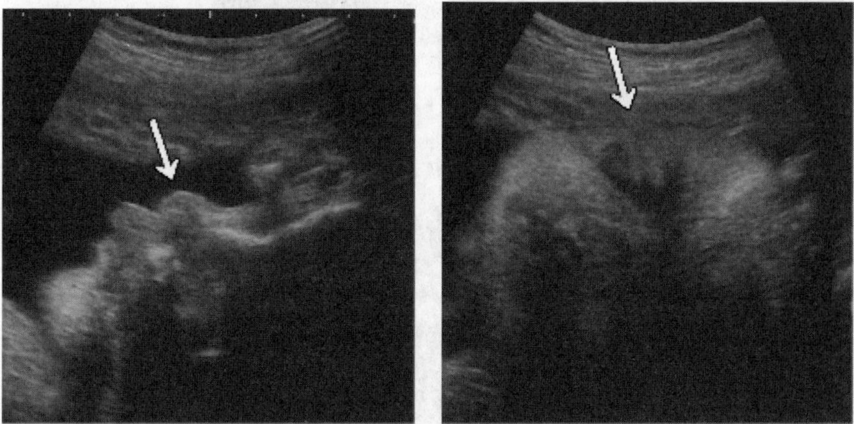

Fig. 2.8. Two-dimensional ultrasound images of a fetal face in the third trimester. *Left:* a longitudinal section showing the nose (*arrow*) and the mouth. *Right:* a coronal section showing a eye (*arrow*)

appropriate section for biometry cannot be always obtained because probe manipulation is limited, especially in the transvaginal approach.

Morphological abnormalities of a fetus can usually be diagnosed by transabdominal 2D ultrasound after 15 weeks of gestation. Figures 2.8, 2.9, and 2.10 show examples of transabdominal 2D ultrasound images. Although a

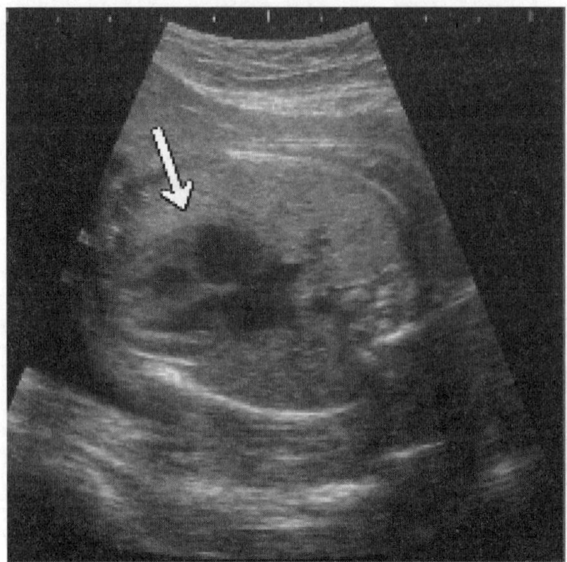

Fig. 2.9. A 2D ultrasound image of a fetal thorax, showing the heart (*arrow*) in the second trimester

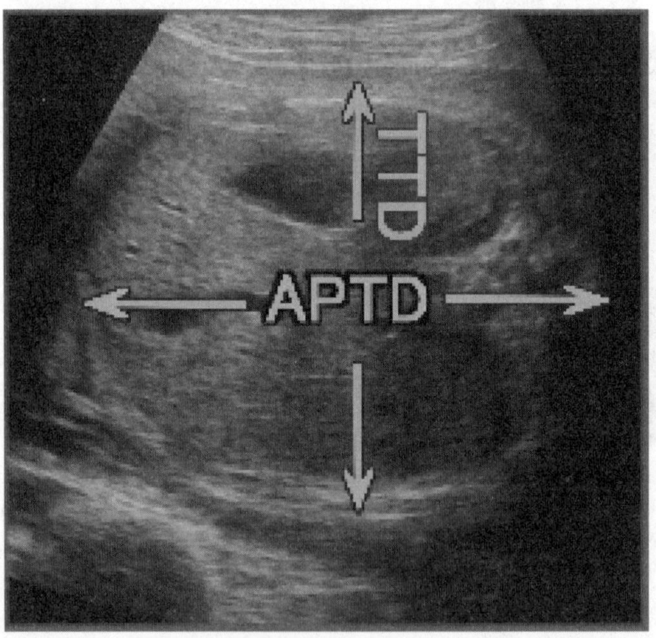

Fig. 2.10. Fetal body weight is estimated by fetal biometry. *TTD:* transverse trunk diameter. *APTD:* antero-posterior trunk diameter

fetus has a three-dimensional (3D) structure, 2D ultrasound can depict only sectional images. This means that 2D ultrasound cannot depict the eyes, nose, and mouth on an image (Fig. 2.8), for example. An examiner has to imagine the 3D structure of the fetus in his or her mind for an accurate diagnosis. However, it is not easy for most examiners to imagine the 3D structure from 2D ultrasound images, and many abnormalities are overlooked in daily practice. Fetal growth is evaluated by estimating fetal body weight. However, the estimation is sometimes inaccurate because it is based on only several lengths, biparietal diameter (BPD), antero-posterior trunk diameter (APTT), transverse trunk diameter (TTD) (Fig. 2.10), and so on.

2.2.2 The Development of Three-Dimensional Ultrasound

Three-dimensional ultrasound for a fetus was introduced in the 1980s [10–12] to overcome the limitations of conventional 2D ultrasound. Many technologies for 3D untrasound were developed in the 1990s. [13,14]. The following 3D ultrasound imaging methods are available at present:

1. Computer processing
   - Section reconstruction
   - Surface rendering
   - Volume rendering
2. Defocusing lens method
3. Real-time ultrasonic beam tracing

Computer Processing

Computer processing is the oldest and the most popular method for 3D ultrasound. Three steps (3D data acquisition, 3D data set construction, and projection/display) are required in this method. Three-dimensional data are usually acquired through movement of a probe. A 3D probe contains a convex probe of conventional 2D ultrasound, and the convex probe swings in the 3D probe to acquire 3D data. A 3D data set is then constructed in the computer. In 3D ultrasound by computer processing, many kinds of images can be generated from the 3D data set. Surface rendering was initially introduced for a 3D fetal image in the 1980s [10–12]. Good 3D images can be obtained in X-ray CT and MRI by surface rendering, because a boundary between different tissues can be clearly identified. But in ultrasonography, the boundary is often obscure and 3D images by surface rendering are often insufficient. Volume rendering usually makes a better 3D image (Fig. 2.11) and is used more often nowadays than surface rendering. Not only a surface image but also a skeletal image (Fig. 2.12) can be obtained by both surface rendering and volume rendering by selecting parameters in their processes. Any arbitrary section can be obtained by section reconstruction from the 3D data set. In practice, three-orthogonal sectional images are usually displayed simultaneously (Fig. 2.13).

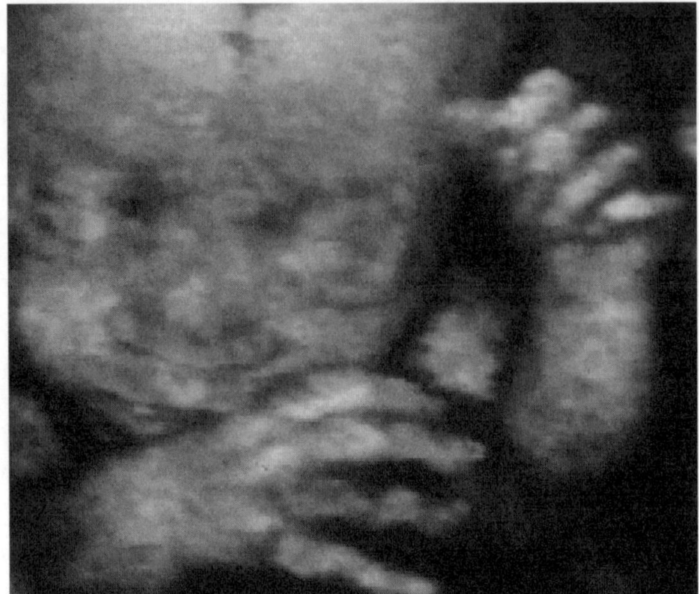

Fig. 2.11. A 3D surface image of a fetus by 3D ultrasound

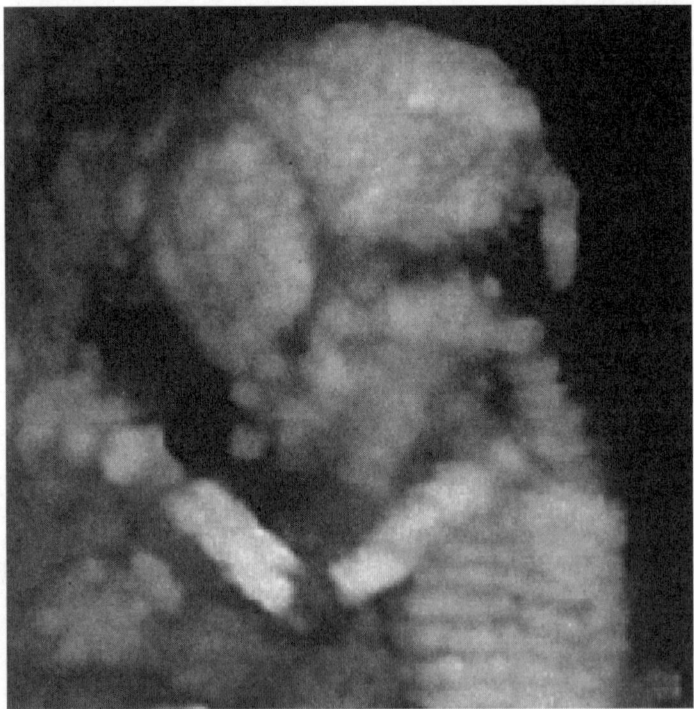

Fig. 2.12. A 3D skeletal image of a fetus by 3D ultrasound

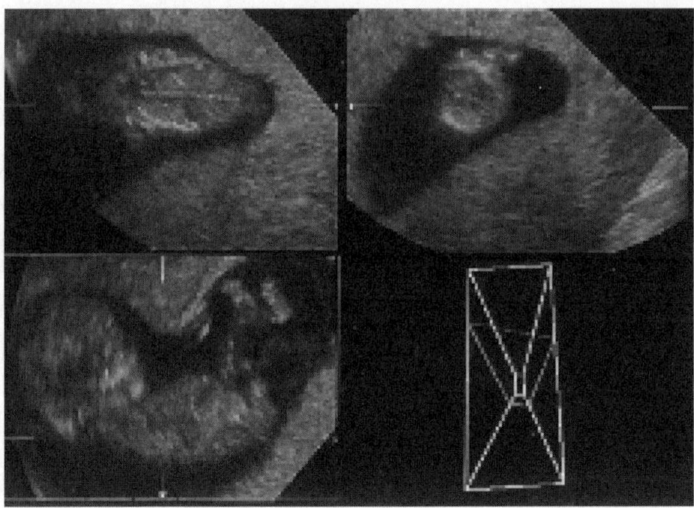

Fig. 2.13. Three orthogonal sectional images of a fetus in the first trimester by 3D ultrasound

Defocusing Lens Method

This method is basically tomography with a thick slice [13,15], but an image just like a transparent 3D image can be obtained. Even a surface image can be obtained under some conditions. The image is recognized three-dimensionally by its observer when the probe is rotating. Real-time observation is possible, but the applications are very limited because the structure surrounding the fetus makes the fetal image obscure.

Real-Time Ultrasonic Beam Tracing

This is the latest method in 3D ultrasound [16,17]. Generation of a 3D image and 3D data acquisition are done simultaneously in this method, and a complete 3D image is displayed on the monitor when a 3D scan is completed. This is the most promising method for real-time 3D ultrasound [18]. By selecting parameters, a 3D fetal skeletal image can be obtained as well as a 3D fetal surface image.

2.2.3 Clinical Applications of Three-Dimensional Ultrasound in Obstetrics

Three-dimensional ultrasound has not been popularized. But its significance in obstetrics has been recognized widely [19].

First Trimester

Three orthogonal sectional displays by 3D ultrasound enable the examiner to locate a gestational sac in the uterus more accurately and to measure crown–rump length (CRL) for gestational age confirmation and nuchal translucency for screening for chromosomal abnormalities more accurately than 2D ultrasound. A 3D image can make complicated fetal abnormalities clear and can demonstrate all fetuses in multiple pregnancy.

Second and Third Trimesters
Morphological abnormalities of a fetus, and their severity, especially on the face, ears, extremities and skeleton, can be depicted clearly on 3D images by transabdominal 3D ultrasound [19,20]. A 3D fetal surface image assures worrying parents that their fetus is normal, and may aid the parents–fetus bonding [21]. Three orthogonal sectional displays are useful for analyzing a complicated structure step by step.

2.2.4 The Advantages and Limitations of Three-Dimensional Ultrasound

Advantages
It is often difficult to imagine the 3D structure of a fetus from 2D images, even for an expert on ultrasonography. Three-dimensional fetal images can make the 3D structure clear and are useful for an accurate diagnosis of fetal abnormalities which may be overlooked by conventional 2D ultrasound. An appropriate section cannot be always obtained by conventional 2D ultrasound, because of the limitation of probe manipulation, especially in the transvaginal approach. In 3D ultrasound, any desirable section for diagnosis or biometry can be obtained by section reconstruction. This results in accurate diagnosis and shortens examination time.

Reexamination is possible without the presence of the patient by using the 3D data set. This means that reexamination or tele-examination by an expert is possible by just sending the 3D data set via the Internet, for example.

Volume measurement is another prominent function of 3D ultrasound. The difference between estimated fetal body weight by 2D ultrasound and actual weight is sometimes not negligible in critical cases, such as a fetus with a severe intrauterine growth restriction or a fetus in preterm labor. Volume measurement by 3D ultrasound is expected to provide more accurate fetal body weight estimation. Volume measurement by 3D ultrasound is also applicable to fetal organs such as the lungs and liver.

Limitations
It takes some time to scan a 3D space. A 3D ultrasound scanner generates a 3D image on the assumption that the object (the fetus) does not move during the scan. When the fetus moves during the scan, motion artifacts will appear on a 3D image. Even maternal aortic pulsation sometimes causes pulsatile movements in the uterus and motion artifacts on a 3D fetal image. Although the fetal heart is an important organ for prenatal diagnosis, it is not a suitable object for 3D ultrasound because of its beating motion.

Three-dimensional fetal surface images are very attractive. But they are not always obtained due to lack of amniotic fluid around the fetus or an inadequate fetal position.

There are some artifacts in 2D ultrasound [22]. Multiple reverberation causes a band-like artifact image under the abdominal wall. The side lobes of an ultrasonic beam causes artifact images from one side of a strong reflector to the other. An acoustic shadow appears under a strong reflector. These artifacts also exist in 3D ultrasound and sometimes hide a fetal image or cause a false deformity on a 3D fetal image.

2.2.5 The Future Development of Three-Dimensional Ultrasound

Three-dimensional ultrasound is still immature and has to be developed much more. Two main developments are expected in future 3D ultrasound. One is real time and the other is tissue characterization.

Real-time 3D ultrasound will resolve the motion artifact problem, one of the major problems in 3D ultrasound. Figure 2.14 shows 3D fetal images obtained at a frame rate of 3.2 frame/s by a specially designed high-speed 3D ultrasound scanner with a real-time ultrasonic beam tracing method. The fetus opened its mouth for a split second in conjunction with its finger bending. Fetal movement or fetal behavior can be observed in this way. Fetal behavior is expected to be used as the basis of a method for evaluating the development of the motor and nervous system of the fetus. It is necessary to

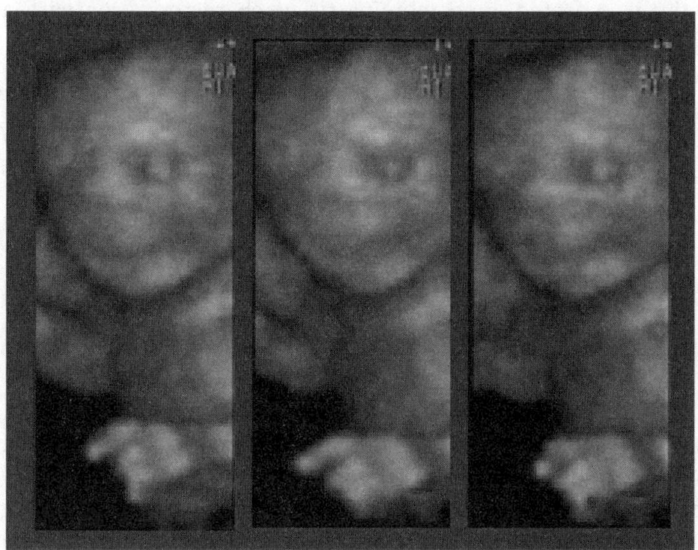

Fig. 2.14. 3D fetal images obtained at the frame rate of 3.2 fps (*left* to *right*) by a specially designed high-speed 3D ultrasound scanner. Note the mouth opening (*center*) and an index finger bending (*right*)

increase the frame rate up to 15 fps or more to observe rapid movement of the fetus in real time. However, an ultrasonic speed does have some limitations. Ultrasound travels at a speed of 1540 m/s (mean) in soft tissue. This speed limits the number of ultrasonic beams available per second. When a diagnostic range is 15 cm, about 5000 ultrasonic beams per second are available at the maximum. This means that only 18×18 pixels can be used for each 3D image when the frame rate is 15 fps. Under these conditions, the resolution of the image will be extremely low or the field of view will be extremely narrow if the ultrasonic beam density is kept high enough. New technologies are needed to realize real-time 3D ultrasound with high resolution and a wide field of view. One of the promising technologies is parallel receiving processing, in which a broad ultrasonic pulse is transmitted and ultrasonic echoes are received as if many ultrasonic beams had come back from different directions. A 2D array probe, in which transducers are arrayed two-dimensionally, can scan a 3D space fast enough if parallel receiving technology is applied. Real-time 3D ultrasound will be used not only for observation but also in treatment, such as fetal surgery, as a noninvasive eye of the operator.

Tissue characterization by ultrasound will differentiate two kinds of tissue and identify the border clearly. A 3D fetal surface image will be possible by tissue characterization even when the fetus touches the uterine wall or the placenta. Tissue characterization will realize automated fetal organ measurement, which will be a new diagnostic method for some fetal conditions such as hypoplastic lung, which is one of the most critical conditions for a newborn baby and can be hardly diagnosed before birth. Even automated diagnosis of fetal abnormalities is expected, by combining tissue characterization with a computerized recognition technology.

Further development of 3D ultrasound will lead to many more fetuses in the world being examined and treated. Three-dimensional ultrasound will be used not only in obstetrics but in most fields of medicine, since the technologies being developed in 3D ultrasound for a fetus are applicable in all fields of medicine.

2.3 Imaging by a Spherical Ultrasound Wave

Ultrasound imaging is widely used in various fields because it visualizes not only the exterior structure of a measurement object but also the interior structure that cannot be visualized with an optical measurement method, such as the inside of a human body. Therefore, ultrasound imaging is an essential technique for the medical field in diagnosis of human organs like the heart, liver, and so on. Furthermore, intravascular ultrasound imaging (IVUS) [23] has been developed to show the cross-section of a blood vessel and reveal morphological information of a blood vessel, such as stenosis, aneurysm, and so on in the clinical field of cardiovascular medicine. This imaging method, however, cannot give the frontal image of a blood vessel, which is useful for the treatment of a blood vessel, such as in percutaneous transluminal coronary angioplasty (PTCA) [24].

We have been studying the ultrasound frontal imaging system [25]. This imaging method requires one transmission of spherical ultrasound to reconstruct a three-dimensional (3D) image. Therefore, this method enables us to show a 3D frontal image with a very high frame rate, above 1000 fps.

In this section, imaging by a spherical ultrasound wave is described. This imaging is an instantaneous imaging method [25] and is based on the synthetic aperture method [26] and a spherical wave that is generated by laser-induced breakdown (LIB).

2.3.1 An Instantaneous Imaging Method

In the conventional ultrasound imaging method, a pulsed ultrasound beam scans the object electrically to reconstruct a two-dimensional (2D) image [27]: this is called a "sequential imaging method". The image has a frame rate of about 10–100 fps, but it is difficult to increase the frame rate to about 1000 fps because ultrasound has a finite velocity in the human body, typically 1500 m/s. In the final decade of the last century, the sequential imaging method was often extended to 3D imaging [28]. However, the sequential imaging method takes a much longer time to reconstruct a 3D image because of the finite velocity of ultrasound. The sequential imaging method, especially 3D imaging, is not suitable for use in a treatment that requires rapid imaging.

To overcome this difficulty, we have proposed an imaging method that we call an "instantaneous imaging method" [25], using the synthetic aperture method and a spherical wave. In the experiment using this method, a 3D object was imaged as follows:

1. A sound source, for example an ultrasound transmitter, a measurement object, and several ultrasound receivers were positioned in water.
2. A limited measurement region was determined and divided into small elements (voxels).
3. The ultrasound flight time from a sound source to a receiver by way of an element was calculated for all receivers and elements.
4. Pulsed and spherical ultrasound were transmitted once.
5. The wave reflected from a measurement object was received at several receivers.
6. The magnitudes of the echo signals from each element were summed for all receivers.
7. The summed values were converted to gray levels in a 255-level scheme.
8. The 3D image of a measurement object was reconstructed.

As spherical ultrasound is transmitted, this imaging method enables us to obtain a 3D image with a frontally wide view. Moreover, the 3D image has a very high frame rate above 1000 fps because the imaging method needs only one transmission of ultrasound. However, the following problems are encountered:

Table 2.1. Laser parameters for generating laser-induced breakdown (LIB)

Pulsed laser	Nd:YAG
Wavelength	1064 nm
Pulse repetition frequency	10 Hz
Pulse energy	10 mJ (typ.)
	0.2×10^9 J/m^2
	1.25×10^6 W
Pulse width	8 ns
Laser beam radius	16 mm
Focused laser beam radius	16 µm
Focal length of the lens	30 mm

- The transmitted waveform does not have a single peak but many peaks, owing to the transducer ringing on transmitting ultrasound (the ringing phenomenon) [25], which results in blurring of the reconstructed image.
- Ultrasound with low directivity cannot be transmitted from a transducer of a piezoelectric device, for example, made of ceramics [29]. The ultrasound enables us to visualize only a very narrow region.

Judging from the matters mentioned above, the instantaneous ultrasound imaging technique has been awaiting the development of ultrasound with low ringing and low directivity.

2.3.2 Laser-Induced Breakdown

When a pulsed laser beam is radiated to atoms or molecules in a very small region, some atoms or molecules absorb light, which induces an increase in their energy levels and leads to ionization of the atoms or molecules. This ionization causes a cascade ionization of other atoms or molecules. Then, ionized atoms or molecules are re-combined with an electron. Immediately, light emission, a shock wave, and cavitation follow. Consequently, light, heat, and sound are generated from the laser focal point. This phenomenon is called "laser induced breakdown" (LIB) [30]. Here, we performed an experiment to create LIB in water. In this experiment, the pulsed laser beam was focused in a water bath using an optical lens with a focal length of 30 mm. Table 2.1 summarizes the parameters of the pulsed laser.

Figure 2.15 is a photograph of LIB recorded with a digital video camera. In this figure, the LIB occurred at the bright spot and heat was also generated because bubbles were thought to be produced by water boiling. Moreover, the click-like sound was heard and was expected to have low directivity because the sound is generated by a very small source.

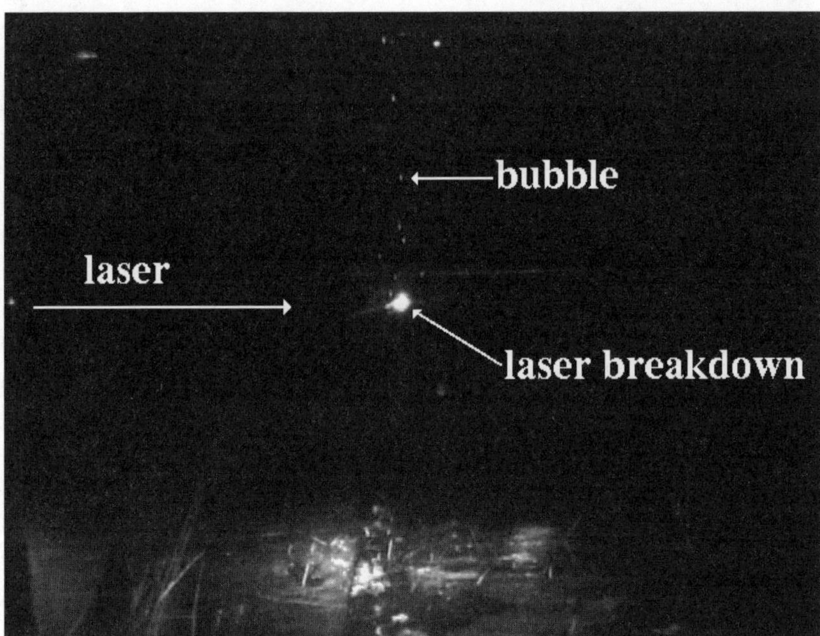

Fig. 2.15. The laser-induced breakdown (LIB) phenomenon in a water bath. The LIB occurred at the bright spot and heat was thought to be generated because of bubble generation

2.3.3 Ultrasound Generated by Laser-Induced Breakdown

We created LIB in a very small region in water. In order to examine the characteristics of ultrasound generated by LIB, we carried out an experiment to measure the waveform and directivity of the ultrasound. Figure 2.16 shows the experimental arrangement, where the z direction is set along the laser beam and the other directions (x and y) are normal to the z-axis; x is horizontal and y is vertical. This time, ultrasound was received with a needle hydrophone made of polymer, because it has a higher sensitivity than a receiver made of ceramics. The received signal was amplified and then stored in a digital storage oscilloscope (DSO) at a sampling frequency of 500 MHz.

Figure 2.17 is an example of an ultrasound waveform. The wave has a large and sharp peak at 30 µs after LIB occurs, and is a sinusoidal wave with a period of 200 ns. Besides, the small peaks appear in the range of 30–50 µs. The former peak is the wave received from LIB directly. As the pulsed laser was radiated in this measurement, this figure shows the impulse response of the measurement system including the hydrophone and the amplifier. This result demonstrates that ultrasound by LIB has little ringing. On the other hand, the latter peaks are the waves reflected at the walls of a water bath or the water surface, and the period of the wave becomes larger as reflected waves from various points are superimposed.

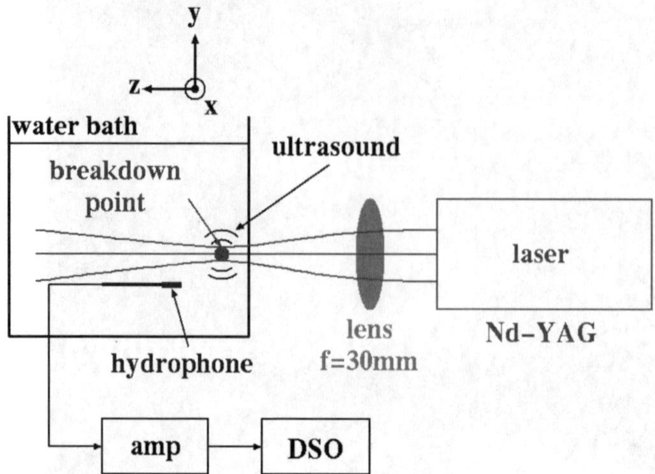

Fig. 2.16. The measurement arrangement for ultrasound by LIB. The z-axis was set along the laser beam and the other axes were normal to z. Ultrasound was received with a needle hydrophone, amplified, and stored in a digital storage oscilloscope (DSO)

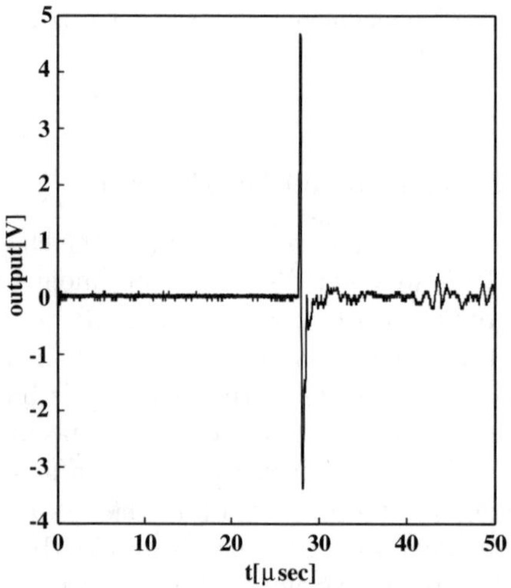

Fig. 2.17. The waveform of the ultrasound by LIB, which has a large peak at 30 µs and small ones at 30–50 µs. The former is the wave received by LIB directly, while the latter is reflected at the walls of a water bath or the water surface

Using the same measurement system as seen in Fig. 2.16, we investigated the directivity of the ultrasound generated by LIB. In this experiment, the origin of the coordinates was set at the LIB point. The measurement points were varied from -70 to 70 mm in the x direction at a constant step of 10 mm, and from 0 to 70 mm in the y and y directions at the same step owing to the finite volume of the water bath. On the z-axis, the measurement point was not set, to avoid damage by the laser. We assumed that the ultrasound propagation is symmetric in the y and z directions.

Figure 2.18 shows the propagation of ultrasound at (a) 200, (b) 250, and (c) 300 µs after LIB, where the nodes of the grid correspond to the measurement points. In this figure, the grids were distorted as the nodes were shifted in proportion to the ultrasound magnitude received at the measurement point, and the nodes were drawn brightly.

This figure reveals that the ultrasound propagated in any direction and that the ultrasound has little directivity. Therefore, we succeeded in transmitting a spherical wave.

2.3.4 Imaging Using Laser-Induced Breakdown

We can obtain ultrasound with little ringing and little directivity by LIB. Using this ultrasound, we constructed an instantaneous imaging system and performed two experiments. The first experiment was performed to observe an echo signal by LIB and to examine whether it can be used for imaging. Figure 2.19 shows the measurement arrangement, where LIB was generated in a water bath and a measurement object and a hydrophone were set in the same bath. The reflected wave from a measurement object was received with 25 hydrophones with a constant step of 10 mm on the x–z plane. A steel wrench was selected as a measurement object, so that a signal with a larger amplitude could be received.

Figure 2.20 shows an example of the reflected waveform. In comparison with Fig. 2.17, this waveform has more peaks, which indicates that echo signals were received from various portions of the measurement object.

Figure 2.21 shows an example of a 3D image with a voxel size of 5 mm × 5 mm × 5 mm, where (a) and (b) are images observed from different directions. The insets in this figure are photographs of the measurement object viewed from the same direction as the reconstructed image. Comparing the 3D images with the insets in Fig. 2.21, the 3D image can be reconstructed correctly: in particular, the corner of the measurement object can be visualized, as indicated by the arrow shown in Fig. 2.21b. These results indicate that ultrasound generated by LIB can be used for instantaneous imaging.

The second experiment was carried out to confirm whether ultrasound by LIB is suitable for frontally wide view imaging. Figure 2.22 shows a measurement arrangement where LIB was generated in a water bath and a measurement object and a hydrophone were set in the same bath. Styrene balls were set at positions 1–3. The reflected wave from a measurement object was received at 12 hydrophones, which were located in a circle with a radius of 15 mm.

Figure 2.23 shows an example of a 3D image, which indicates that every object could be observed wherever it was positioned. It is noted that the

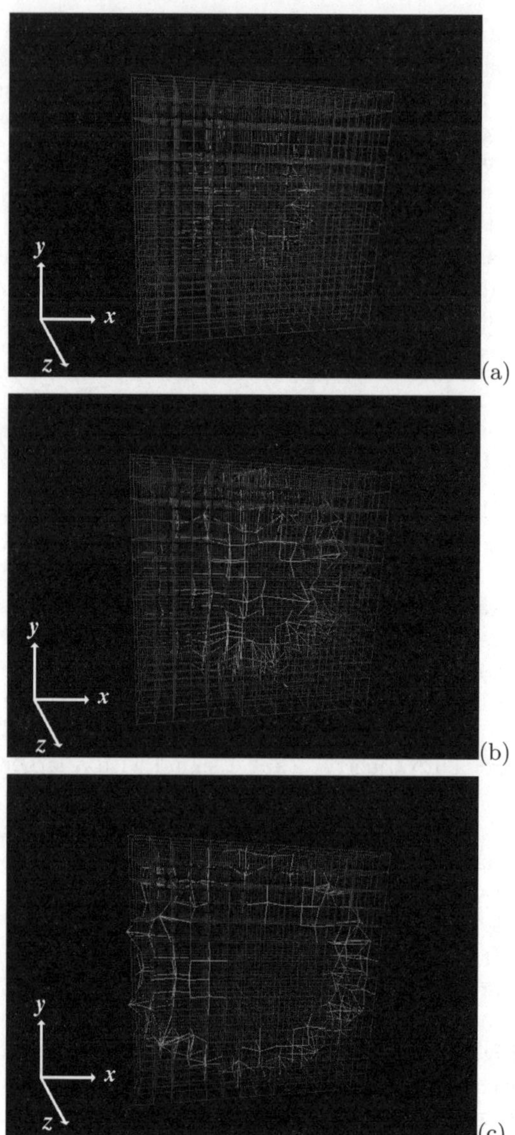

Fig. 2.18. The propagation of ultrasound at 200 (**a**), 250 (**b**), and 300 µs (**c**) after LIB occurred, where the nodes of the grid correspond to the measurement points. The grids were distorted as the nodes were shifted in proportion to the ultrasound magnitude

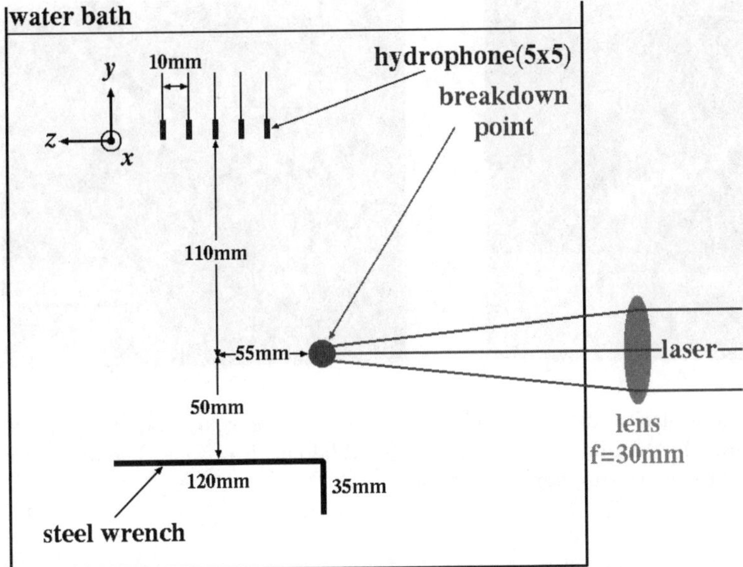

Fig. 2.19. The arrangement for measurement to observe an echo signal by LIB and to examine whether it can be used for the imaging. The measurement was done in water and the waves were measured at 25 points

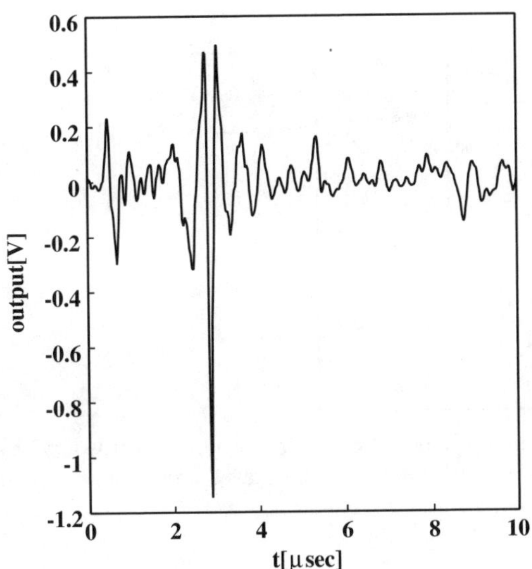

Fig. 2.20. An example of a reflected wave. This waveform has more peaks, which indicates that echo signals were received from various portions of the measurement object

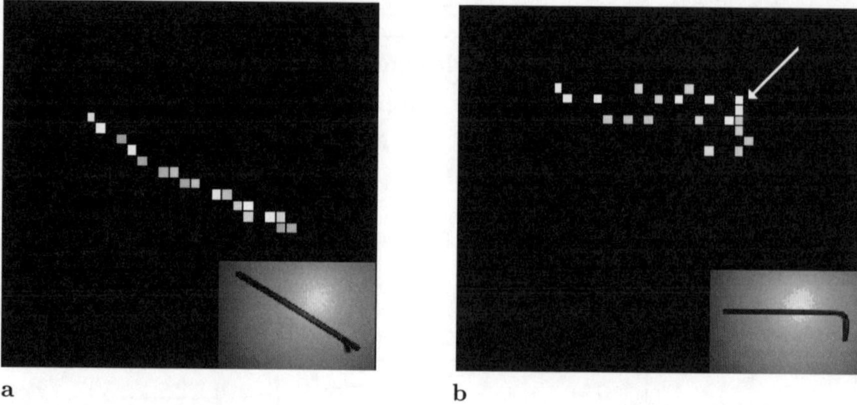

Fig. 2.21. Examples of 3D reconstructed images with a voxel size of 5 mm × 5 mm × 5 mm, where **a** and **b** are observed from different directions. The insets are photographs of the measurement object

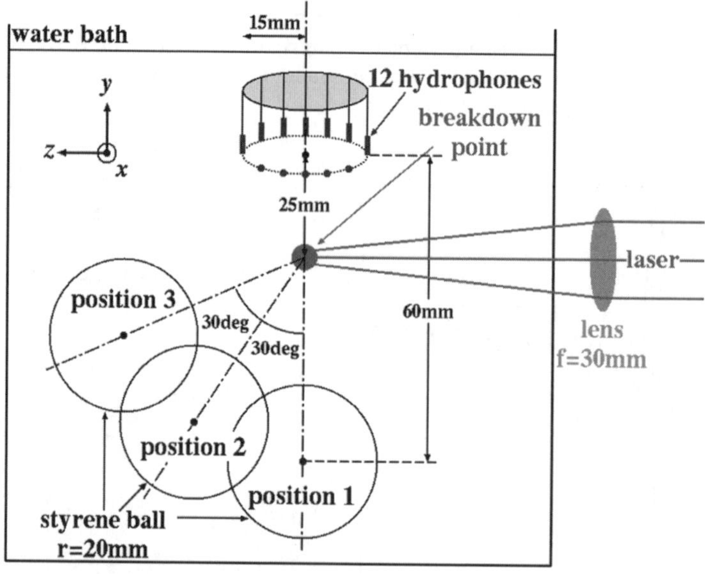

Fig. 2.22. The arrangement for measurement to confirm whether ultrasound by LIB is suitable for frontally wide view imaging. The waves were measured at 12 points on a circle with a radius of 15 mm

object at position 3 could be visualized, although it was located outside the aperture. These results demonstrate that ultrasound generated by LIB is fit not only for frontal imaging but also for frontally wide view imaging.

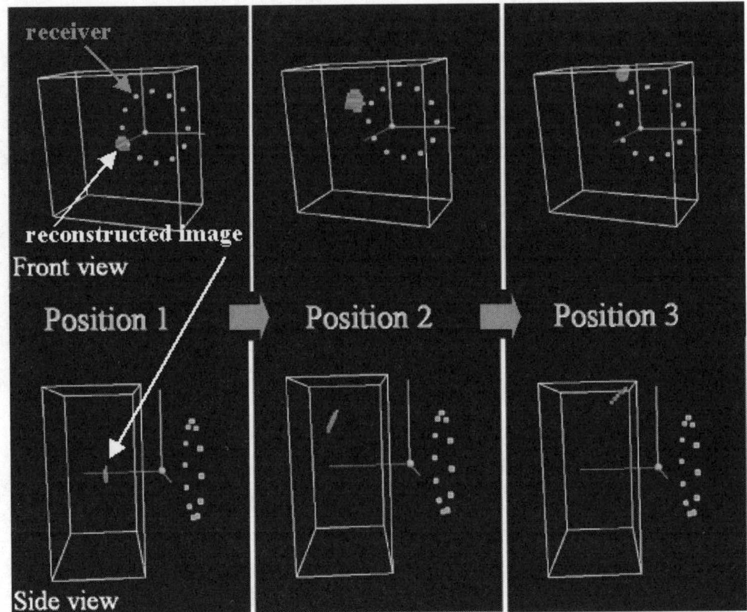

Fig. 2.23. Examples of 3D reconstructed images, which indicate that every object could be observed wherever it was positioned

2.3.5 Summary

The instantaneous imaging method was developed using the synthetic aperture method and a spherical wave generated by laser-induced breakdown. LIB was created by a focused laser, and we succeeded in transmitting the ultrasound with little ringing and little directivity. The imaging experiments demonstrate that spherical ultrasound enables us to reconstruct a 3D image, and that ultrasound can obtain not only a frontal image but also frontally wide view imaging; that is, a panorama image. Therefore, with this imaging it should be possible to obtain various morphological information from a blood vessel. In the future, this system will be implemented with a catheter, where laser energy is imported into a blood vessel through an optical fiber.

2.4 Imaging Tissues Using an Ultrasound Wave and Light

2.4.1 An Ultrasound Wave and Light for Tissues

The imaging of biological tissues by ultrasound wave and light suggests the possibility of noninvasive, informative observation for thick tissues in vivo. Ultrasound wave has the characteristic of high penetration (centimeter order) of tissues at a MHz frequency, although light is strongly scattered in the tissues and cannot propagate over a long distance in a straight line. However, the light interacts with the atoms and molecules, of which the tissue consists through the energy level and/or the vibration energy level, and informs us about the live state of the tissues. By combining both of these advantageous characteristics, new techniques for medical diagnostics, surgical operations, and biopsy can be pioneered.

In aiming at these applications, optoacoustic tomography and sonoluminescence imaging have been applied to the observation of tissue for the past several decades. In optoacoustic tomography, the modulating light irradiated to the tissues is converted to ultrasound waves at the location of the optical absorbers, because the heated regions due to light absorption produce expansion and shrinking of the local volumes, which generates compression waves. Observing these waves, the optical information, such as the complex refractive index, in the tissue can be obtained. In contrast with tomography, light is generated locally at the tissues by means of irradiating with focused ultrasound in sonoluminescence, and the resultant light intensity is observed. Among the research projects using ultrasound wave and light, new ultrasound-assisted optical imaging has been developed recently as a good candidate observing the inside of the tissues [31–35]. In this section, one of these new imaging methods is described [31].

2.4.2 Ultrasound-Assisted Optical Imaging

In ultrasound-assisted optical imaging, the light scattered in the tissues is tagged locally with an ultrasound wave to distinguish it from the other materials. A converging pulsed-ultrasound wave and a continuous light are used to irradiate a tissue. The ultrasound wave, which interacts with light via the density change in the tissue, modulates the density of the material at the instantaneous location: meanwhile, the light is strongly scattered by the tissue, consisting of the sum of small components in a cell. The modulation changes the optical scattering efficiency according to the dependence of the optical property, because the path of the scattering for the incident light is changed by the change of the efficiency. As the pulsed ultrasound wave propagates through the sample, the temporal intensity of the light transmitted through the tissues is changed. Through the observation of this change, optical information such as the complex refractive index can be obtained.

2.4.3 The Experimental Setup

Figure 2.24 shows the experimental setup of the ultrasound-assisted cross-sectional imaging system. A specially shaped ceramic ultrasound transducer generates converging pulsed ultrasound waves. A laser beam from a He–Ne laser (632.8 nm wavelength, 0.5 mW output power, and 1.0 mm beam diameter) is irradiated to the sample in the focal region of the converging ultrasound wave, which means that the ultrasound wave and the light are

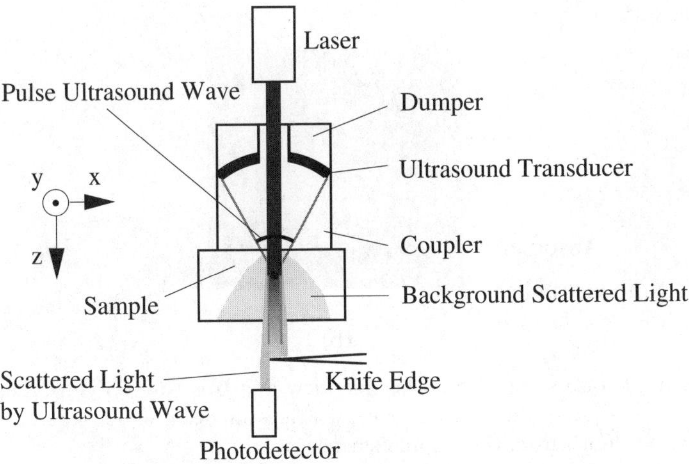

Fig. 2.24. The experimental setup of the optical cross-sectional imaging system with pulsed ultrasound wave assistance

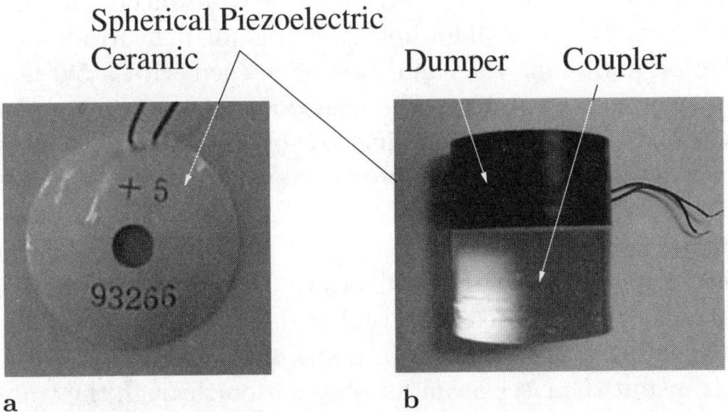

Fig. 2.25. A ultrasound transducer head. **a** is a spherical piezoelectric ultrasound transducer. **b** is an ultrasound transducer head, which consists of a piezoelectric transducer attached by rubber and a damper

on the same axis and in the same same direction. The pulsed ultrasound wave travels through the sample and scatters some of the incident beam. The scattered light is detected on the other side of the sample with a PIN photo-detector, positioned behind a knife edge that blocks light that is not scattered by the ultrasound wave. The signal from the detector is recorded with a digital storage oscilloscope up to a 20 MHz bandwidth. The time-dependent scattered light intensity is synchronized from the start pulse of the ultrasound wave.

Figure 2.25 shows a piezoelectric ceramic and an ultrasound transducer head designed specially for use in this experiment. The piezoelectric ceramic ultrasound transducer in Fig. 2.2a has a spherical shape with a 25.0 mm ra-

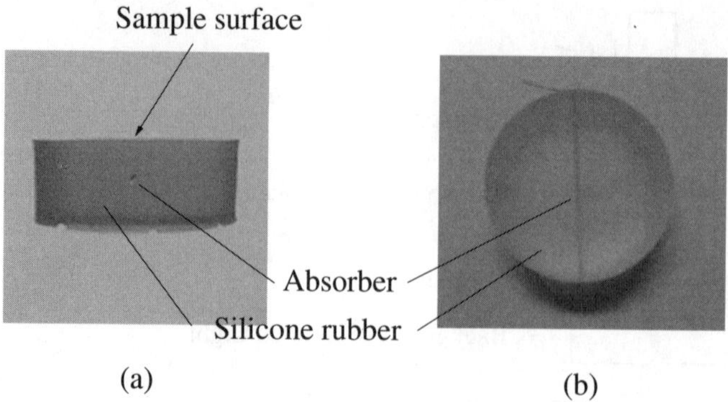

Fig. 2.26. An example of a sample. **a** is the side view and **b** is the top view. The thickness of the sample is 10 mm. An absorber which is made of a silk thread is embedded at a 5-mm depth from the sample surface

dius of curvature, so that the pulsed ultrasound wave converges to a point located at a distance of 25.0 mm from the surface of the transducer. The diameter of the transducer is 25.0 mm and a hole, which is for light irradiation from the same direction and along the same axis, is opened with a 5.0 mm diameter at the center of the transducer. The transducer contacts a coupler made of silicone rubber to reduce the coupling loss of the ultrasound wave at the surface of the sample. The silicone rubber is transparent in the visible region, to transmit the laser beam efficiently to the sample. The distance from the transducer to the surface of the coupler is 20 mm, and the focal point of the ultrasound wave is located at a depth of 5.0 mm in the sample. The rear of the transducer is mounted on a rubber damper to generate the ultrasound wave efficiently. A hole 5.0 mm in diameter is also made in the rubber damper, to transmit the laser beam for observation through the transducer. In generating an ultrasound wave, the voltage of 12.0 V peak applied to the ceramic, which produces peak pressure of approximately 2.7×10^5 Pa at the focal point of the ultrasound wave, is a rectangular wave from a pulse generator. The pulse width is around 0.5 μs and the repetition rate of the pulses is 1 kHz. The repetition rate was set at a suitably low value to wait for a sufficient time for the next pulse after the sample had been irradiated with pulsed ultrasound. The ultrasound velocity of the silicone rubber in the experiment was around 1000 m/s. With these parameters, the diameter of the focal spot of the ultrasound wave is 0.3 mm in the full width at half maximum (FWHM) and the depth of the focal area of the ultrasound wave is 2.5 mm, also at FWHM.

Figure 2.26 shows an example of the samples used in this experiment. Figures 2.26a,b are the side and top views of the sample, respectively. The sample is made from an absorber and its surrounding material is made from silicone rubber, which is the same material as the coupler in the transducer head. The absorber is made of a silk thread stained with victorial blue dye, which has a high absorbance around 633 nm. The diameter of the thread is 0.3 mm and the length is around 10 mm along the y-axis. In our experiment, the location of the thread in the depth is changed. The scattering coefficient of the surrounding material is adjusted according to the amount of Intralipid® (Teruno, Tokyo, Japan) resolved in the sample. The thickness of the sample is 10 mm. Experiments are performed in an almost clear sample and in a scattering sample with absorbers.

2.4.4 An Experiment for Weak-Scattering Samples

Observation of an Absorber in the Medium. A sample with an absorbing object in an almost transparent medium is observed. The thread was located at a depth of 5 mm, so that the position of the thread could be set at the focal point of the ultrasound. Figure 2.27a shows an example of the axial cross-sectional signal obtained by the apparatus. The observed signal contains an alternating component (ac) that describes the modulation of the scattered light intensity, induced by the pulsed ultrasound wave and a direct component (dc) that represents almost constant intensity of the scattered light, which is caused by the silicone rubber. The signal is an accumulation of 256 data sets, with each set containing 500 data points along the z-axis. The longitudinal axis shows the time delay after pulse initiation from the piezoelectric ceramic. The value of 20.0 µs on the axis corresponds to the position at the surface of the sample, estimated from the velocity of the ultrasound wave. After the pulsed ultrasound wave reaches the surface of the sample, the scattered signal is detected for a period of 10.0 µs. At the 25.0 µs

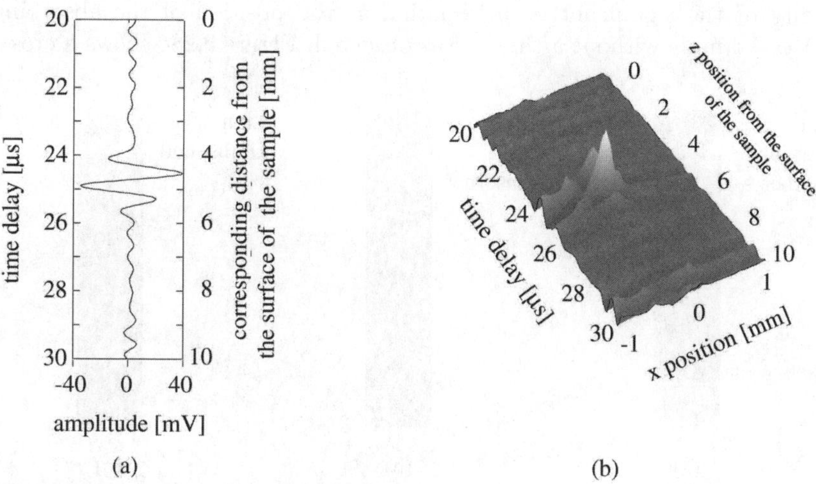

Fig. 2.27. The observed signal of an absorber in a silicone medium. **a** is a graph of the temporal light intensity scattered by an ultrasound wave. **b** is a cross-sectional image of an absorbing object embedded in silicone rubber

delay the signal changes dramatically. This time delay corresponds to a location of 5.0 mm from the surface of the sample when using a value of 1000 m/s as the ultrasound velocity in the sample. This is the same position at which the thread was located at a depth of 5.0-mm. This shows that the detected signal is changed only when the pulsed ultrasound wave is located at the absorber. The shape of the signal was mainly affected by the pulse shape of the ultrasound wave. Because the damping of the pulse ultrasound wave was not optimized, the pulsed wave had some side lobes caused by ringing.

By scanning the sample against the ultrasound transducer head laterally, the cross-sectional image of the sample can be obtained. Figure 2.27b shows a surface plot of the cross-sectional image of the same sample. The corresponding size of the image is 10.0 mm in depth and 2.0 mm in a lateral direction. The height of the image is equivalent to the intensity of the scattered light. The height level shows an ac signal that corresponds to the range between −40.0 mV and 40.0 mV. The center of the image shows one major peak that corresponds to the location of the embedded thread.

The Effect of a Pulsed Ultrasound Wave and an Absorber. The effect of the pulsed ultrasound and an absorber is confirmed. Figure 2.28a is the result which is observed the same sample by the same setup in Fig. 2.27b. Figure 2.28b is an image resulting from an ac signal measurement in which the same sample is observed by the same setup but without applying the pulsed ultrasound wave to confirm the effect of the pulse ultrasound wave. The ac signal was reduced to less than 1% compared with the ac signal that was measured with the pulsed ultrasound wave. Consequently, the image without the pulsed ultrasound wave had almost no structure. To verify the validity of the significant ac modulation at the position of the absorbing object, a sample without a thread was observed. Figure 2.28c shows a cross-

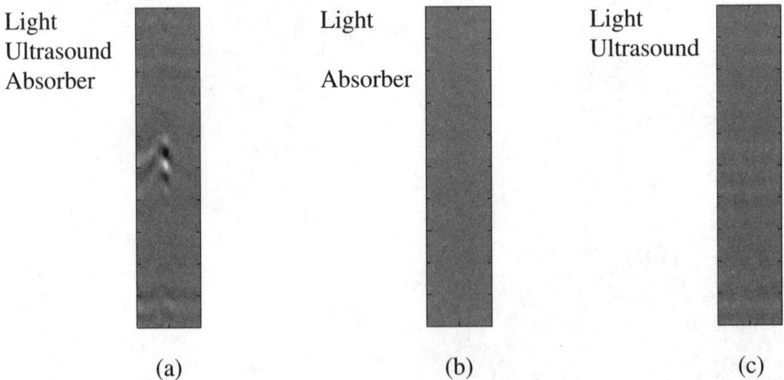

Fig. 2.28. A cross-sectional image of an absorbing object embedded in silicone rubber and two reference images taken at various conditions for confirmation of the effect of ultrasound and light

sectional image of the sample while a pulsed ultrasound wave was applied. There is no significant modulation that may be attributed to an absorbing object. Only small fluctuations can be observed along the optical axis. These fluctuations are caused by the electronic noise generated by the ultrasound transducer. The sensitivity of the system is derived as 0.005 from the ratio of the modulation amplitude in Fig. 2.28a and the fluctuation amplitude in Fig. 2.28c.

Observation of Absorbers at Different Depths. A series of samples that contains the blue-stained threads embedded in silicone rubber at different depths were investigated. The scattered light intensity is expressed as a height. Both light and ultrasound are radiated in the $+z$ direction. The position of the sample surface is on the x-axis. The z-axis corresponds to the depth of the absorbers. The locations of seven threads were 2.0, 3.0, 4.0, 5.0, 6.0, 7.0, and 8.0 mm, respectively. Figure 2.29 shows the two-dimensional cross-sectional image obtained from each sample. In sample (a), the modulation of the signal caused by the absorber is seen at a time delay of 22.0 µs. The corresponding distance of the time delay of the signal estimated from the ultrasound velocity is 2.0 mm from the surface of the sample. This value is equal to the distance from the surface of the sample to the absorbing object. From this series of results, the modulations of the signal caused by the thread in each result appeared at time delays of around 23.0, 24.0, 25.0, 26.0, 27.0, and 28.0 µs, respectively. These values are in accordance with the delay time expected from the thread positions.

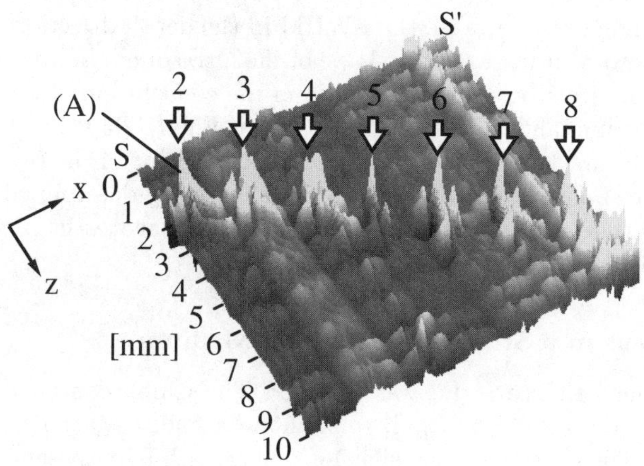

Fig. 2.29. A cross-sectional image of a series of samples which contain absorbing objects at different depths. The depth of the objects are 2, 3, 4, 5, 6, 7, and 8 mm. Light and focused pulsed ultrasound are irradiated in the z direction

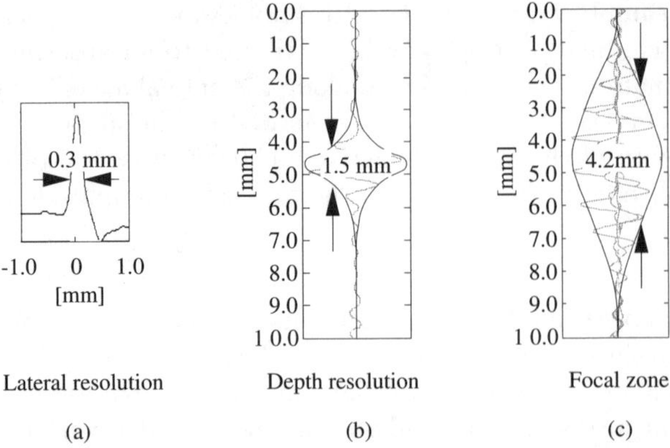

Fig. 2.30. The system performance of the apparatus. **a** shows the lateral resolution of 0.3 mm. **b** shows the axial resolution of 1.5 mm. **c** shows the focal zone of the focusing ultrasound wave along the z-axis

The Performance of the Apparatus. The performance of the developed system has been estimated. Figure 2.30a shows the lateral cross-section of Fig. 2.28a at the 5 mm location. In defining the lateral resolution as a FWHM of the cross-section, the value is obtained as 0.3 mm. This value is reasonable from the expected diameter of the ultrasound wave. Figure 2.30b shows the resolution along the z-axis, which is estimated as 1.5 mm from the measurement at FWHM of the envelope of the ac signal. The envelope is fitted by the Gaussian distribution. The shape of the signal was mainly affected by the pulse shape of the ultrasound wave, because the damping of the pulsed ultrasound wave was not optimized, so that the pulsed wave had some side lobes caused by ringing. To decrease the FWHM in the depth direction, the observed signal is deconvoluted with the shape of the ultrasound, because the FWHM in the depth direction is determined from the convolution of the pulse shape of the ultrasound and the size of the absorber. Figure 2.30c is the superposed trace of each raw data value in Fig. 2.29. The envelope of the raw data exhibits the axial response of the apparatus, which is mainly determined by the change in amplitude of the ultrasound wave as it propagates in the axial direction. The FWHM of the envelope was 4.2 mm.

2.4.5 An Experiment in a Strongly Scattering Medium

A sample that gave strong light scattering was observed. The sample consisted of silicone rubber that contained Intralipid, to simulate a highly scattering medium. Intralipid was dispersed into the silicone rubber with a surfactant. The scattering coefficient of the silicone rubber was measured at 2.0 cm^{-1} by using the Beer–Lambert law. An absorber was embedded at a depth of 4.5-mm from the surface of the sample. Figure 2.31a shows the axial cross-sectional signal of the sample. The lateral axis shows an ac signal level, which

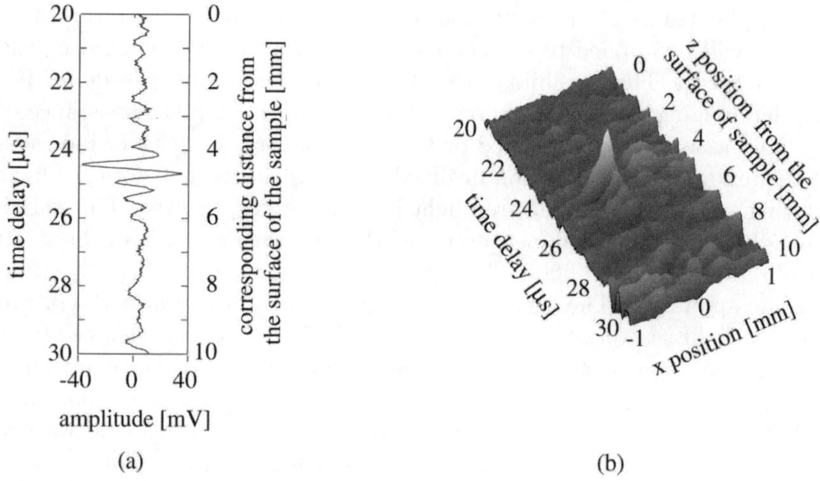

Fig. 2.31. The observed signal for a scattering medium. **a** is a graph of the temporal light intensity scattered by an ultrasound wave. **b** is a surface plot of the cross-sectional image of an absorbing object embedded in silicone rubber

ranged from −3.5 mV to 3.5 mV. At 4.5 mm from the sample surface, the modulated signal caused by the absorbing object appeared strongly. The amplitude of the ac signal is smaller compared to the signal obtained from the sample without Intralipid in Fig. 2.27a. This is due to the fact that the scattered light modulation generated by the pulsed ultrasound wave decrease as the scattered light propagates from the absorber through the sample. Figure 2.31b shows a cross-sectional image of the sample. The height indicates the ac signal level of the scattered light. In this result, a peak of the signal has been observed, which corresponds to the location of the absorber.

2.4.6 Conclusions

In this section, a detailed description of the imaging of tissues by an ultrasound wave and light is given. The basic physical mechanism leading to the change of the optical scattering pattern due to a pulsed ultrasound wave has been explained. A new transmission system for optical measurement has been developed, and the system performance has been assessed. Furthermore, this system have shown the cross-sectional image of an optical absorber embedded in scattering medium. From the results, this system shows the possibility of depth discrimination of an absorbing.

The technique described is able to be extended to the imaging of blood vessels under a scattering medium such as skin tissue, and is therefore an aid in the study of biological structures. This system limits the thickness of the sample at around a few centimeters, because the incoming light needs to be transmitted through the tissue sample. The proposed technique may possibly be applicable as a diagnostic tool in mammography. In the future, this technique will be applied to a reflection system, which is not limited by the size of the target. The possibility of a reflection system can be considered, because the imaging mechanism in optical measurement with the assistance of ultrasound assistance is explained by scattering theory. The theory indicates that the main kind of scattering involved in biological tissue is forward light scattering, and that backscattered light is only slightly involved. This shows the possibility to reflection measurement, through the detection of the small amount of backscattered light. The reflection system can be realized by replacing some apparatus. The light source is replaced by a near-infrared light, to reduce the light scattering effects. The photodetector is also exchanged for a highly sensitive one, such as an avalanche photodiode and a photomultiplier. Furthermore, a high voltage is applied to the ultrasound transducer produces a high-power ultrasound field, and the pulse shape of the ultrasound wave is optimized. In extending the application to biological tissue, a coupler is needed in which the acoustic impedance of the ultrasound transducer is matched to that of biological tissue. The specification also demands a material that is clear in the visible region so that light can be transferred efficiently into the tissue, and the power of the pulsed ultrasound and light should be increased, while adhering to safety standards.

2.5 An Ultrasonic Drug Delivery System Using Microcapsules

2.5.1 The Requirement for a Drug Delivery System

We have developed an ultrasonic drug delivery system (DDS) using air-filled microcapsules. We have confirmed the physical possibility of the resonance design for the drug carriers, depending on the shell structure of the microcapsule. In addition, we have developed a novel echographic imaging system to locate microcapsules precisely. The path lines of microcapsules in blood flow could be clearly visualized.

Cancer, which ranks as one of the principal causes of death, and arteriosclerotic thrombosis, are both local lesions. However, it is often inevitable that drugs having that are strongly effective against such diseases will be used, even though they have seriously adverse effects on the whole body. As a result, there has been a great demand for a method that will release the drug only to the diseased region.

To meet such a demand, various DDSs for realizing the most effective cure by drugs while suppressing the side effects have been proposed with regard to the desired absorption passage, degradation rate, metabolic rate, and other factors. Nevertheless, none of the proposed methods have satisfied all of the following important conditions that are necessary in clinical application:

1. Drug delivery should be controlled so that only a specific organ or diseased region is targeted, or is subjected to concentrated administration of the drug.
2. The release or activation of a drug should be able to be specified for a certain time following administration.
3. A method for observing the distribution state and the drug concentration should be available.
4. The drug delivery system should applicable to any type of drug.
5. The drug delivery system should be noninvasive.

There are some techniques that satisfy some of these conditions. However, if any one of these conditions is not satisfied, the DDS cannot be regarded as a complete method for medication.

To eliminate these problems, a DDS should be a microrobotic system; that is, controllable DDS system. It should be able to carry any type of drug and should be measurable and controllable using intelligent diagnostic decision-making.

Therefore, we have proposed a new concept of an intelligent microrobotic system in vivo, as a DDS in which drug carriers of air-filled microcapsules are measured using a novel high-resolution echography system [36–38] and are noninvasively actuated [39,40] using resonant ultrasound from the skin surface of a patient.

2.5.2 The Acoustic Characteristics of Microcapsules as Drug Carriers

In this system, the micro-encapsulated drug carriers are microrobots within the blood vessels.

The substantial requisite for these microrobots is to be micro-balloons or to be microcapsules containing harmless gas. A significant difference between acoustic impedances of the water and the gas is essential to maintain suitable ultrasonic backscatter and resonance.

The microrobots – i.e., microcapsules – are visualized and measured using nonresonant ultrasound. In this regard, it is well known that microbubbles in the blood are a good contrast agent in the field of ultrasonic diagnosis.

The control of the microrobots is based on the principle that an air-filled microcapsule in liquid absorbs ultrasound energy at the resonant frequency [41]. In other words, it can be excited with resonant ultrasound. Fortunately, the resonant frequency of a microcapsule having the same, or a smaller, diameter as a red blood cell coincides with the frequency of medical ultrasound.

As active microrobots, the micro-encapsulated drug carriers are noninvasively actuated and controlled by resonant ultrasound from outside the patient's body. Equation (2.1) represents the resonant frequency of a microbubble in water [42]:

$$f = \frac{1}{\pi d}\sqrt{\frac{3kP}{P}}, \qquad (2.1)$$

where f is the resonant frequency; d is the diameter of the microbubble or microcapsule; k is the specific heat ratio of gas, a constant; P is the pressure of the water; and p is the density of the water, a constant.

The resonant frequency depends mainly on the bubble diameter and the pressure of the surrounding water.

Although this formula does not describe the structure of a microcapsule shell, it is valid for the order estimation of microcapsule resonance with thin shell walls.

Equation (2.1) assumes that the microcapsule is a perfect sphere. Microscopic photographs show perfect spherical microcapsules made of polymethylmethacrylate (PMMA). Newton rings, which interfere between the microcapsule membrane and the stage glass, demonstrate the perfect smoothness of the microcapsule surface. They all contain harmless gas. These data confirm the validity of (2.1).

In addition, an ultrasonically excited drug matrix releases the drug component into the surrounding liquid [43]. If the shell of a microcapsule was made of drug matrix, then the release of the drug component from the microrobotic DDS could be controlled by resonant ultrasound.

To confirm the physical possibility of resonance control based on shell design, three series of microcapsules were made from one series of PMMA and two series of vinylidenedichloride VCl, with diameters ranging from 5 to 90 µm. The experimental results showed that the resonant frequencies of the microcapsules made of hard VCl with mean diameters of 82, 54, and 26 µm were 110, 220, and 330 kHz, respectively.

Thus, the smaller the diameter, the higher is the resonant frequency. The resonant frequencies of the microcapsules almost coincided with the theoretical values calculated by (2.1) [44].

In the case of microcapsules having the same diameter, the resonant frequency depends on the material. The harder the material, the higher is the resonant frequency. The resonance of the soft VCl capsules was lower than that of the hard VCl capsules. Rigid PMMA microcapsules showed the highest resonance.

These findings show that we can design microcapsules of sufficient resonance to be controlled by ultrasound. Microcapsules smaller than red blood cells can provide a feasible resonant frequency for clinical use through adjustment of their diameter and material.

2.5.3 A Noninvasive Measurement System for DDS

For adequate control of the microrobotic system, a measurement method is necessary. Information concerning the accurate location and velocity of a microcapsule is essential to achieve measurement and control of the system in the blood flow [45].

However, with a conventional imaging technique, echography must be used, which provides an insufficient spatial resolution of 1–3 mm. The Doppler method only provides the flow velocity component projected into the ultrasonic beam.

To alleviate these constraints, a two-dimensional moving target indication (2D-MTI) method has been developed to enable direct visualization of the 2D vector and path line (streamline) of a microcapsule in the blood flow.

The principle of the 2D-MTI method is serial image accumulation of subtracted images (high-speed DSE images) of echoic particles in blood like optical flow method. In the DDS system, the high-speed DSE system [37] executes image subtraction between successive frames of high-speed B-mode echograms of the blood flow containing highly echoic particles, such as microcapsules.

The digital subtraction unit reveals the new location of an echoic microcapsule as a positive intensity (white), on its 2D high-speed DSE image and displays the vanishing location as a negative intensity (black). Finally, the accumulated high-speed DSE image shows the streamline; that is, the trace of rapid motion of the microcapsule.

Through the 2D-MTI method, the path lines of rapid motion in vitro were demonstrated with highly echoic particles, air-filled microcapsules made of PMMA. The current 2D-MTI system was able to visualize the path of microcapsules smaller than the lung capillaries of 8 µm diameter without the restraints of Doppler echography.

2.5.4 Collapse Monitoring of Microcapsules

The brightness of the microcapsule suspension captured on the echogram decreases after ultrasound emission, because microcapsules collapse at their resonant frequency. Therefore, by comparing two successive echograms, the location and degree of microcapsule collapse can be indicated [46].

To evaluate the degree of microcapsule collapse, we applied a method to measure the density of capsules from the B-mode echogram. This method elucidated the relation between the density of the capsules and in the brightness of the echogram. The variation in the density of the capsules is calculated throughout digital processing of successive B-mode echograms. We constructed in vitro phantom with fixed microcapsules and calculated the density of the capsules. Figure 2.32 indicates the experimental setup to capture the echogram of capsules inside a phantom [47].

Figure 2.33 shows an echogram of a cross-section of a phantom. Inside the phantom, capsules can be seen as speckles. From the brightness variation in proportion to the depth, the density of the capsules can be estimated. The brightness B depends on the depth x by calculation of the constant

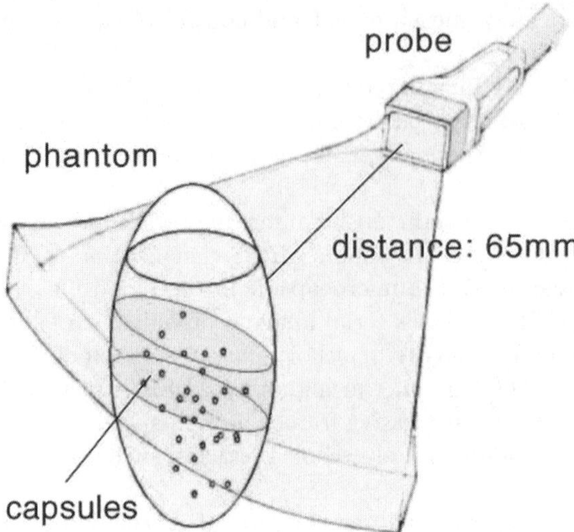

Fig. 2.32. The experimental setup to capture an echogram of capsules inside a phantom

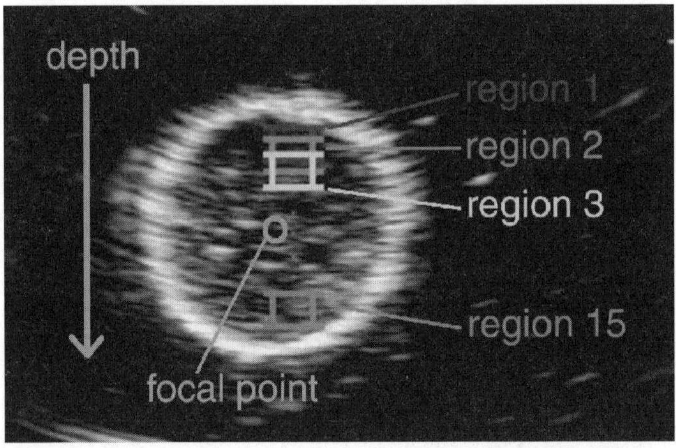

Fig. 2.33. An echogram of a cross-section of a phantom

coefficients c and d using an approximate logarithmic function as follows [46]:

$$B = c \ln x + d\,. \tag{2.2}$$

Figure 2.34 is a result of brightness variation measurement. The measured variation in the density of the capsules is classified and superimposed on the original gray-scale echogram using color, to assist in the measurement of the effect of ultrasound emission. By using this method, we could easily evaluate not only the collapse phenomena of microcapsules, but also the optimum

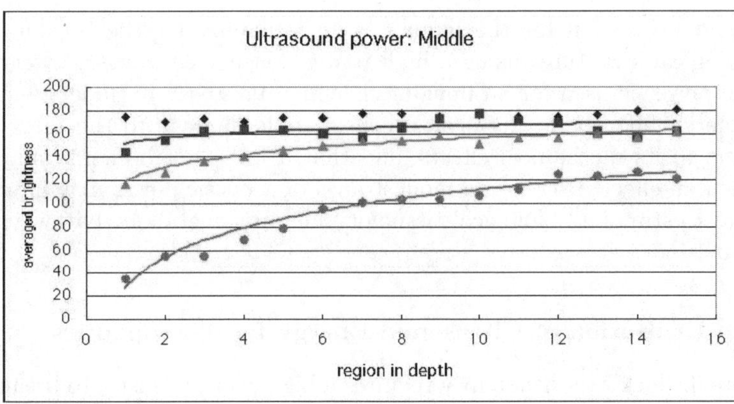

Fig. 2.34. Brightness variation depending on the depth of the phantom

emission power and period of ultrasound for the collapse of microcapsules in vivo.

2.5.5 Conclusion

We can conclude that the microrobotic DDS can be measured and controlled using ultrasound. Determining the exact location with the 2D-MTI system, the microrobot can be activated anywhere and at any time with resonant ultrasound.

If the system requires much more complex control, the current microrobotic DDS could provide an intelligent DDS system, with a microcomputer or a man–machine interface installed within the sensing and control feedback loop.

This system could form the basis of an intelligent noninvasive in vivo drug delivery system, although no conventional DDS has an active feedback loop for measurement and activation drugs. Using this method, any type of drug can be delivered anywhere in a patient. This technique will play an important role in medical microrobot systems.

2.6 The Biological Effects of Ultrasound

Ultrasound has been utilized as a source of vibration and cavitation in the industrial field. The term "sonication" means the use of ultrasound for washing small fine parts of a machine, homogenizing suspensions and emulsions such as milk, destroying structure, and many other purposes. In the medical field, the ultrasound scalpel is very useful for soft tissue surgery. Ultrasound (shockwave) lithotripsy has become a common therapy for stones in the urinary tract calculi. In the production process of medicine, ultrasonic homogenizing and standing wave formation can be used for homogenizing and particle separation.

Ultrasound utilization for the human body is mainly for the breaking down of biological structures using a high power density. Otherwise, a very low acoustic power is used for an imaging machine. It is easy to understand that high-energy ultrasound may cause considerable damage to the target structure due to its mechanical effects, and that a low enough energy may not have such an effect. However, a modest level of acoustic power may have the potential to stimulate biological tissue or may alter metabolic pathways and tissuefunctions.

2.6.1 The Utilization of Ultrasound Energy for Therapeutics

Extracorporeal shockwave lithotripsy requires a high energy density to break up a hard stone. The peak-to-peak pressure change is up to 30 MPa, 32 kW/cm^2, at the focal point in water. This high-energy shock wave is problematic in that it causes damage to skin and tissue, such as subcutaneous hemorrhages, by a simple mechanical effect even far from the focus. Another therapeutic usage of acoustic power is as a heat source. Hyperthermia, in the form of ultrasonic thermotherapy for cancer treatment, is one such major application. The use of low-intensity ultrasound in the field has been limited, but minute mechanical and thermal effects are expected, which may help to resolve tendonitis or to relieve pain from a shoulder.

The effects of mechanical and thermal damage on biological tissue vary depending on the acoustic power and its absorption coefficient. The absorption coefficient basically depends on the protein content and density. For example, tendons and bones have a relatively higher absorption coefficient than other tissues. The nervous system also has a higher absorption coefficient than other organs. The theory of ultrasound imaging is based on the fact that ultrasound energy can pass through muscles and organs. Those ultrasound-transparent tissues are relatively insensitive to the acoustic power. If an appropriate level of acoustic power is chosen, ultrasound imaging should be applicable without any anxiety concerning side-effects on biological tissue.

2.6.2 The Biological Effects of Diagnostic Ultrasound

Accordingly, imaging devices use an acoustic power that is low enough to avoid apparent damage to human organs and skin. Ultrasound power for diagnostic usage may reach as high as 2 W/cm^2 even in continuous-wave Doppler examination, which requires the most intensive ultrasound power. Early studies showed that biological effects were subtle at this acoustic power level, except for the thermal effect on bony tissue and collagen-rich tissue. Because there is no mechanical and thermal damage, clinical ultrasound examination may be considered completely safe. The diagnostic usage of ultrasound, including imaging, geometry measurement, and backscatter examination, has become common today.

These are continuing experiments that report the biological effects of diagnostic ultrasound [48]. In their official statements, The World Federation for Ultrasound in Medicine and Biology (WFUMB) and the American Institute of Ultrasound in Medicine (AIUM) have mentioned the biological effects and adverse effects of diagnostic ultrasound [49,50]. There is some evidence suggesting that the heating effect causes a biological response [51]. Ultrasound-induced heating of the embryo or fetus should not be neglected, because a small volume of developing bone tissue and nerve system is heated significantly by ultrasound at diagnostic power. The lens of the human eye is highly sensitive to heat, with the risk of the development of cataracts. Cavitation effects on the gas/fluid border and gas-saturated fluid produce free radicals at diagnostic power [52]. Lung tissue bleeding has been reported as a result of cavitation-related effects. Free radicals have chemical activity such as oxidation, and disrupt chemical bonds and structures of biological molecules [53]. Such effects cannot be explained by the thermal effects of ultrasound.

2.6.3 The Effect of Ultrasound on the Cell Membrane

Progress in molecular biology provides us with a lot of information that helps us to understand machinery of cell and sub-cellular structure on a molecular basis. Proteins and other organic compounds have their biochemical roles, including interactions that determine biological responses. Ultrasound may affect protein particle aggregation and may change its chemical properties. The fluid mosaic model of the cellular membrane indicates that the lipid fluidity of the membrane and the membrane proteins change their compositions due to local pressure variation such as ultrasound exposure. Phase transition in a lipid bilayer changes membrane stability and permeability. Effects on membrane proteins, such as receptors, can trigger a biochemical signaling cascade followed by a variety of cellular responses. In recent years, interest in the biological effects of ultrasound has been rekindled, as it has become a method of manipulating living phenomena.

Biological tissue is constructed from cell and interstitial matrix and connective tissue. Each cell has a membrane at its outer boundary. The basic structure of the cell membrane is a lipid bilayer that has a hydrophobic layer in the middle. Hydrophilic, water-soluble dye cannot penetrate the intact membrane. Ethidium bromide is one fluorescent hydrophilic dye that does not stain the inside of an intact cell. However, after 5 s of ultrasound exposure (2 MHz, 200 mW/cm^2, SPTA), an ethidium bromide stain was found (Fig. 2.35).

This result shows us that ultrasound exposure at the same acoustic power as diagnostic equipment has the effect of increasing membrane permeability. The cell membrane works as a protective barrier when attempts are made to transport materials into a cell. Severe disruption of the integrity of the membrane may cause drastic changes in the internal environment of the cell that results in immediate cell death.

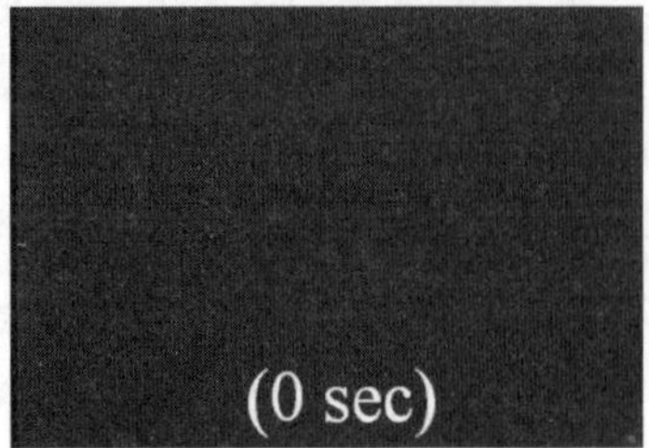

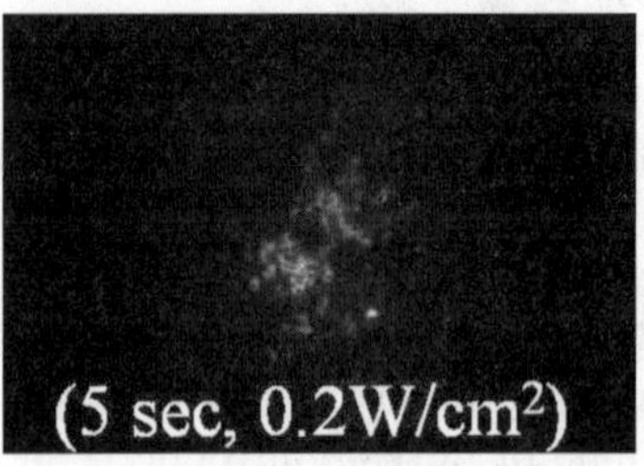

Fig. 2.35. Hydrophilic fluorescent dye staining showed an increased membrane permeability due to ultrasound exposure. *Left panel:* No staining without ultrasound exposure. *Right panel:* Staining (*white dots*) by 5 s of ultrasound (0.2 W/cm^2, 2 MHz) exposure

2.6.4 Ultrasound for Gene Therapy

Current research is making great efforts to use high molecular weight compounds as medicines. Because those molecules are primary functional units in cells, their replacement or repair means the direct treatment of diseases based on molecular functions, not on symptoms. A simple strategy is the insertion of defective or missing functional genes. An introduced gene molecule, deoxyribonucleotide (DNA), would be transcribed as a template for protein synthesis.

This kind of therapy requires the insertion of a functional form of those molecules into a cell to obtain a therapeutic effect. One problem is that such a high molecular weight, "macro" molecule is very difficult to introduce into a cell. As gene therapy, several methods have been contrived. Electroporation uses a electric pulse to make pin holes in a cell membrane. A particle delivery system uses accelerated micro-particles that fire DNA molecules into a cell like a gunshot. A gene transfection technique uses a cationic liposome, a positively charged lipid particle, as a carry molecule that makes a DNA carrier complex.

Increased cell membrane permeability might help gene transportation into a cell. The efficiency of gene transfection can be assessed by the amount of gene products, which is usually measured by enzyme activity. Ultrasound exposure, hypothetically, makes a perforation in the cell membrane, so this technique is called "sono-poration". The activity of the gene product luciferase was 3.4-fold higher after 30 s of ultrasound exposure (1 MHz, 800 mW/cm^2, SPTA). However, the amount of introduced gene vector was 16-fold greater in the ultrasound exposed group than in the control group. Because more than 30 s of ultrasound exposure caused cell damage, the total gene activity was not increased from the condition of 30 s exposure. Here, we saw two phenomena. First, there was an optimal power of ultrasound to increase the efficiency of gene transfection. In other words, too much exposure to ultrasound increases the chance of seeing adverse effects. Second, there was a discrepancy between the increasing ratio of the gene activity and the net

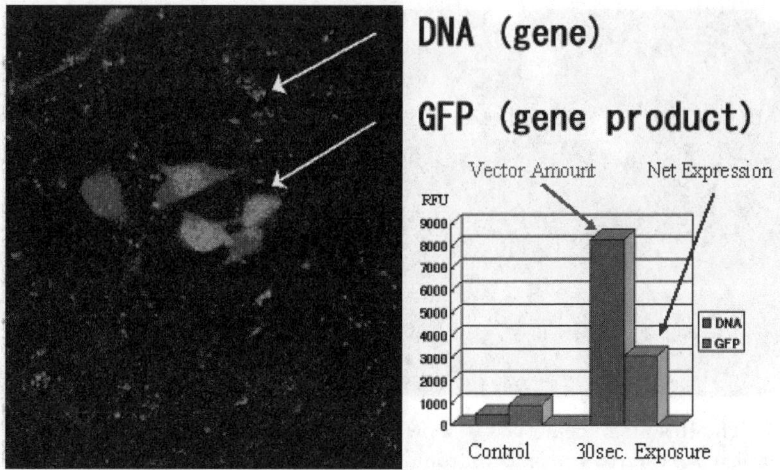

Fig. 2.36. Fluoresence labeled vectors (DNA) and their products (GFP). The vectors were introduced into the cell by ultrasound. Then cells transcripted them and synthesized fluorescent protein

amount of introduced gene. This suggests that ultrasound exposure has effects on cellular properties in addition to changing the permeability of the membrane (Fig. 2.36).

2.6.5 Direct Effects on Cell Components

The mutagenic activity of radiation is well known and is mainly explained by damage to DNA. Any kind of energy may lead to the same effect on biological materials as radiation. Fragmentation of genomic DNA was assayed by gel electrophoresis study. Smaller DNA fragments that have a smaller molecular size move greater distances than unfragmented DNA. The electrophoresis pattern of the cell exposed to ultrasound showed a fragmented DNA tail, while the pattern of the untreated cell did not (Fig. 2.37).

This result indicates that ultrasound exposure causes DNA damage. Free radicals and hydrogen peroxide that are generated by the cavitation effect of ultrasound may cause DNA damage just as radiation does. Nevertheless, this hypothesis is likely to be supported: the lifetime of the free radicals is too short to allow them to reach the nucleus DNA from the cell membrane. There is no clear evidence that cavitation may occur inside a cell at the acoustic power used in diagnosis today. Thus, we should consider another hypothesis, that DNA damage results from cellular activity.

2.6.6 The Stress-Induced Cellular Response

A mammalian cell has a sophisticated system to repair damaged DNA. Severe DNA damage does not allow expression of any gene, even for repair, which

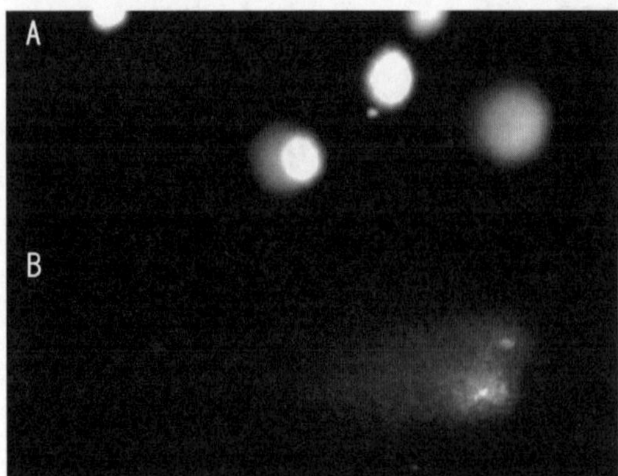

Fig. 2.37. The Results of comet assay show that fragmented DNA that has a small molecular size is easier to translocate from its original position than the intact and larger DNA molecules. **A:** The absence of comet tail like trace of DNA without ultrasound exposure meant that DNA was intact in the control condition. **B:** A comet tail like DNA trace (the blurred image) from the position of the nucleus means the existence of fragmented DNA in an ultrasound-treated cell

means immediate arrest of cell function, and cell death. If complete repair is achieved, the damaged cell can survive as before. At the intermediate level of ultrasound exposure, not severe enough to cause immediate death but too severe for complete restoration, cells go into apoptosis. Necrosis means a status of cell death. Apoptosis is another route to cell death, that can be explained as programmed cell death.

In apoptosis, cell activity initiates a process of degrading the genomic DNA into fragments. The sequence of apoptosis consists of three phases: damage sensing, the initiation of degradation activity, and cell death with morphological change. Initially, apoptosis has been found to be induced via the stimulation of several cell surface receptors. Cell membrane alternations, activation of enzymes, disruption of mitochondria, and DNA damage are other possible key triggers of apoptosis.

A protein, p53, was initially introduced as a cancer suppressive factor. Its function is regulation of the cell cycle. DNA damage results in phosphorylation of p53, which protects p53 from degradation. Accumulation of p53 arrests the cell cycle at the G1 phase, which is the pre-phase of DNA synthesis for mitosis. This regulation of the cell cycle plays a role in the time taken to repair the DNA, before DNA duplication in preparation for cell proliferation. Further accumulation of p53 molecules triggers the induction of apoptosis. This means the abandonment of any effort to repair.

The accumulation of p53 was monitored by the expression of a manipulated gene which is a fused gene of p53 and green fluorescent protein (GFP) into a cultured cell. Then, we monitored p53 accumulation by measuring the

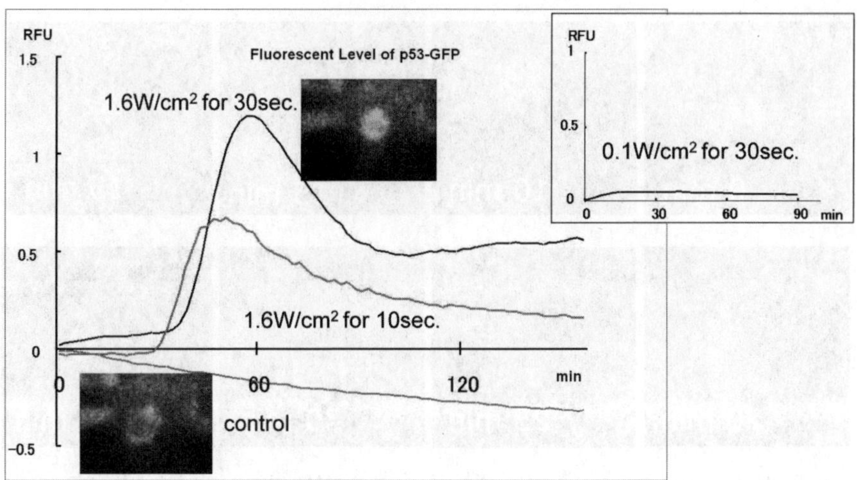

Fig. 2.38. The Accumulation of p53-GFP fusion protein by ultrasound exposure. The graph of time versus the relative fluorescence level shows that p53 accumulation begins approximately 30 minutes from the start of ultrasound exposure. The right upper panel shows the result of 0.1 W/cm^2 acoustic power. There was no significant increase of the fluorescent level, which meant that P53 accumulation by ultrasound had exposed the dose dependency

intensity of the green fluorescent. Exposure to ultrasound at 1600 mW/cm^2 acoustic power showed its increase, but at 100 mW/cm^2 there was no significant change (Fig. 2.38).

At a lower acoustic power, the different aspect of the response was seen. Cellular function is regulated by changing protein characteristics via the activation and inactivation of enzymes. Protein kinase C is an enzyme protein that has signaling roles of in the regulation of cellular function. It is well known as an upstream factor of the cell proliferation sequence in several cell lines.

Protein kinase C (PKC) is located at cytosol when it is in an inactive form, and it is activated by stimulation. The activated form of PKC is located in the cell membrane. Therefore, we may assess the activation of PKC by observation of its location. A fusion of PKC and green fluorescent protein (GFP) was introduced into a culture cell as a marker of the PKC location. An accumulation of PKC-GFP on the cell membrane was seen as a result of continuous exposure to low acoustic power ultrasound (Fig. 2.39).

2.6.7 Potential Applications of Low-Power Ultrasound

We have seen several adverse effects of ultrasound for biological tissue. Recently, another aspect of the biological effects of echo image contrast agents has become known. Gas-filled particles for an echo image contrast enhancer

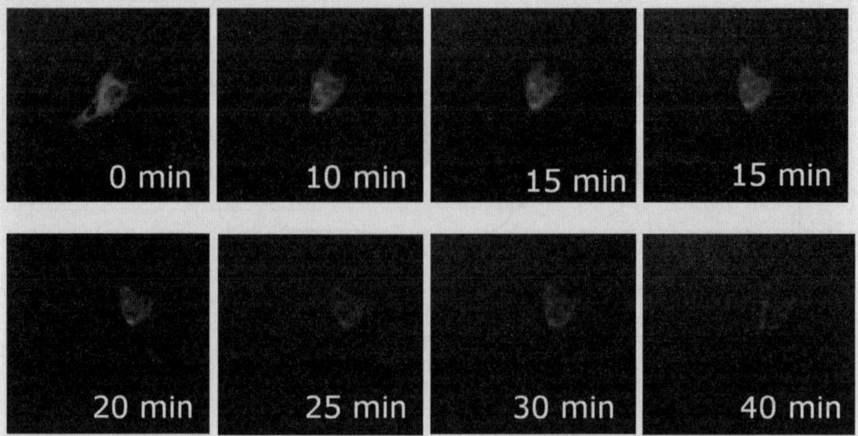

Fig. 2.39. Continuous exposure to 0.1 W/cm^2, 2 MHz ultrasound resulted in translocation of green fluorescent dye from cytosol to the cell membrane, which demonstrated the ultrasound medicated activation of protein kinase C

also enhance the cavitation effect [54]. Thus, today, biological effects should not be ignored even if ultrasound is used for imaging purpose. Risk–benefit oriented decision-making might be required even for ultrasound exposure. Ultrasound examination should not be performed lightly, without any need for clinical benefit. On the other hand, an osteogenesis acceleration device which features daily exposure to low acoustic power ultrasound has been introduced. Ultrasound-induced apoptosis is potentially applicable to the treatment of leukemia [55]. Low acoustic power ultrasound, at 1 W/cm$^2{}_{SPTA}$ or lower, does not suffer from problems of adverse effects, and has the potential for cellular function control. Further investigations on the biological effects of ultrasound, especially exposure to an acoustic power lower than the power of an imaging device, may lead to mechano-chemical manipulation of biological functions.

References

1. J.G. Miller, J.E. Perez, B.E. Sobel: Prog. Cardiovasc. Dis. **28**, 85 (1985)
2. J.W. Mimbs et al.: Circ. Res. **48**, 49 (1980)
3. K.K. Shung, J.M. Reid: Proc. IEEE Ultrason. Symp. **CH12364-ISU**, 230 (1977)
4. M. O'Donnell, J.W. Mimbs, J.G. Miller: J. Acoust. Soc. Am. **69**, 580 (1981)
5. J.G. Mottley, J.G. Miller: J. Acoust. Soc. Am. **83**, 755 (1988)
6. J.W. Mimbs et al.: Circ. Res. **49**, 89 (1981)
7. E.I. Madaras et al.: Ultrason. Imag. **5**, 229 (1983)
8. S.A. Wickline et al.: Circulation **72**, 183 (1985)
9. K.A. Wear, T.A. Shoup, R.L. Popp: IEEE Trans. Ultrason. Ferroelectric Freq. Control **33**, 347 (1986)

10. K. Baba, K. Satoh: Acta Obstet. Gyn. Japon. **38**, 1385 (1986)
11. K. Baba et al.: "Non-invasive three-dimensional imaging system for the fetus in utero", in: *The Fetus as a Patient '87*, ed. by K. Maeda (Excerpta Medica, Amsterdam, 1987)
12. K. Baba et al.: J. Perinat. Med. **17**, 19 (1989)
13. K. Baba, T. Okai: "Basis and principles of three-dimensional ultrasound", in: *Three-Dimensional Ultrasound in Obstetrics and Gynecology*, ed. by K. Baba and D. Jurkovic, Chap. 1 (Parthenon Publishing, Carnforth, 1997)
14. K. Baba: "Development of three-dimensional ultrasound in obstetrics and gynecology". in: *3-D Ultrasound in Obstetrics and Gynecology*, ed. by E. Merz, Chap. 1 (Lippincott Williams & Wilkins, Philadelphia, 1998)
15. G. Kossoff et al.: "Principles of three-dimensional volume imaging in sonography", in: *Three-Dimensional Ultrasound in Obstetrics and Gynecology*, ed. by K. Baba and D. Jurkovic, Chap. 2 (Parthenon Publishing, Carnforth, 1997)
16. K. Baba, T. Okai, S. Kozuma: Lancet **348**, 1307 (1996)
17. K. Baba et al.: Radiology **203**, 571 (1997)
18. S. Kozuma et al., Ultrasound Obstet. Gynecol. **13**, 283 (1999)
19. K. Baba, T. Okai: "Clinical applications of three-dimensional ultrasound", in *Three-Dimensional Ultrasound in Obstetrics and Gynecology*, ed. by K. Baba and D. Jurkovic, Chap. 3 (Parthenon Publishing, Carnforth, 1997)
20. K. Baba et al.: Radiology **211**, 441 (1999)
21. B. Maier et al.: "The psychological impact of three-dimensional fetal imaging on the fetomaternal relationship", in: *Three-Dimensional Ultrasound in Obstetrics and Gynecology* ed. by K. Baba and D. Jurkovic, Chap. 7 (Parthenon Publishing, Carnforth, 1997)
22. K. Baba, K. Kinoshita: *The Physics and Equipment of Ultrasound Imaging for Accurate Diagnosis* (Medical View, Tokyo, 1993)
23. N. Bom et al.: Int. J. Cardiac Imaging **4**, 79 (1989)
24. K. Mitsuto: Clinic All-Round **45**, No. 9, 2159 (1996)
25. O. Oshiro, H. Tojo, K. Chihara: Trans. Inst. Syst. Control Inf. Eng. **8**, No. 8, 344 (1995)
26. D.H. Johnson, D.E. Dudgeon: "Apertures and arrays", in: *Array Signal Processing*, Chap. 3 (Prentice Hall, Englewood Cliffs, NJ, 1993)
27. K.K. Shung, M.B. Smith, B. Tsui: "Ultrasound", in: *Principles of Medical Imaging*, Chap. 2 (Academic Press, California, 1992)
28. A. Fenster, D.B. Downey: IEEE Eng. Med. Biol. Soc.**15**, No. 6, 41 (1996)
29. O. Oshiro et al.: Trans. Inst. Syst. Cont. Inform. Eng. **13**, No. 5, 244 (2000)
30. P.M. Kennedy, D.X. Hammer, B.A. Rockwell: Prog. Quant. Electr. **21**, No 3, 155 (1997)
31. M. Hisaka, T. Sugiura, S. Kawata: J. Opt. Soc. Am. A., in print
32. M. Hisaka, T. Sugiura, S. Kawata: Jpn. J. Appl. Phys. **38**, L1478 (1999)
33. L. Wang, S. Jacques, X. Zhao: Opt. Lett. **20**, 629 (1995)
34. L. Wang, X. Zhao: Appl. Opt. **36**, 7277 (1997)
35. S. Lévêque et al.: Opt. Lett. **24**, 181 (1999)
36. K. Ishihara et al.: "High-speed digital subtraction echography: principle and preliminary application to arteriosclerosis, arrhythmia and blood flow visualization", Proc. 1990 IEEE Ultrasonics Symposium, p. 1473 (Honolulu, HI, Dec. 1990)
37. K. Ishihara et al.: "Principle of high-speed digital subtraction echography and the potential for clinical applications", Jpn. J. Appl. Phys. **30**, Suppl. 30-1, 322

38. K. Ishihara: "Development of high-speed digital subtraction echography and clinical application in the field of circulation", Osaka Daigaku Igakuzasshi (in Jpn) **43**, No. 5, 61 (1991)
39. K. Ishihara et al.: "Microballoon as ultrasonic sensor-actuator in vivo", Jap. Soc. Prof. Eng. (in Jpn) **56**, No. 12, 2152 (1990)
40. K. Ishihara et al.: "Drug delivery system controlled by resonant ultrasonics", Med. Biol. Eng. Comp. **29**, Suppl. Part 1, 151 (1991)
41. K. Ishihara et al.: "New approach to noninvasive manometry based on pressure dependent resonant shift of elastic microcapsules in ultrasonic frequency characteristics", Jpn. J. Appl. Phys. **27**, Suppl. 27-1, 125 (1988)
42. M. Minnaert: "On musical air-bubbles and the sounds of running water", Phil. Mag. **16**, 235 (1933)
43. N. Negishi et al.: "Ultrasonic control of drug releasing", Jpn. J. Artificial 0rgans (in Jpn) **13**, No. 3, 1205 (1984)
44. K. Ishihara et al.: "Experimental study of noninvasive pressure measurement method based on pressure dependent resonant shift in ultrasonic frequency characteristics", Jpn. J. Med. Ultrasonics (in Jpn) **15**, No. 2, 107 (1988)
45. K. Ishihara et al.: "Noninvasive and precise motion detection for micromachines using high-seed digital subtraction echography", Proc. IEEE Micro Electro Mechanical Systems, Jan.–Feb. 1991, p. 176, Nara, Japan (1991)
46. K. Masuda, K. Ishihara: "Estimation of collapsed microcapsules for drug delivery system using successive echograms", Proc. of World Cong. on Med. Physics and Biomed. Eng., Jul. 2000, Chicago, No. 5004–59481 (CD-ROM)
47. K. Masuda, K. Ishihara: "Collapse monitoring of microcapsules and its quantitative evaluation from successive echograms", Proc. of 1st International Symposium on Ultrasound Contrast Imaging, 1999, Kyoto, p. 120
48. D.L. Miller: Ultrasound Med. Biol. **13**, 443 (1987)
49. S.B. Barnett et al.: Ultrasound Med. Biol. **23**, 805 (1997)
50. S.B. Barnett et al.: Ultrasound Med. Biol. **26**, 355 (2000)
51. S.B. Barnett, G. Kossoff, M.J. Edwards: Med. J. Aust. **160**, 33 (1994)
52. D.L. Miller, R.M. Thomas: J. Acoust. Soc. Am. **93**, 3475 (1993)
53. M.W. Miller, D.L. Miller, A.A. Brayman: Ultrasound Med. Biol. **22**, 1134 (1996)
54. N. Nanda, R. Schlief, B. Goldberg: *Advances in Echo Imaging Using Contrast Enhancement*, 2nd edn. (Kluwer, Dordrecht, 1997)
55. L. Lagneaux et al.: Exp. Hematol. **28**, 1503 (2000)

3 The Imaging of a Magnetic Source

H. Kado, H. Ogata, Y. Haruta, M. Higuchi, M. Shimogawara, J. Kawai, Y. Adachi, C. Bertrand, and G. Uehara

3.1 The Principle of Magnetic Field Measurement

The purpose of this section is to provide information about magnetic field measurement, especially of very weak magnetic signals, and about the analysis of the source of the measured signal. It is not our intention to provide full information on magneticism and the related area, which is too wide and beyond our capability. To realize the purpose, limited discussions will be provided on topics from basic principles of measurement to several examples of the state-of-the-art issue. The discussions will include the definition of a magnetic field, an example of a magnetic field source, the concept of measurement of the field, practical technology and measuring devices and an example of source analysis. Most of those items are also described in the following subsections in more detail. Some of the details are basic knowledge, while others have to do with contemporary studies on applications of the technology. We have tried to arrange the description as self-contained as possible and to keep reference citations to the necessary minimum.

A knowledge of magnetism is assumed to be already established at a level such as a university undergraduate course on electromagnetism, including basic vector concepts. If you are not familiar with these subjects, we strongly recommend that you refer to textbooks on electromagnetism that are accessible at most public or university libraries. Two physical quantities in magnetism – the magnetic field and magnetic flux – are important concepts in the context of this chapter. Two systems – SI and cgs – are commonly used to express electrical and magnetic quantities. In this chapter, we use the SI system.

3.1.1 The Magnetic Field

A magnetic field is a result of the existence of magnetic sources. One simple magnetic source would be a magnet. To observe the magnetic field generated by a magnet, small compass magnets, which are one of the most simple magnetic measurement devices, can be used. Figure 3.1 shows the pattern of the compass magnets placed around the magnetic source. The pattern indicates the existence of an invisible force distribution between the compass magnets and the magnetic source. The strength and direction of the force

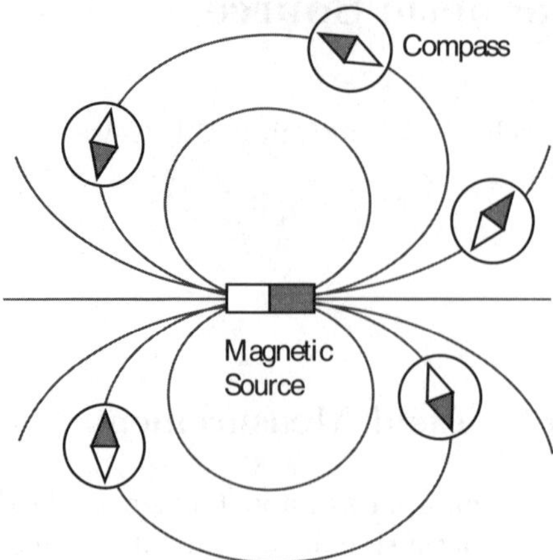

Fig. 3.1. Compass magnets and a magnetic source

differ from place to place around the source magnet. The concept of the pattern is one of the operational definitions of a magnetic field. Actually, electromagnetism gives us various definitions of a magnetic field. If we need to quantify the field, we could calculate the strength of the field by (3.1), which describes the relationship between the magnet and the magnetic field.

$$\boldsymbol{B}(\boldsymbol{r}) = \frac{\mu}{4\pi} \left\{ \frac{1}{|\boldsymbol{R}-\boldsymbol{r}|^3} \left(3\,\boldsymbol{r} \frac{\boldsymbol{r}\cdot\boldsymbol{M}}{|\boldsymbol{R}-\boldsymbol{r}|^2} - \boldsymbol{M} \right) \right\} , \tag{3.1}$$

where μ is the relative magnetic permeability, $\boldsymbol{M}$ is the magnetic moment of the source magnet, $\boldsymbol{R}$ is the location of the magnet, and $\boldsymbol{r}$ is the measurement location in the magnetic field. In the SI system, the magnetic field is expressed in Tesla. Assuming that a number of iron particles instead of compass magnets are distributed around the magnetic source, we could see the same pattern, which could be described as many lines in a pattern. These lines give us the concept of the magnetic force line in electromagneticism. We can define the idea of the density of lines, and this density varies from place to place. The density of the lines expresses the strength of the magnetic field.

3.1.2 The Magnetic Dipole

To be more precise, the length of the magnet is relatively small in comparison with the distance between $\boldsymbol{r}$ and $\boldsymbol{R}$ (say, the length is 1 cm, and the distance is 1 m). The source magnet is regarded as a single particle, which is called a magnetic dipole. Equation (3.1) is the equation of the magnetic dipole.

3.1.3 Magnetic Flux

Assuming that there is a small area around the magnetic source, we could accumulate the magnetic force lines that pass through this small area. In other words, the magnetic field integrated over the area is defined as the magnetic flux. The magnetic flux is expressed in webers (Wb) in the SI system. A quantitative expression for the magnetic flux is also given in (3.2).

$$\Phi = \int_S B(\boldsymbol{r})\mathrm{d}S\,. \tag{3.2}$$

From the discussion on the magnetic field and flux using the concept of the magnetic force line, the magnetic field is also called the magnetic flux density. In physics, the terms magnetic field and magnetic flux density are equivalent. Both expressions are used in the following text, depending on the context.

3.1.4 The Electromagnetic Coil

Instead of a source magnet, the elementary electromagneticism teaches us about a device that generates an equivalent magnetic flux density by an electrical current, using a coil. Figure 3.2 is an electrical coil that generates a magnetic field. The relation between the coil parameters and the magnetic field is defined in the following equation:

$$\boldsymbol{B}(\boldsymbol{r}) = \boldsymbol{F}(I, A, \boldsymbol{R}, \boldsymbol{N} \mid \boldsymbol{r})\,, \tag{3.3}$$

where I is the electric current in ampere, A is the diameter of the coil, $\boldsymbol{R}$ is the position of the coil, $\boldsymbol{N}$ is the direction of the coil, and $\boldsymbol{r}$ is the measurement

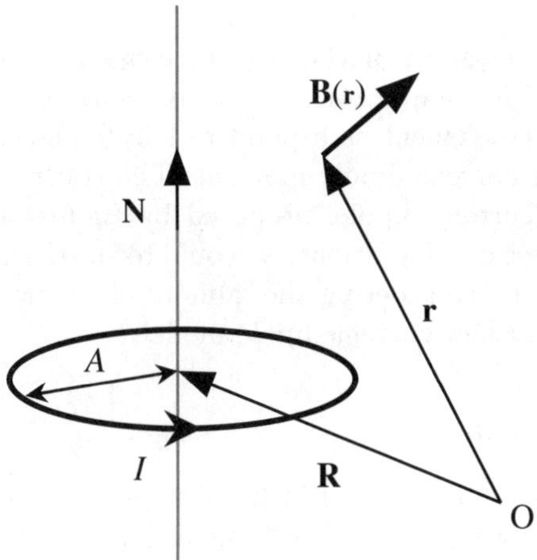

Fig. 3.2. The magnetic field generated by a coil

location in the magnetic field. These definitions are shown in Fig. 3.2. When the diameter of the coil is small, the magnetic field becomes similar to that of a small magnet.

3.1.5 The Current Dipole

To divide the coil into small current elements, the magnetic field can be approximated by the contribution of all of the current elements. The magnetic field $\mathrm{d}\boldsymbol{B}(\boldsymbol{r})$ generated by a current element is expressed by the following equation:

$$\mathrm{d}\boldsymbol{B}(\boldsymbol{r}) = \frac{\mu}{4\pi}\frac{I\,\mathrm{d}L\,\boldsymbol{N}\times(\boldsymbol{R}-\boldsymbol{r})}{|\boldsymbol{R}-\boldsymbol{r}|^3}\,, \tag{3.4}$$

where I is the electric current of the element, $\mathrm{d}L$ is the length of the element, $\boldsymbol{N}$ is the direction of the element, $\boldsymbol{R}$ is the position of the element, and $\boldsymbol{r}$ is the measurement location in the magnetic field. If we use this idea, the magnetic field can be expressed by the integration of $\mathrm{d}\boldsymbol{B}(\boldsymbol{r})$ along the coil:

$$\boldsymbol{B}(r) = \int_c \mathrm{d}\boldsymbol{B}(\boldsymbol{r})\,\mathrm{d}L\,. \tag{3.5}$$

As you can imagine from this example, a general magnetic source can be approximated by the summation of current elements with the length being small. Therefore, such a small current element itself is one of the most fundamental magnetic sources. We can rewrite (3.4) as the magnetic field expression of a small current element:

$$\boldsymbol{B}_{\mathrm{cd}}(r) = \frac{\mu}{4\pi}\frac{i\,\boldsymbol{N}\times(\boldsymbol{R}-\boldsymbol{r})}{|\boldsymbol{R}-\boldsymbol{r}|^3}\,, \tag{3.6}$$

where $i = I\ \mathrm{d}L$. From the physics viewpoint of the current element, the length of the element is relatively small in comparison with the distance from $\boldsymbol{R}$ to $\boldsymbol{r}$. In this case, we can regard the element as a point of current flow. The parameter i is called the electric current dipole moment. The concept of current flow is called the electric current dipole, proposed by Biot and Savart (the Biot–Savart law). From these considerations, we could reconstruct the magnetic field of a source magnet by considering the number of electric current dipoles. This is called the equivalent current dipole model.

3.1.6 Time-Varying Magnetic Fields

Magnetic fields sometimes vary. This is mostly caused by variation of the magnetic source parameter. If we are measuring the ambient magnetic field at a certain point in the street, and an automobile passes by the magnetic sensor, we can observe the time-varying magnetic field intensity. In this case, the automobile could be regarded as a magnet and the movement of the automobile is modeled by the motion of a magnet. The parameter of the magnetic source varies over time, and the magnetic field of the source varies over time. The time-varying magnetic field is expressed by a function of time. For example, in the case of a current dipole as a magnetic source, (3.6) is modified by the time parameter as follows:

$$\boldsymbol{B}_{\mathrm{cd}}(t, \boldsymbol{r}) = \frac{\mu}{4\pi} \frac{i(t)\ \boldsymbol{N}(t) \times (\boldsymbol{R}(t) - \boldsymbol{r})}{|\boldsymbol{R}(t) - \boldsymbol{r}|^3} \,. \tag{3.7}$$

The magnetic field at location $\boldsymbol{r}$ which is generated by a current dipole as a function of $i(t)$, $\boldsymbol{N}(t)$, and $\boldsymbol{R}(t)$. In a practical application of a magnetic field measurement, when we observe a time-varying magnetic field signal with a measurement instrument, this means the possibility of a time-varying magnitude and/or direction and/or position of the magnetic source. Most of the applications of magnetic field measurement are to identify the characteristics of the source of the magnetic field. This study is in the field of inverse problems. In particular, magnetic source analysis is called an inverse source problem. We will address the basic considerations of magnetic source analysis in the later part of this section, and an example of the analysis ins a subsection of Sect. 3.3 (Magnetic Source Analysis). In the issue of magnetic source analysis, one single-point observation of a time-varying magnetic field data does not give us enough information on whether the intensity and/or the direction and/or the position of the signal source contribute to the data. This is the so-called ill-posed problem of an inverse problem. Before addressing magnetic source analysis, some of the technologies on how to measure the magnetic field will be discussed.

3.1.7 The Search Coil Magnetometer

One of the simple methods to measure a magnetic field is to use a search coil. If one places a coil in a varying magnetic field, a voltage is induced the coil. The induced voltage versus the intensity of magnetic field depend on the following equations:

$$\begin{aligned} V(t) &= \frac{\mathrm{d}\Phi(t)}{\mathrm{d}t}, \qquad \Phi(t) = kSB(t)\,, \\ \Rightarrow V(t) &= kS\frac{\mathrm{d}B(t)}{\mathrm{d}t}\,. \end{aligned} \tag{3.8}$$

By integrating $V(t)$ with an electronic circuit, we could obtain the value of $B(t)$. Figure 3.3 shows the electronic integration circuit of a magnetometer. It is sometimes called a search coil. The search coil measures the normal component of the magnetic field vector facing the diameter of the coil. If we define the direction of the search coil as $\boldsymbol{N}$ and note the magnetic field

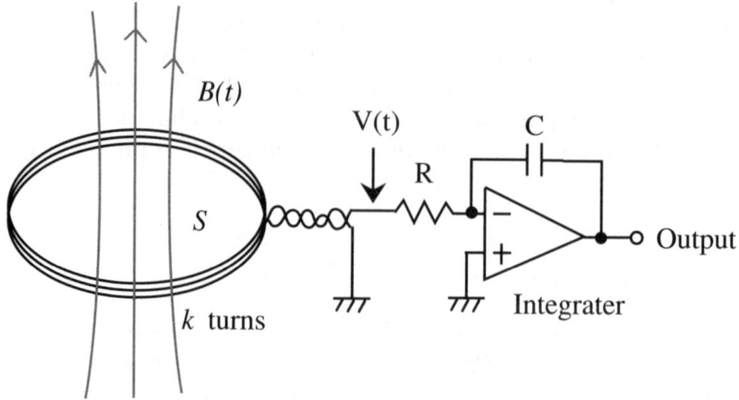

Fig. 3.3. the configuration of a search coil

vector at the search coil, the value measured by the magnetometer is the inner product of two vectors $\boldsymbol{B}(t)$ and $\boldsymbol{N}$:

$$B(t) = \boldsymbol{B}(t) \cdot \boldsymbol{N} . \tag{3.9}$$

Most of the magnetometers describe later, such as the search coil, the fluxgate magnetometer, and the SQUID magnetometer, are needed to measure a component of the vector magnetic field as an inner product of two vectors $\boldsymbol{B}(t)$ and $\boldsymbol{N}$. As important information for further analysis, not only the sensitivity of the magnetometer, but also the position and direction of the sensor should be recorded.

3.1.8 The Proton Magnetometer and Other Magnetometers

On the other hand, a proton magnetometer measures the direct value of the intensity of the magnetic field; in other words, the absolute value of the vector magnetic field. The proton magnetometer uses the phenomenon of nuclear magnetic resonance (NMR). A proton in water has spin, and if the proton is superposed on an artificial dc magnetic field, the spin shifts to a certain angle, which is determined by the summation of the ambient and the artificial magnetic fields. If we erase the artificial magnetic field stepwise, the shifted spin comes back to the initial state: this is called a relaxation process. During the process, the proton emits an electromagnetic wave whose frequency is proportional to the intensity of the ambient magnetic field. Therefore, if we count the wave of the relaxation process, we could calculate the intensity of the magnetic field. This is called a proton magnetometer. The concept of the configuration of the proton magnetometer is shown in Fig. 3.4.

There are many magnetic measurement devices: the compass magnetometer, the fluxgate magnetometer (Fig. 3.5), the optically pumped magnetometer, the Hall effect magnetometer, and the SQUID magnetometer. The

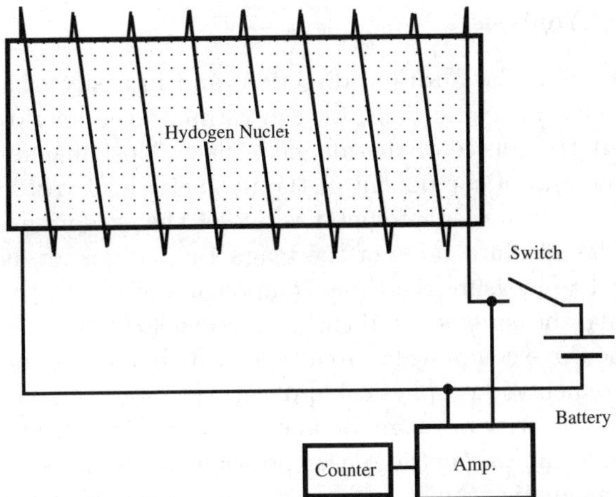

Fig. 3.4. The proton magnetometer

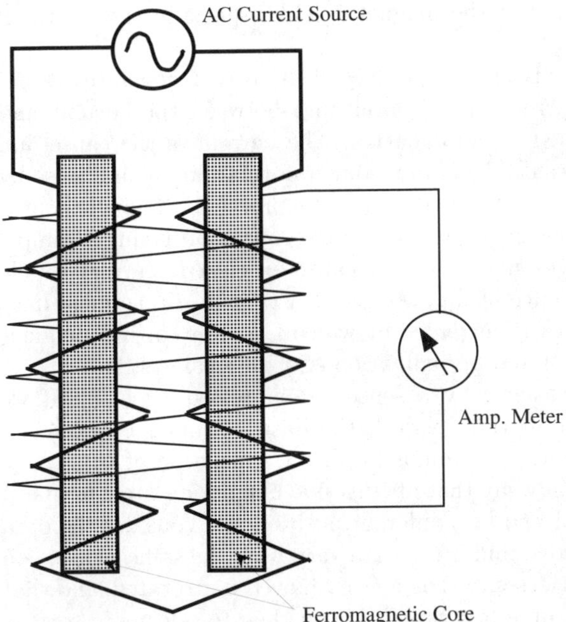

Fig. 3.5. the fluxgate magnetometer

most sensitive magnetometer is the SQUID, which is introduced in a later section for very weak magnetic signal measurement.

3.1.9 Magnetic Source Analysis

As mentioned above, if we know the position, direction, and intensity of a certain component of a vector magnetic field, we can estimate the spatio-temporal information about the source of the magnetic field. Most magnetometers are designed to measure a component of the magnetic field vector in the sensor direction. The proton magnetometer is one of the exceptions. If we would like to know the absolute value of the magnetic field vector at the sensor position, we need to measure the three components of the vector by independent perpendicular measurements. If the magnetic field varies over time, we should measure the three components simultaneously. In most of the studies on magnetic measurements in geophysical applications – for example, magnetic storms due to solar activity or magnetic anomaly detection for mining – the magnetic field information at the sensor position has been used. On the other hand, in biomagnetic studies such as magnetoencephalogram (MEG), the main target of the measurement is not only to get the magnetic field value at the measurement point but also to analyze the obtained value by localization of the source of the magnetic field, especially the electrical activity in the object.

The Biot–Savart law tells us that an electrical current generates a magnetic field around the current. The relationship between the current and the magnetic field is defined by an equation. The current is the cause and the magnetic field is the result. To obtain some information on the cause by observing the result is what is known as an inverse problem. Magnetic source analysis is once such inverse problem, or inverse source problem. A simple example of magnetic source analysis is to localize an electric current dipole (strictly speaking, no true current dipole exists). A current dipole must have a current source and a return current, but we can assume an independent current dipole as a model of an approximation of a physical system.

The magnetic field measured by a sensor is expressed by (3.7). If we observe the field in a certain time period, the data obtained is the time-varying series. From the data, we cannot localize the position of the source and, furthermore, we can only say that at least one magnetic source exists. If you add one more sensor placed at a different position and consider the data from the two different sensors, and if the data vary with the same phase, you can conclude that at least the source has neither moved nor rotated, and that only the intensity of the source has varied. If you then to add more sensors placed at yet more different positions, the signals of all of the channels would also vary with the same phase, and you would obtain a distribution pattern on the magnetic field which would not change during the time period of observation. You could estimate the position and direction of the source from this pattern. If the source does not vary in intensity but varies in position and/or direction, the observed data are assumed to be influenced by the actions. For example, if there are two sensors at different positions and the source is moving between these two positions, the observed data from the

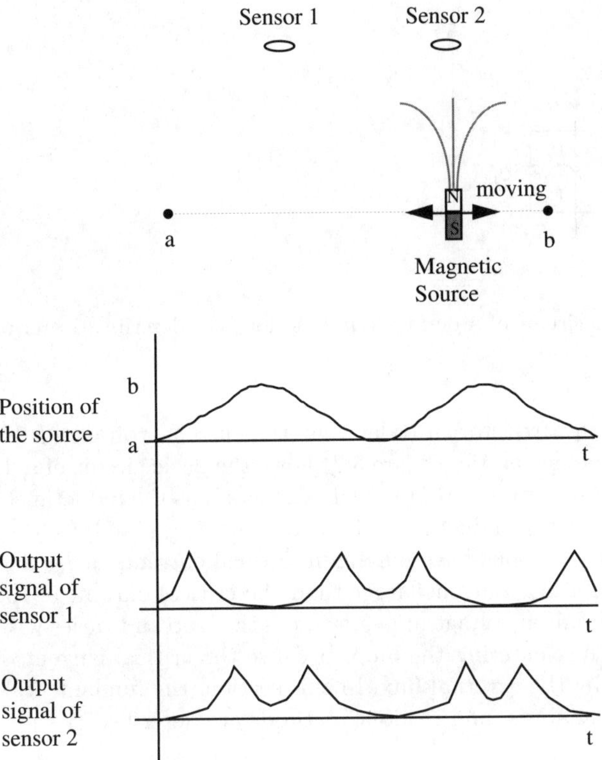

Fig. 3.6. A moving magnetic source

two sensors would vary with different phases, as shown in Fig. 3.6. The phase difference depends on the relation of the positions of and directions between the sensors. If the sensors are close together, the phase difference becomes very small and it is difficult to make a decision as to whether or not the source is moving. In this case we need more sensors to learn about the source. From this simple example, we can see the possibility of knowing the dynamics of the source, such as movment, rotation, or a combination of both, if we have more than two sensors. To analyze such data, we need to consider not only the distribution of the magnetic field but also the time variation. Therefore, this can be called this spatiotemporal analysis. Practical data and methods about this will be described in later sections.

3.2 A High-Sensitivity Magnetic Field Sensor

3.2.1 The SQUID

The fundamentals of a dc SQUID are described in this section. A dc SQUID consists of a superconducting loop with two Josephson tunnel junctions. Each

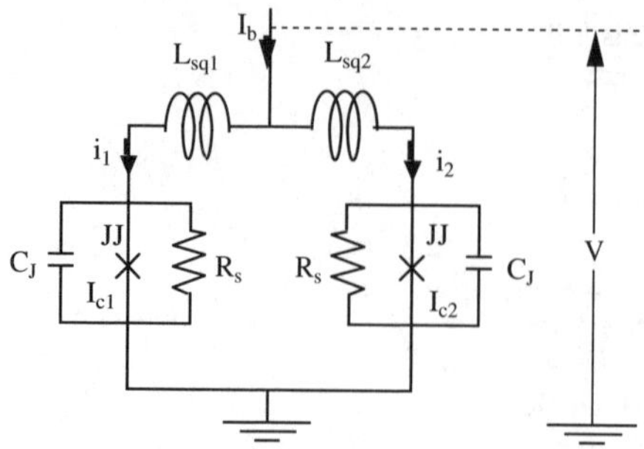

Fig. 3.7. The equivalent circuit of a dc SQUID. Junctions are described in terms of the RSJ model

junction is shunted with a resistor in order that the current–voltage (I–V) characteristics cannot be hysteretic. Figure 3.7 shows the basic circuit of a dc SQUID, which is analyzed with the RSJ model. The loop has the inductances L_{sq}, which is the sum of the inductance of each branch L_{sq1} and L_{sq2}, or $L_{\mathrm{sq}} = L_{\mathrm{sq1}} + L_{\mathrm{sq2}}$. The junctions have maximum critical currents of I_{c1} and I_{c2}, respectively. When a bias current larger than the critical current of the loop is applied to the circuit, a voltage appears across the loop and varies with the external magnetic flux entering the loop, because the critical current of the loop is modulated by the external flux. In this section, the fundamentals of dc SQUIDs, a noise analysis, and readout methods are described [1–5].

Critical Current Modulation by Applied Magnetic Flux. If an external magnetic flux Φ_{ex} enters the loop, there is an exclusion flux Φ_{sc} caused by a screening current $\mathrm{I_s}$ in the loop. The total magnetic flux Φ_{t} in the loop is

$$\Phi_{\mathrm{t}} = \Phi_{\mathrm{ex}} + \Phi_{\mathrm{sc}} \,. \tag{3.10}$$

In terms of L_{sq1}, L_{sq2}, I_{c1}, I_{c2}, and the phase differences of the two junctions ϕ_1 and ϕ_2, Φ_{sc} is given by

$$\Phi_{\mathrm{sc}} = L_{\mathrm{sq1}} I_{\mathrm{c1}} \sin\phi_1 + L_{\mathrm{sq2}} I_{\mathrm{c2}} \sin\phi_2 \,. \tag{3.11}$$

According to the flux quantization condition, ϕ_1 and ϕ_2 satisfy the following relation:

$$2n\pi = 2\pi + \frac{\Phi_{\mathrm{ex}} + \Phi_{\mathrm{s}}}{\Phi_0} + \phi_1 - \phi_2 \,. \tag{3.12}$$

Combining (3.11) with (3.12),

$$\frac{\Phi_{ex}}{\Phi_0} = n - \frac{\Phi_{ex} + \Phi_s}{2\pi} - \beta_{L1} \sin\phi_1 + \beta_{L2} \sin\phi_2 \tag{3.13}$$

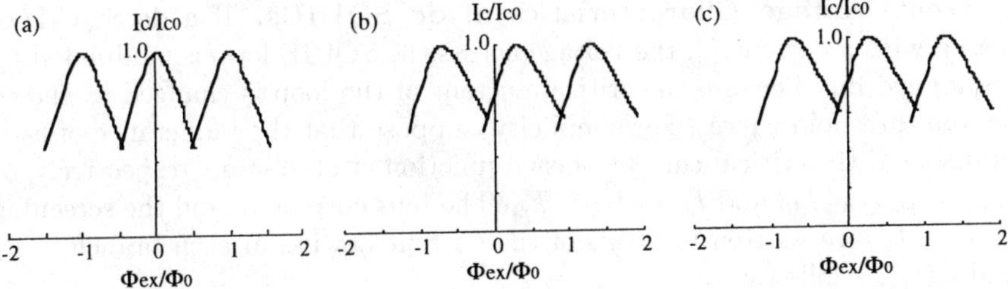

Fig. 3.8. Threshold curves of a dc SQUID numerically calculated for **a** $I_{c1} = I_{c2}$, $L_{sq1} = L_{sq2}$, **b** $I_{c1} = I_{c2}$, $L_{sq1} = 0.25L_{sq2}$, and **c** $I_{c1} = 0.25I_{c2}$, $L_{sq1} = L_{sq2}$

is derived, where

$$\beta_{L1} = \frac{L_{sq1}I_{c1}}{\Phi_0}, \qquad \beta_{L2} = \frac{L_{sq2}I_{c2}}{\Phi_0}. \tag{3.14}$$

Since the critical current of the loop I_c is the sum of the superconducting currents of the two junctions, I_c is written as

$$I_c = I_{c1}\sin\phi_1 + I_{c2}\sin\phi_2 . \tag{3.15}$$

The relationship between Φ_{ex} and I_c is obtained with the maximum I_c for various ϕ_1 and ϕ_2 under the condition indicated as (3.13), using Lagrange's method of undetermined multipliers. The solutions are

$$\frac{\Phi_{ex}}{\Phi_0} = \frac{\phi_2^* - \phi_1^*}{2\pi} + \beta_{L2}\sin\phi_2^* - \beta_{L1}\sin\phi_1^* , \tag{3.16}$$

$$\phi_2^* = \cos^{-1}\left(-\frac{I_{c1}\cos\phi_1^*}{I_{c2} + 2\pi I_{c1}\beta_L^*\cos\phi_1^*}\right), \tag{3.17}$$

where ϕ_1^* and ϕ_2^* are the phase differences for the maximum I_c, and β_L^* is given by

$$\beta_L^* = \beta_{L2} + \beta_{L1}\frac{I_{c2}}{I_{c1}} . \tag{3.18}$$

The calculated Φ_{ex}–I_c characteristics from the above equations, which are called threshold curves, are shown in Fig. 3.8. In the case that the SQUID is symmetric, or $L_{sq1} = L_{sq2}$ and $I_{c1} = I_{c2}$, the threshold curve is also symmetric, as shown in Fig. 3.8a. If the SQUID is asymmetric, the threshold curve becomes asymmetric, as shown in Fig. 3.8b,c in the case of $L_{sq1} \neq L_{sq2}$ and $I_{c1} \neq I_{c2}$, respectively. The vertical axis is normalized with the maximum critical current of the loop I_{c0} ($I_{c0} = I_{c1} + I_{c2}$).

Current–Voltage Characteristics of dc SQUIDs. If a dc SQUID is biased with a current I_{b}, the voltage across the SQUID loop is modulated by an applied flux because the critical current of the loop is changed as shown for the threshold curves. For simplicity, suppose that the inductance of each branch and the critical current of each junction are the same, respectively, or $L_{\mathrm{sq1}} = L_{\mathrm{sq2}} = L_0$, and $I_{\mathrm{c1}} = I_{\mathrm{c2}} = I_{\mathrm{c0}}$. The bias current I_{b} and the screening current I_{sc} are written in terms of the current flowing in each branch, $i_1(t)$ and $i_2(t)$, as follows:

$$I_b = i_1(t) + i_2(t)\,, \tag{3.19}$$

$$I_{sc} = \frac{i_2(t) - i_1(t)}{2}\,. \tag{3.20}$$

$i_1(t)$ and $i_2(t)$ are expressed using the voltages across each junction, v_1 and v_2, namely:

$$i_1(t) = I_{\mathrm{c0}} \sin\phi_1(t) + \frac{v_1}{R_{\mathrm{s}}}\,, \qquad i_2(t) = I_{\mathrm{c0}} \sin\phi_2(t) + \frac{v_2}{R_{\mathrm{s}}}\,, \tag{3.21}$$

where the phase differences ϕ_1 and ϕ_2 are related to the Josephson equation:

$$\frac{\mathrm{d}\phi_1}{\mathrm{d}t} = \frac{4\pi e}{h} v_1\,, \qquad \frac{\mathrm{d}\phi_2}{\mathrm{d}t} = \frac{4\pi e}{h} v_2\,. \tag{3.22}$$

The voltage across the SQUID loop V is the sum of the voltages across the junctions and the inductance:

$$V = v_1 + L_0 \frac{\mathrm{d}i_1}{\mathrm{d}t} = v_2 + L_0 \frac{\mathrm{d}i_2}{\mathrm{d}t}\,, \tag{3.23}$$

where the mutual inductance between two branches is neglected. Rearranging (3.19), (3.21), and (3.23), for a constant bias current,

$$V(t) = \frac{h}{4\pi e} \frac{\mathrm{d}\phi(t)}{\mathrm{d}t} \tag{3.24}$$

is obtained, where $\phi(t)$ is written as

$$\phi(t) = \frac{\phi_1(t) + \phi_2(t)}{2}\,. \tag{3.25}$$

Equation (3.24) indicates an "ac Josephson effect", which means that the phase difference varying in time across the junction results in a voltage output.

Using (3.21) and (3.23), (3.19) and (3.20) are rewritten as

$$I_b = 2I_{\mathrm{c0}} \cos\left[\frac{\phi_1(t) + \phi_2(t)}{2}\right] \cos\left[\frac{\phi_1(t) - \phi_2(t)}{2}\right] + \frac{2V(t)}{R_{\mathrm{s}}}\,, \tag{3.26}$$

$$I_{\rm sc} = 2I_{\rm c0} \cos\left[\frac{\phi_1(t)+\phi_2(t)}{2}\right] \sin\left[\frac{\phi_1(t)-\phi_2(t)}{2}\right], \tag{3.27}$$

where the second term on the right-hand side of (3.27), which is the contribution of quasi-particles, is neglected for simplicity.

Considering the flux quantization condition,

$$\frac{\phi_1(t)-\phi_2(t)}{2} = -\pi\frac{\Phi_t(t)}{\Phi_0} \tag{3.28}$$

is true. Substituting (3.28) into (3.26),

$$I_b = 2I_{\rm c0}\sin[\phi(t)]\cos\left[\pi\frac{\Phi_{\rm t}(t)}{\Phi_0}\right] + \frac{2V(t)}{R_{\rm s}} \tag{3.29}$$

is derived. Since the flux caused by the screening current $\Phi_{\rm sc}$ is written as

$$\Phi_{\rm s} = L_{\rm sq}I_{\rm sc}\,, \tag{3.30}$$

the following equation is yielded by rearranging (3.19), (3.26), and (3.28):

$$\frac{\Phi_{\rm ex}-\Phi_{\rm t}}{\Phi_0} = \beta_{\rm L}\sin\left[\pi\frac{\Phi_{\rm t}(t)}{\Phi_0}\right]\cos[\phi(t)]\,, \tag{3.31}$$

where $2L_0$ is replaced by $L_{\rm sq}$, and the parameter

$$\beta_{\rm L} = \frac{2I_{\rm c0}L_{\rm sq}}{\Phi_0} \tag{3.32}$$

is introduced. If the screening current $I_{\rm sc}$ is neglected, $\Phi_{\rm t}(t) \sim \Phi_{\rm ex}(t)$ because $\beta_{\rm L} \longrightarrow 0$. Equation (3.29) is simplified as

$$I_{\rm b} = I_{\rm c}(\Phi_{\rm ex})\sin[\phi(t)] + \frac{2}{R_{\rm s}}\frac{h}{4\pi e}\frac{{\rm d}\phi(t)}{{\rm d}t}\,, \tag{3.33}$$

where

$$I_{\rm c}(\Phi_{\rm ex}) = \cos\left[\pi\frac{\Phi_{\rm ex}}{\Phi_0}\right]. \tag{3.34}$$

The I–V characteristics of SQUIDs are obtained by solving these nonlinear equations with changing $I_{\rm b}$ for a constant flux. The Φ–V characteristics are also obtained with changing Φ for a constant $I_{\rm b}$. However, a simple approximation has been proposed for the detailed analysis, as follows:

$$0 \geq |I_{\rm b}| \geq I_{\rm c}(\Phi_{\rm ex}) \Rightarrow \overline{V(t)} = 0\,, \tag{3.35}$$

$$|I_{\rm b}| \geq I_{\rm c}(\Phi_{\rm ex}) \Rightarrow \overline{V(t)} = \frac{R_{\rm s}}{2}\sqrt{I_{\rm b}^2 - [I_{\rm c}(\Phi_{\rm ex})]^2}\,. \tag{3.36}$$

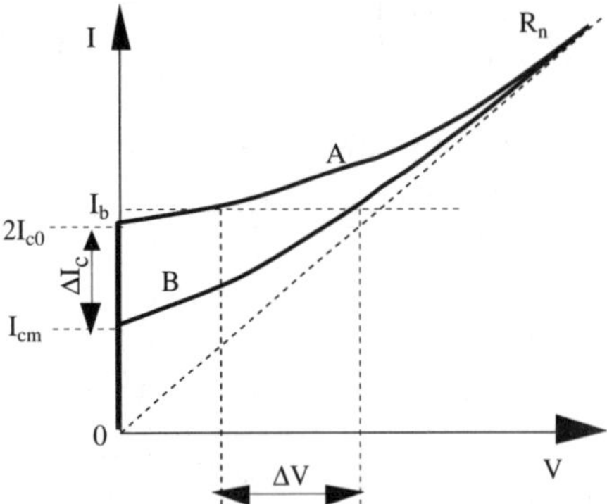

Fig. 3.9. A schematic drawing of the I–V characteristics obtained with the hyperbolic approximation. The curves A and B are drawn for the applied flux of $n\Phi_0$ and $(n+1/2)\Phi_0$

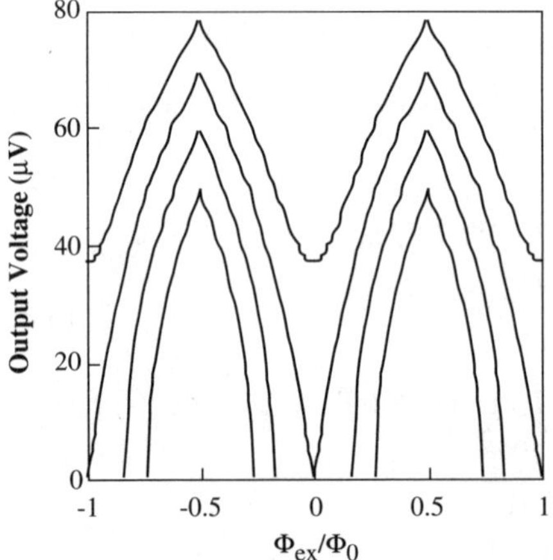

Fig. 3.10. Φ–V characteristics calculated numerically with $2I_0 = 40$ µA, $2L_0 = 50$ pH, $R_s = 4\ \Omega$, and $I_b = 32$, 36, 40, and 44 µA

Equation (3.36) indicates that the output voltage of a SQUID is approximated by a hyperbolic function of I_c, which is practically useful for the analysis of the operation of SQUIDs. The schematic I–V characteristics with the hyperbolic approximation are shown in Fig. 3.9, where two curves are drawn for $\Phi = n\Phi_0$ and $\Phi = (n+1/2)\Phi_0$. Φ–V curves calculated numerically with an increase in the bias current are shown in Fig. 3.10, where $I_{c0} = 20\ \mu$A, $L_0 = 25$ pH, $R_s = 4\ \Omega$, and $I_b = 32$, 36, 40, and 44 µA. The Φ–V curve shifts upward with an increase in the bias current and the maximum peak-to-peak voltage, which is called the "maximum voltage modulation", is obtained at a bias current equal to the critical current of the SQUID, or $I_b = 2I_{c0}$.

In this calculation, no noise is taken into account. Therefore, theoretically, the I–V curve intersects with I-axis at $V = 0$ with zero slope. The corresponding Φ–V curve has infinite slope near $V = 0$ for $I_b < 2I_{c0}$. But in practical SQUIDs, the I–V curve and the Φ–V curve are rounded by noises, which consist of the thermal fluctuation of the critical current and some other noises. The noise attributed to the thermal fluctuation at a finite temperature T is characterized in terms of the noise parameter Γ, defined as the ratio of the thermal energy $k_B T$ and the coupling energy of the junction $I_{c0}\Phi_0/2\pi$, where Γ is

$$\Gamma = \frac{2\pi k_B T}{I_{c0}\Phi_0} . \tag{3.37}$$

It is necessary for stable operation of a SQUID that the coupling energy is much larger than the thermal energy, or $\Gamma \ll 1$. According to the study by Tesche and Clarke [6], Γ should be smaller than 0.05, from which a SQUID requires a minimum critical current larger than 3.5 μA at 4.2 K.

Noise in dc SQUIDs. In practical SQUIDs, noise should be discussed in order to evaluate the field sensitivity. After thermal fluctuation of the critical current, the main source of noise is the Johnson noise generated in the shunt resistors. In this section, noise estimation of dc SQUIDs is outlined, including the attribution of preamplifier noise for practical use [4–6].

A dc SQUID provides the maximum voltage modulation, ΔV_{max}, if the bias current I_b is equal to the maximum critical current of the SQUID, $2I_{c0}$. From (3.36), ΔV_{max} is given by

$$\Delta V_{max} = \frac{R_s}{2}\sqrt{(2I_{c0})^2 - I_{cm}^2} = R_s I_{c0}\sqrt{1 - (\frac{I_{cm}}{2I_{c0}})^2} , \tag{3.38}$$

where $2I_{cm}$ is the minimum critical current of the SQUID obtained with the flux of $(n + 1/2)\Phi_0$. The difference of $2I_{c0}$ and $2I_{cm}$ is called the modulation depth ΔI_c, which is approximately given by

$$\frac{2I_{c0} - I_{cm}}{2I_{c0}} = \frac{\Delta I_c}{2I_{c0}} = \frac{1}{1 + \beta_L} . \tag{3.39}$$

From (3.38) and (3.39), ΔV_{max} is written as

$$\Delta V_{max} = \frac{R_s I_{c0}}{1 + \beta_L}\sqrt{1 + 2\beta_L} . \tag{3.40}$$

A SQUID has the maximum sensitivity if it is operated at the steepest slope on the Φ–V curve, which provides the largest differential coefficient $\partial V/\partial\Phi$. $\partial V/\partial\Phi$ is called the "flux-to-voltage transfer coefficient", or simply "transfer coefficient", with the units of $[V/\Phi_0]$ normalized by Φ_0. Using (3.40),

$\partial V/\partial \Phi$ is approximately given by

$$\frac{\partial V}{\partial \Phi} \simeq \frac{\Delta V_{\max}}{1/2\Phi_0} = \frac{2I_{c0}R_s}{\Phi_0}\frac{\sqrt{1+2\beta_L}}{1+\beta_L}. \tag{3.41}$$

Johnson noise generated in the shunt resistors appears not only as voltage noise across the SQUID, but also as current noise in the SQUID loop, which induces flux noise through the SQUID inductance. Both noises are converted into flux noises and discussed in general terms. The spectral density of the voltage noise S_Φ^{v} is

$$S_\Phi^{\mathrm{v}} = 8k_BTR_s\left(\frac{\partial V}{\partial \Phi}\right)^{-2}, \tag{3.42}$$

while the spectral density of the current noise S_Φ^{i} is

$$S_\Phi^{i} = \frac{8k_BT}{R_s}L_{sq}^2. \tag{3.43}$$

Accordingly, the total spectral density $S_{\Phi s}$ is the sum of S_Φ^{v} and S_Φ^{i}:

$$S_{\Phi s} = 8k_BTR_s\left[\left(\frac{\partial V}{\partial \Phi}\right)^{-2} + \left(\frac{L_{sq}}{R_s}\right)^2\right]. \tag{3.44}$$

In normal SQUIDs with a small inductance of ~10 pH, S_Φ^{i} is negligible compared to S_Φ^{v}. In this case, $S_{\Phi s}$ is of the order of ~1 $\mu\Phi_0^2/\mathrm{Hz}$.

According to Tesche and Clarke [6], optimized SQUIDs with $\beta_L = 1$ have the minimum energy resolution ε, which is derived from (3.41) and (3.44):

$$\varepsilon = \frac{S_{\Phi s}}{2L_{sq}} \approx \gamma h\left(\frac{k_BT}{eI_{c0}R_s}\right), \tag{3.45}$$

where γ is generally 2–5 with $\Gamma = 0.05$. Typical SQUIDs with a small inductance have ε values of ~10 h. As indicated in (3.45), the energy resolution becomes small with a large product of I_{c0} and R_s in keeping $\beta_c < 1$. This means that the capacitance of a junction should be as small as possible to obtain higher energy resolution.

In practical operation of a SQUID, preamplifier noise must be taken into consideration. The voltage noise of the amplifier V_{na} is converted into flux noise through the transfer coefficient. The spectral density $S_{\Phi a}$ is

$$S_{\Phi a} = V_{na}^2\left(\frac{\partial V}{\partial \Phi}\right)^{-2}. \tag{3.46}$$

The total noise in the SQUID, S_Φ, is given by the sum of the intrinsic noise and the amplifier-attributable noise:

$$S_\Phi = S_{\Phi s} + S_{\Phi a} = 8k_BTR_s\left[\left(\frac{\partial V}{\partial \Phi}\right)^{-2} + \left(\frac{L_{sq}}{R_s}\right)^2\right] + V_{na}^2\left(\frac{\partial V}{\partial \Phi}\right)^{-2}. \tag{3.47}$$

Commercially available amplifiers have an input voltage noise of about 0.4 nV/Hz$^{1/2}$ at best at room temperature, which is normally larger than the intrinsic noise level of SQUIDs. Therefore, larger $\partial V/\partial \Phi$ should be required to reduce the amplifier-attributable noise. Some methods have been proposed to obtain a large $\partial V/\partial \Phi$, which are mentioned later.

The above theory is valid in the white region. On the other hand, $1/f$ noise exists in the low-frequency region below ~1.0 Hz. Various mechanisms of $1/f$ noise are predicted, such as the fluctuation of the tunneling probability in a junction caused by trapped electrons [7–9] and fluctuation of trapped flux in superconducting films [10]. However, a clear theory is lacking at present.

Readout Methods: Flux-Locked Loop (FLL). The output voltage from SQUIDs is a nonlinear function with a periodicity of Φ_0. The linear relationship between the flux and the voltage is achieved with a method called a "flux-locked loop (FLL)" as a null detector, which provides high resolution, a large dynamic range, and a high slew rate. Two popular FLL methods are described in this section.

FLL with Flux Modulation. Figure 3.11 shows a typical circuit of an FLL with flux modulation. A bias flux Φ_b and modulated flux Φ_m with a peak-to-peak amplitude of $\Phi_0/2$ and a frequency of f_m are applied to the SQUID loop. A rectangular wave or a sine wave is used as the modulation signal. With a quasistatic flux in the SQUID of $n\Phi_0$, the resulting output voltage V is a rectified signal with $2f_m$, as shown in Fig. 3.12a. If this signal is lock-in detected with reference to f_m, with a phase-sensitive detector (PSD), the output voltage from the lock-in detector V_L will be zero. On the other hand, if the quasistatic flux is $(n+1/4)\Phi_0$, V is a modulated signal with f_m, as shown in Fig. 3.12b, and V_L will be a maximum. Thus the linearized voltage for a signal flux can be obtained, as shown in Fig. 3.12c. The lock-in detected signal is integrated with an integrator and fed back to the SQUID loop through a feedback coil L_f, via a feedback resistor R_f. If there is an external flux $\Delta\Phi$

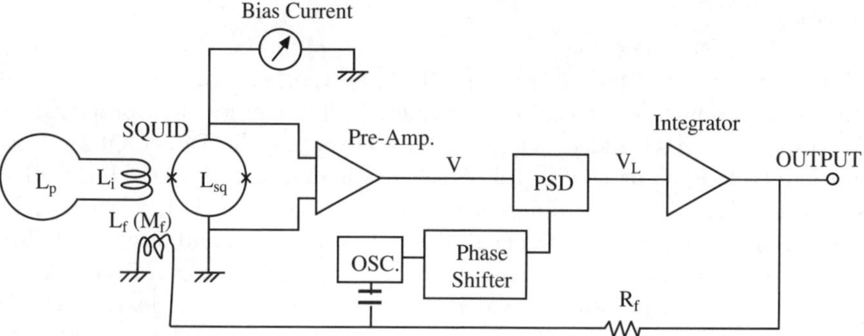

Fig. 3.11. A schematic diagram of an FLL with flux modulation

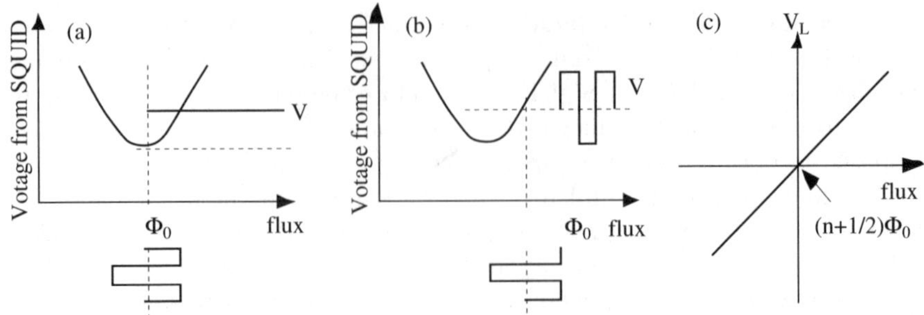

Fig. 3.12. A flux modulation scheme showing the voltage from a SQUID V for **a** $\Phi = n\Phi_0$ and **b** $\Phi = (n+1/4)\Phi_0$. **c** The output voltage from a lock-in detector V_L vs. flux [4]

into the SQUID, the feedback circuit applies an opposite flux of $-\Delta\Phi$; that is, the net change in flux in the SQUID is zero, or locked at the bias flux Φ_b. Therefore, one can know the change in the flux with the voltage that appears across the feedback resistor proportional to $\Delta\Phi$. The maximum feedback flux Φ_{max} to lock the SQUID – in other words, the maximum flux that one can measure – is restricted by the mutual inductance M_f between the SQUID and the feedback coil, R_f, and the voltage of the power supply V_0, namely given by

$$\Phi_{max} = \frac{R_f}{M_f} V_0 \,. \tag{3.48}$$

The slew rate s in the circuit depends on the modulation frequency f_m:

$$s_\Phi \simeq \frac{\pi^2}{16} f_m \,. \tag{3.49}$$

With this technique, detection of flux much larger than Φ_0 is possible. In addition, the use of modulating flux has the advantage of eliminating $1/f$ noise. This method is not useful for SQUIDs with asymmetric Φ–V characteristics.

Direct Offset Integration Technique (DOIT). Another FLL method without flux modulation has been adopted recently. This method was proposed by McWane et al. [11] and developed by D. Drung as the Direct Offset Integration Technique, or DOIT [12]. A schematic diagram of a DOIT circuit is shown in Fig. 3.13a. The difference between the voltage V from a SQUID, which is amplified, and a reference voltage V_r is detected and integrated. The output from an integrator is fed back to the SQUID through a feedback coil L_f via a feedback resistor R_f, so that the change in voltage in the SQUID is zero. Consequently, the SQUID is locked at V_r. V_r should be adjusted to the voltage at which $\partial V/\partial\Phi$ is a maximum on the Φ-V curve, usually for

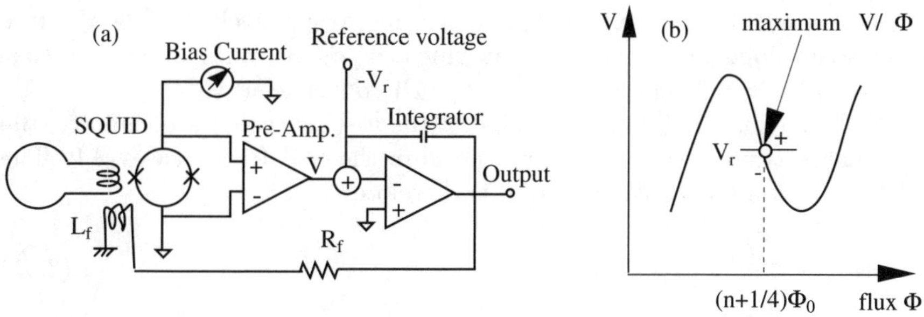

Fig. 3.13. A schematic diagram of a DOIT circuit (**a**), and reference voltage adjustment on a Φ–V curve (**b**)

$(n + 1/4)\Phi_0$, as shown in Fig. 3.13b. The slew rate of a DOIT circuit is limited by the time constant of the integrator.

The advantage of a DOIT is that the electronics is quite simple compared to that of a flux modulation type. There is, however, a disadvantage, in that that amplifier noise affects a SQUID directly. Especially at low frequency, the $1/f$ noise of the amplifier degrades the performance of the SQUID. DOIT is useful for SQUIDs with a large $\partial V/\partial\Phi$, which eliminates the amplifier-attributable noise.

Increase of $\partial V/\partial\Phi$: Additional Positive Feedback (APF). As mentioned in a previous section, the resolution of a practical SQUID is limited mainly by preamplifier noise rather than the intrinsic noise of the SQUID. However, if the transfer coefficient $\partial V/\partial\Phi$ is increased, the amplifier-attributable noise $S_{\Phi a}$ given by (3.46) can be reduced. Especially for SQUIDs operated with DOIT, amplifier noise is dominant – if $\partial V/\partial\Phi$ is small. D. Drung et al. proposed that additional positive feedback (APF) is effective to obtain Φ–V characteristics with high $\partial V/\partial\Phi$ [13]. An equivalent circuit of APF is shown in Fig. 3.14, where a resistor R_a and a inductor L_a are connected in parallel to the SQUID loop as a APF circuit. Because a part of the bias current I_b flows in the APF circuit, the inductor L_a generates a magnetic flux, which couples to the SQUID through the mutual inductance M_a.

The flux induced in the APF circuit makes the Φ–V curve of the SQUID asymmetric. The asymmetric Φ–V curve is calculated numerically, adding the flux by APF Φ_a to the sets of equations indicated in a previous section, where Φ_a is given by

$$\Phi_a = M_a \frac{V}{R_a} . \tag{3.50}$$

Figure 3.15 shows the calculation results of the Φ–V curve of a SQUID with an APF effect for various bias currents. The calculation parameters are $I_{c0} = 0.1\ \mu A$, $L_{sq} = 20$ pH, $R_s = 1\ \Omega$, $R_a = 100\ \Omega$, and $M_a = 300$ pH. The bias currents are 0.101 mA, 0.11 mA, and 0.2 mA, respectively. The Φ–V curve has a steep slope on the left-hand, at which $\partial V/\partial\Phi$ becomes larger than that of a symmetric Φ–V curve. Thus, a large $\partial V/\partial\Phi$ is achieved.

If Φ_a is equal to $1/2\Phi_0$, the slope of the Φ–V curve goes to infinity, the Φ–V curve is no longer stable, and noise in the SQUID increases [14]. The APF condition for stable operation is therefore

$$M_a \frac{V}{R_a} < \frac{\Phi_0}{2} . \quad (3.51)$$

When the SQUID has the maximum output voltage modulation ΔV_{max}, the following condition should be satisfied for stable operations:

$$\frac{R_s}{R_a}\frac{M_a}{L_{sq}} < \frac{1+\beta_L}{\beta_L\sqrt{1+2\beta_L}} , \quad (3.52)$$

where L_{sq} is the inductance of the SQUID.

The transfer coefficient $\partial V/\partial\Phi$ is exactly the differential coefficient of the curve. The effective $(\partial V/\partial\Phi)a$ is, however, approximately characterized with a parameter β_a:

$$\left(\frac{\partial V}{\partial\Phi}\right)_a = \frac{(\frac{\partial V}{\partial\Phi})_{sym}}{1-\beta_a} , \quad (3.53)$$

where $(\partial V/\partial\Phi)_{sym}$ is the transfer coefficient of the symmetric Φ–V curve without APF, and β_a is given by

$$\beta_a = \frac{M_a}{R_a}\left(\frac{\partial V}{\partial\Phi}\right)_{sym} . \quad (3.54)$$

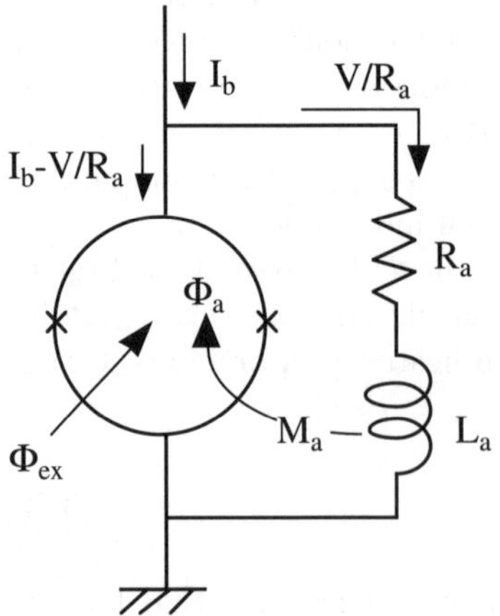

Fig. 3.14. The equivalent circuit of a SQUID with APF

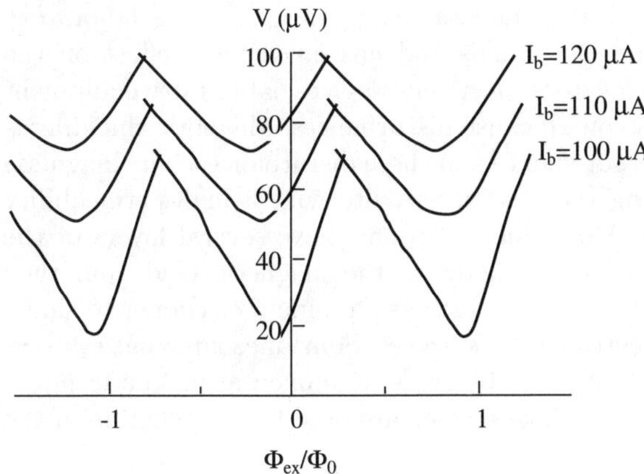

Fig. 3.15. Calculated asmmetric Φ–V characteristics with the APF effect. $I_{c0} =$ 0.1 mA, $L_{sq} = 20$ pH, $R_s = 1\ \Omega$, $R_a = 100\ \Omega$, and $M_a = 300$ pH

$(\partial V/\partial \Phi)a$ becomes larger with increase of β_a up to unity. But if $\beta_a > 1$, the Φ–V curve is no longer stable, which provides the same condition as indicated in (3.52).

With a large APF gain, that is, with large M_a/R_a the output voltage is reduced, if R_a is too small. While with large L_a, the current noise induced in the resistor R_a affects the SQUID through large M_a [15]. APF circuit should be designed to take into consideration the reduction of the output voltage and the attribution of the noise induced in the APF circuit.

3.2.2 A System for Biomagnetic Measurement

Configuration. In the previous section, SQUID sensors and their driving electronics were described. The system configuration for high-sensitivity magnetic measurement can be accomplished by highly integrated digital and analog signal processing technology. All the signals from the SQUID electronics are acquired digitally for further numerical analysis. The analog signals must be filtered and amplified before recording the signals as digital data. The dynamic range of the amplifier and the cutoff frequency of the analog filters are important for high resolution of the digitized data.

SQUID sensors must be kept in a refrigerant such as liquid helium. Therefore, cryogenics are indispensable in SQUID magnetometry. The cryostat specially designed for biomagnetic measurement is used in a biomagnetic field. The cryostats for different measurement objects have different structures. For example, the tail of the cryostat of the multi-channel system for magnetoencephalography has a helmet-like shape in order to allocate the sensors around the whole head at high density.

The magnitude of the biomagnetic field to be measured is extremely low. The magnitude of the geomagnetic field is from one to ten billion times larger than the biomagnetic field. Environmental magnetic noise in a laboratory is several orders higher than the signals and has much more effect on the measurement. Power lines, elevators, electronic devices, laboratory equipment (metal doors, chairs), and so on all cause disturbances. Magnetic shielding is necessary to protect the measurement from these disturbances. The magnetic shield is formed by enclosing the observation site with a high-permeability material such as mu-metal. Most shielded rooms have several layers of the mu-metal to increase the shielding factor of the magnetic field, and they have a layer of high-conductivity metal such as aluminum or copper to shield relatively high-frequency electromagnetic waves. Some measurement systems have reference sensors especially to observe environmental magnetic noise. The signals provided by the reference sensors are used for cancellation of the external magnetic noise.

A computer system is also included in a modern measurement system. The computer system covers control of the electronics, digital data acquisition, signal processing, magnetic source analysis, data visualization, management data and networking, and so on.

Estimation of the source of neuronal electronic activity needs simultaneous magnetic observations at multiple locations. A higher density of sensor location provides higher accuracy of the source estimation. Therefore, every recent biomagnetic measurement system has a multiple SQUID sensor array. Many recent magnetoencephalography systems have more than 100 channels.

Figure 3.16 shows a conceptual drawing of a multi-channel SQUID measurement system for magnetoencephalography. Every piece of magnetic material can cause artifacts when it moves during measurement. Therefore, all mechanical items in the shielded room (the cryostat, the table for the subject, and the stimulators) must be made of nonmagnetic materials.

Data Acquisition. The neuromagnetic signals observed by the SQUID sensors are recorded digitally for further numerical analysis on computers, such as magnetic source localization. Most of the signals have a time dependency and should be recorded as wave data. The analog signals from the SQUID electronics should be amplified and filtered before analog–digital conversion to give an adequate signal-to-noise ratio (SNR) and dynamic range. In this section, analog signal processing and digital data acquisition are described.

In general, when analog wave signals are digitized, we always have to be careful about the sampling rate and quantizing resolution. A higher sampling rate provides a more frequent quantization and more accurate analog information can be maintained. However, high-speed sampling needs a more expensive A/D converter and yields a huge amount of digitized data, which is often hard to treat even in a modern multi-channel SQUID measurement system equipped with more than 100 sensors. The sampling points should be decreased to much as possible to save time and money, and the original analog wave information should not be lost. The basic rule to satisfy this re-

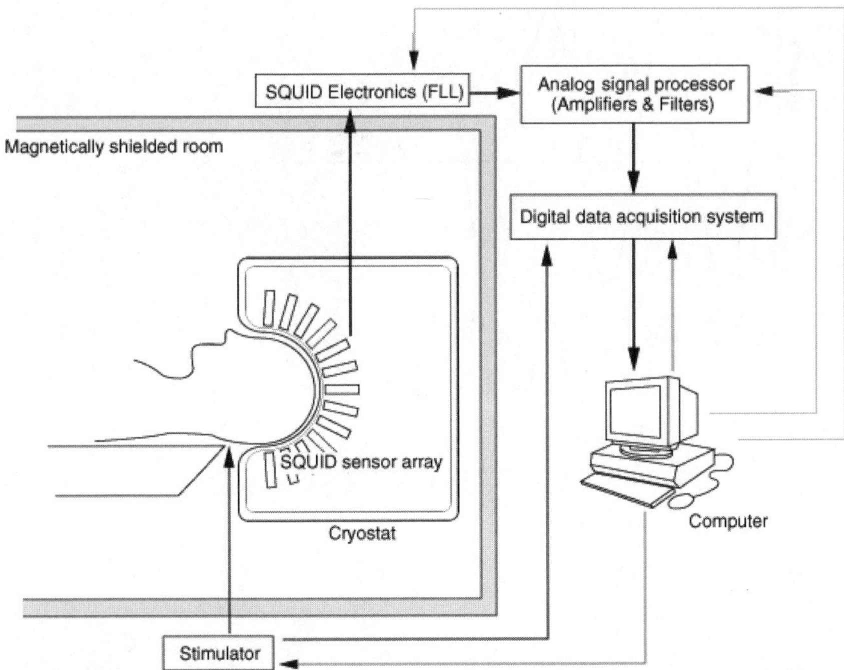

Fig. 3.16. The configuration of a modern magnetoencephalography system. *Bold arrows* represent flows of the signal from the SQUID sensors. *Thin arrows* represent controling signals from the computer

quirement is the sampling theorem. The theorem states that a band-limited analog signal $x(t)$ – that is, a signal in a finite frequency band (e.g. between 0 and f_c Hz) – can be completely reconstructed from its samples $x(n) = x(nT)$, if the sampling frequency is greater than $2f_c$ (the Nyquist rate); expressed in the time domain, this means that the sampling interval T is at most $1/2f_c$ seconds. Sampling below the Nyquist rate can produce serious errors (aliasing) by introducing low-frequency artifacts into the digitized data, as shown in Fig. 3.17.

Any analog signals have their own bandwidths. In the case of the neuromagnetic signal, the maximum limit of the bandwidth is from several hundreds of hertz to several tens of kilohertz at most. The sampling rate should be twice as high as at least the maximum limit. The analog signals include noise in a higher-frequency region than the bandwidth, and the aliasing error caused by the higher-frequency noise can be a problem in the digitized data. Therefore, an analog low-pass filter that reduce the higher-frequency noise has to be applied before digitizing the signals, even if digital noise reduction processing is to be used after data acquisition. This low-pass filter is called an anti-aliasing filter.

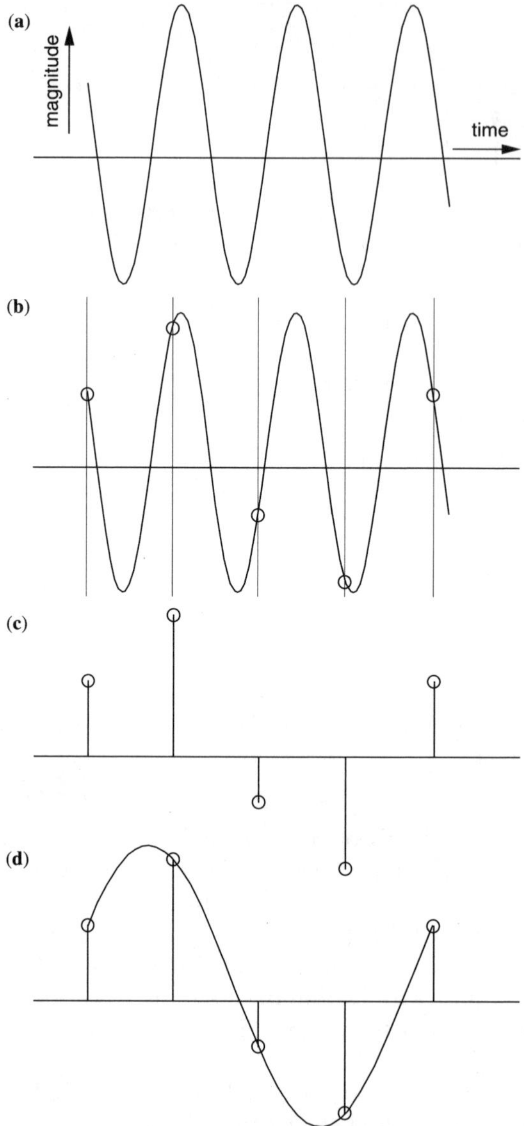

Fig. 3.17. Aliasing error. **a** The original signal. **b** Sampling under the Nyquist rate. **c** The sampled data. **d** The low-frequency artifact

An instantaneous value of the magnitude of the continuous analog signal is quantized by the A/D converter and stored as discrete digital data at each sampling point. The minimum quantization of an A/D converter q is given

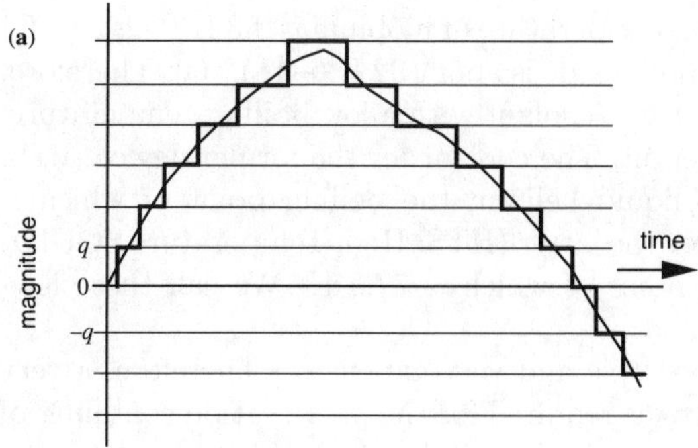

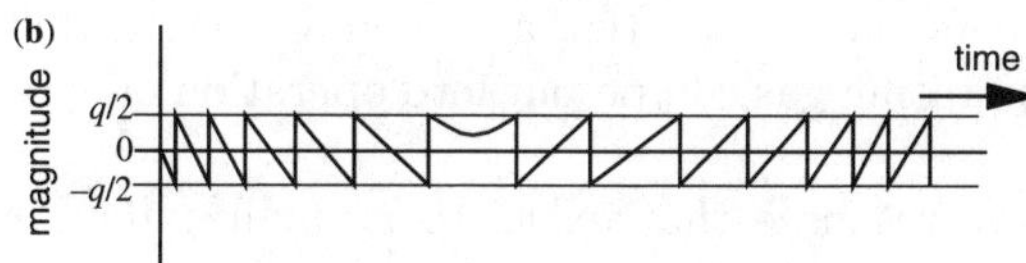

Fig. 3.18. Quantization noise. **a** Quantization of a continuous signal. **b** The difference between the quantized data and the original signal

by

$$q = \frac{S}{2^n}, \tag{3.55}$$

where S is the magnitude of maximum input signal and the A/D converter has n bits of resolution. The value of the quantized data is limited to multiple of q. Therefore, the magnitude of analog signal x, satisfying

$$\left(N - \frac{1}{2}\right) q \leq x < \left(N + \frac{1}{2}\right) q, \tag{3.56}$$

is clipped to Nq, as shown in Fig. 3.18a. The difference between the actual value of the analog signal and the quantized value is called the quantization error. Quantization error is unavoidable in digital data acquisition. Figure 3.18b shows the quantization error in the digitized data. This error acts as white noise added on the original signal. A larger value of n provides less quantization noise. There is a tradeoff between the resolution and speed of an A/D conversion. Therefore, we should compare the noise included in the original analog signal with the quantization noise before applying a high-resolution A/D converter. A higher resolution that gives much less quantization noise than the original analog noise is an overspend in view of the cost performance.

Cryogenics. To operate a SQUID, we have to cool the device to the very low temperature which is necessary to maintain superconductivity. For example, the superconductivity transition temperature of niobium is 9.3 K (−263.9° C) and that of yttrium–barium copper oxide is about 92 K (−181° C). Therefore, we cool a device by immersing it in a coolant with a low boiling-temperature, or cool it directly with a refrigerant. The coolant for the former device (LTS: Low Temperature SQUID) is liquid helium, the boiling point of which is 4.2 K, and the typical coolant for the latter (HTS: High Temperature SQUID) is liquid nitrogen, the boiling point of which is 77.4 K. We call these low-temperature coolants "cryogen".

The cryogen is highly evaporative and very expensive. Therefore, a very high thermal insulation efficiency is required for the preservation container of the cryogen. We can procure a cryogen through local gas sales companies, by having it supplied in a container that is a kind of large-sized thermos flask. If equipped with a cryogen liquefaction apparatus, a recycling system that collects and re-liquefies the evaporating gas can be put into operation on site.

Liquid Helium. Helium is a rare resource that exists to a small extent in air (1/200 000), but in practical use, it is separated from natural gas, which contains helium. A considerable part of the helium consumed in the world is produced in North America. Helium is chemically inactive and harmless. The density of the gas at room temperature is approximately 1/7 that of air. The density of liquid helium is 1/8 that of water (0.125 kg/l), and the liquid becomes a gas with about a 750 times increase in volume at room temperature when vaporized. 1.4 liters of liquid per hour evaporate when 1 W of heat is added.

Liquid Nitrogen. On the other hand, nitrogen is the main element in the atmosphere, the separation is easy, and there are no resource problems. The density of liquid nitrogen is 80% that of the water (0.8 kg/l), and the liquid becomes a gas with about a 700 times increase in volume at room temperature when vaporized. The liquid evaporates at 0.022 liters per hour when 1 W of heat is added. It is easy to handle compared with liquid helium and, moreover, it is economical.

Dewar. Here, the thermos flask that accommodates the SQUIDs and the cryogen is called a "dewar". The general method of cooling used for biomagnetic measurement at present is a method of keeping a constant temperature with the vaporization of the cryogen stored up in the dewar. Therefore, the cryogen must be supplied regularly to compensate for the vaporized portion.

The outline structure of the dewar for multi-channel SQUIDs (LTS) which is used for magnetoencephalography (MEG) is shown in Fig. 3.19. The dewar has a helmet-shaped base that is equipped with a hundred or so SQUIDs and covers much of a human head. The material of the main part of a dewar must be nonmagnetic and nonmetallic, so a plastic resin is selected. As the material should withstand a low temperature and have enough strength, GFRP (Glass Fiber Reinforced Plastics) is optimal.

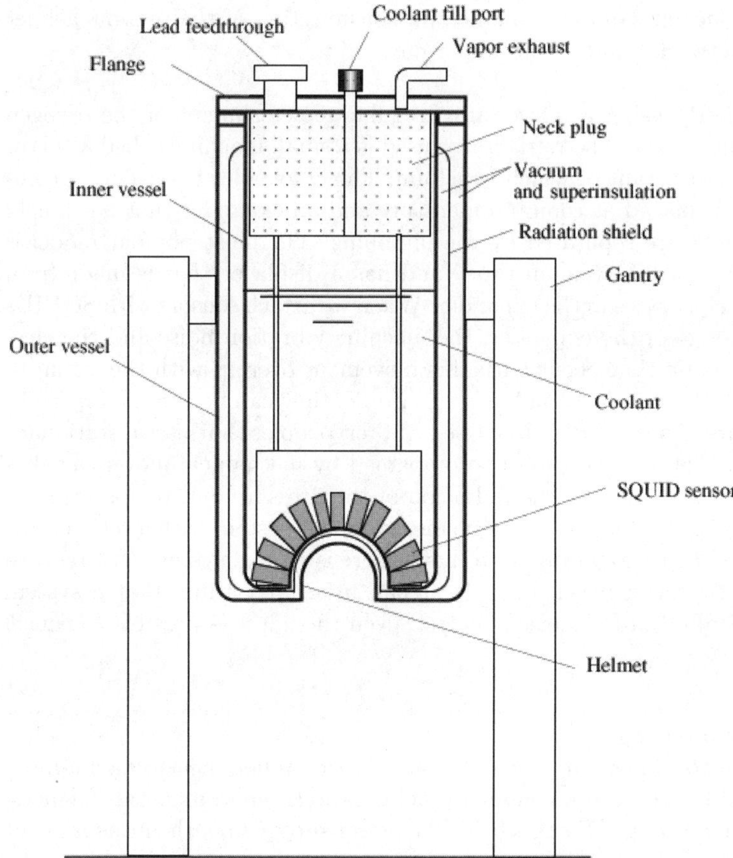

Fig. 3.19. A schematic of a typical dewar and SQUID system used for measuring the magnetic field of the brain

The thermal insulation structure is quite complicated, but basically it is the same as that of a thermos flask, except that a radiation shield and layered insulation sheets fill the vacuum layer. Because a considerable amount of heat enters into liquid helium from the opening in the upper part, called a "neck tube", the structure is taller than it is wide to secure a sufficiently long distance from the room-temperature part. In the opening, a measurement line lead-out port, a supply port for liquid helium, and an exhaust port for vaporized helium are fitted.

The magnetic detection coils coupled to the SQUIDs are attached to the bottom of the inside vessel and the thickness of the heat insulation layer is made as thin as possible to approach a measurement object. However, because such construction becomes a factor in reducing heat insulation efficiency, the optimal structure taking into consideration the measurement efficiency and convenience is required. The helium consumption of the present helmet systems is between 8 and 20 liters per day.

Refrigerator. By using a refrigerator, regular replenishment of the cryogen becomes unnecessary. The refrigerator used is a small machine called a "cryocooler", the refrigerant of which is helium. The cryocooler consists of a compressor that is placed at room temperature and an expander that has a cold stage, and both are separated by the plumbing. The most popular model is called the "Gifford–McMahon type" and has a displacer that is made from metal and reciprocates in the expander. When we attach sensors with SQUIDs to the cold stage of the expander, the machine vibration noise and the electromagnetic noise that occur with this movement overlap with the signal to be measured. Because these noises are periodic, they can be canceled by computer software. There is a "pulse-tube type" cryocooler that uses a stationary displacer. In this case, the noise accompanied by mechanical movement dies down, but noise caused by the pulsation of the pressure and temperature in the displacer is left. The biomagnetic measurement system with a refrigerator has been used experimentally up to now: there is no one system for practical use as yet. To detect a very minute signal, users may think that a system that intrinsically has less noise is better, even though it is possible to cancel this noise.

Magnetic Shielding

The Biomagnetic Field and Environmental Noise. When measuring a minute magnetic field such as a biomagnetic field, various environmental magnetic noises induced from the Earth's field, the power supply line, the measurement equipment, and city traffic such as cars and trains become obstacles. The strength of the Earth's magnetic field is about 10^{-5} to 10^{-4} T (tesla: the strength of the magnetic field in SI units) and changes daily by about 1/1000. The environmental magnetic noise depends on its frequency [16] as shown in Fig. 3.20. As for the magnitude of this noise, the difference between that of the central part of a city and a silent place in the country can be hundredfold.

The strength of the biomagnetic signal is tens of pT (10^{-12} T) for the heart and from tens to hundreds of fT (10^{-15} T) for the brain. Therefore, it becomes necessary to shelter or cancel the environmental magnetic noise to detect these signals. Two ways of sheltering the magnetic noise exist. "Passive shielding" is realized by sheltering the measurement equipment with a ferromagnetic material or a superconductor. "Active shielding" is a method of canceling the magnetic noise with an opposite magnetic field generated by an electric current fed to the compensation coil.

On Magnetic Shielding Using a Ferromagnetic Material. The strength of the magnetic field in air is proportional to the interval of the magnetic force lines. Figure 3.21 shows the distribution of the magnetic force lines when a long circular cylinder which consists of ferromagnetic material is placed in a uniform transverse field. Because the magnetic permeability of the ferromagnetic material is large, the outside magnetic force lines go through the cylinder wall, and the density of the magnetic force lines becomes sparse in the inner air space of the cylinder.

As the material used in the magnetically shielded room (MSR) for the biomagnetic measurement, a metal having excellent magnetization characteristics in a low field is selected. The most popular material with a high permeability is "permalloy", which is an iron alloy containing 78% nickel.

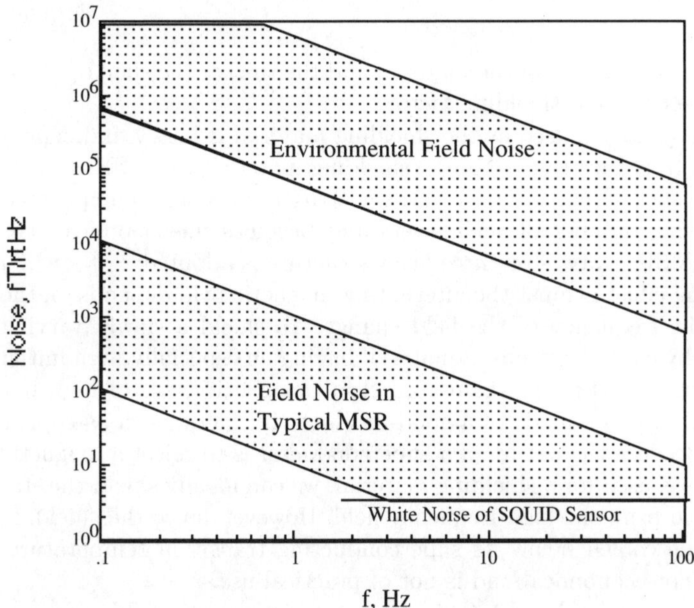

Fig. 3.20. The frequency dependence of magnetic field noise in various environments

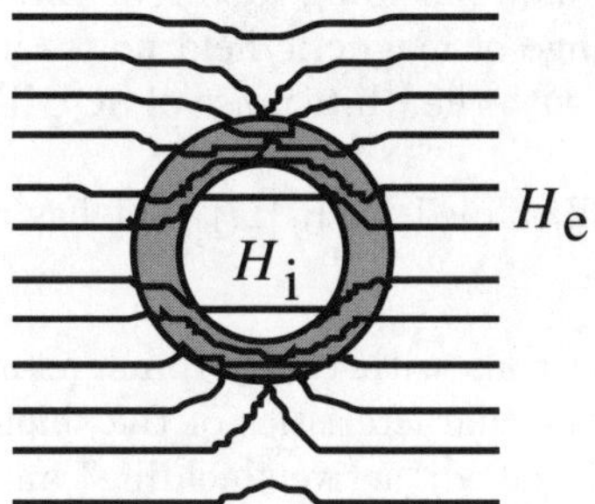

Fig. 3.21. Magnetic force lines for a cylinder with high permeability in a transverse field

The Design of the Magnetically-Shielded Room. The shielding factor is used as the measure that shows the efficiency of the magnetically shielded room (MSR). The shielding factor S is shown as the ratio of the magnetic field outside, H_{e}, and the magnetic field inside, H_{i}, of the MSR:

$$S = \frac{H_{\mathrm{e}}}{H_{\mathrm{i}}} . \tag{3.57}$$

Supposing that the MSR is spherical, with outside diameter D, thickness t, and relative permeability of the material μ, the shielding factor is simply shown by

$$S = \frac{4}{3} \frac{\mu t}{D} . \tag{3.58}$$

The higher the permeability of the material and the thicker the shell of MSR, the higher is the expected shielding factor.

The permalloy that is used as the shielding material is heavy and expensive. Therefore, to reduce the use of the shielding material, the MSR is made with a multiple shell structure. However, the structure becomes complicated with an increasing number of shell layers and becomes uneconomical. For most practical MSRs, a two- or three-layer structure is adopted.

The shielding effect against the alternating magnetic field depends on the material and the frequency of the field change. To shield a slowly varying field generated by cars and trains, a material having an especially high initial permeability must be chosen. Against an electromagnetic wave of high frequency, a shield must be made of a good-conductivity metal such as copper or aluminum. Because the nature of a superconductor is to reject a magnetic force line, if this is used for a shielding material, we can ideally shield the static magnetic field from the high-frequency field. However, since the shielding material must be cooled below its superconducting transition temperature, this method is not economical and is not of practical use.

Most of the magnetically shielded rooms for biomagnetic field measurement consist of a box of side 2–3 m to house the primary equipment and the bed inside. The wall consists of two layers of permalloy plate and a single layer of copper or aluminum plate. The common shielding factor of these MSRs is approximately 1000 at 1 Hz. The range of magnetic field noises in the MSR at various sites is shown in Fig. 3.20, with the white noise of SQUID sensors.

As a special example, there is a soccer ball shaped MSR [17] that has a shielding factor of 100 000 and four layers.

Active Shielding. By combining compensation coils with the ferromagnetic shield, attempts have been made to improve the characteristics of the shielding in the low-frequency area. This method is called "active shielding" and is a way of making the magnetic noise zero by adding an artificial magnetic field that has the same strength in the direction opposite to the detected magnetic noise.

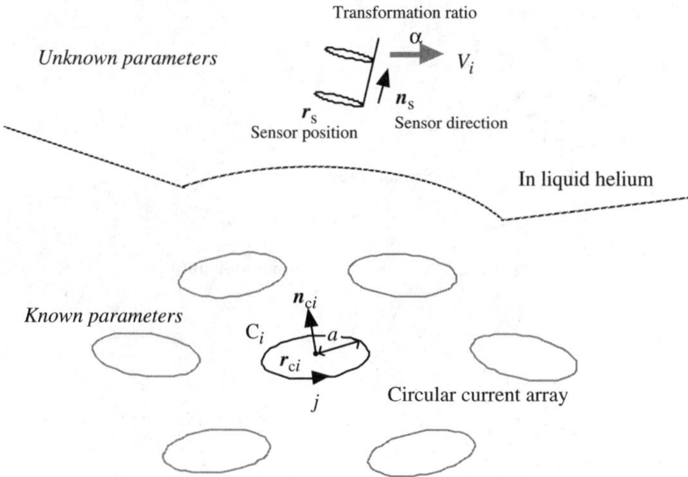

Fig. 3.22. Calibration of sensor parameters by using a circular current array

Calibration and Alignment

Calibration of the Sensitivity and the Position of the Sensors. The biomagnetic measurement system is an instrument to measure a magnetic field, so it is important that the value is correctly measured in the sense of "traceability". The sensitivity, position, and direction are calibrated with a multi-coil set which is machined precisely and fed with current using a generator-of-standard instrument. The coils are fed with a known current and emit a known magnetic field at a known position in a known direction. The output voltage of the FLL circuit is represented by $V = h(\boldsymbol{I}, x, y, z, \boldsymbol{n}, b; g_\mathrm{s}, x_\mathrm{s}, y_\mathrm{s}, z_\mathrm{s}, \boldsymbol{n}_\mathrm{s})$, where $\boldsymbol{I}$ is the current in the calibration coil, x, y, and z are the position of the calibration coil, $\boldsymbol{n}$ is the unit vector (n_x, n_y, n_z) of the direction of the coil, b is the baseline between the sensing coil and the reference coil, g_s is the magnetometer or gradiometer sensitivity, x_s, y_s, and z_s are the position of the sensor, and $\boldsymbol{n}_\mathrm{s}$ is the unit vector $(n_\mathrm{sx}, n_\mathrm{sy}, n_\mathrm{sz})$ of the direction of the sensor (Fig. 3.22). The detailed expression is described by an integral of the Biot–Savar equation. Six variables, g_s, x_s, y_s, z_s, n_sx, and n_sy, are unknown. (The variable n_sz is a dependent value of n_sx and n_sy.) This means that measurements by the five different calibration coils yield five different equations to be solved using the five answers to these variables. It is desirable to do the calibration using 10–20 coils in order to obtain better precision. The procedure is carried out for all sensors installed in the cryostat. With this calibration procedure, the magnetic field can be precisely identified and the source of the field can be estimated through an inverse problem. To demonstrate that the calibration has been carried out correctly, we have done the following. We have prepared another coil and measured the magnetic field produced by this coil (Fig. 3.23). With this measurement result, we have estimated the

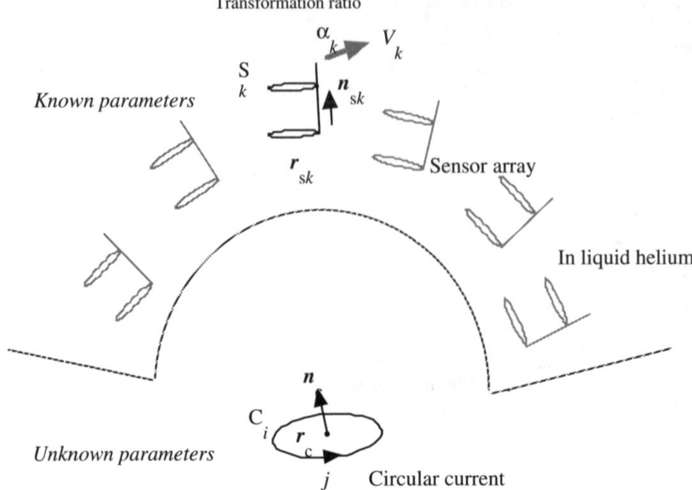

Fig. 3.23. Verification of calibrated sensor parameters

Table 3.1. Examples of calibration error

	x (mm)	y (mm)	z (mm)
Real value	0.000	−75.000	0.000
Estimated value	0.181	−75.302	0.290
	n_x	n_y	n_z
Real value	1.00000	0.00000	0.00000
Estimated value	0.99999	0.00893	−0.00177

coil position (x_c, y_c, z_c) and the coil direction $\boldsymbol{n}_c$. Then we have compared this estimation with the real coil position and direction. Table 3.1 shows the results of this procedure, illustrating that the real position and the estimated current source position are in quite good agreement. For a circular current source, the estimation error is ±0.3 mm as shown in Table 3.1.

Alignment of the Field Distribution and Anatomical Information. The measured biomagnetic response is not of much use unless the data is proposed in reference to the anatomical position of the subject. In case of a magnetoencephalography measurement, the position of the head is calibrated with five markers, five sets of small coils, attached to the heads which is visible to SQUID sensor array. This means that the position of the head is identified in "the coordinate system of the SQUID sensor array", which is calibrated through the procedure described in the previous section. Different five markers visible to MRi (Magnetic Resonance Imaging), which contain water, are attached at the same position to be scanned by MRi after/before the MEG measurement. The information from the MEG markers and the MRi markers is combined, and registration is carried out in order to superimpose the MEG data on the MRi data.

3.3 Magnetic Source Analysis

3.3.1 The Forward Problem

Biological Magnetic Sources. The magnetic field of the human being has been considered to be caused by ionic currents due to the activity of the muscle and neurons. Especially in the case of MEG, the postsynaptic potentials at dendrites in neurons have been considered to be main sources. Figure 3.24 shows the phenomenon of the postsynaptic potential at a synapse near the end of a neuron's dendrite. The local decrease of the transmembrane voltage increases the transmembrane sodium conductivity and allows an inflow of ions. The electrostatic repulsion provided by these ions causes an intracellular current to flow primarily toward the cell body. The intracellular current makes a closed circuit with the return current in the extracellular medium [18].

This is a simple source current model in a microscopic view. But the magnetic field due to this model is very small, so it is impossible to detect this signal outside the head. It is necessary to concentrate a large number of neurons oriented in the same direction to detect the signal outside the head. It is estimated that the collection of 10^4 neurons in 1 mm^2 is needed for a measurable magnetic field. Fortunately, the dendrites and axons in the brain are distributed with a high density and tend to be oriented perpendicular to the cortical surface, as shown in Figs. 3.25 and [19,20].

Such a collection of activated neurons can be assumed to form a current dipole. This is a source current model for the simplest model in MEG. Actually, most biological magnetic sources seem to be more complex, but this model has often been used in practical MEG studies.

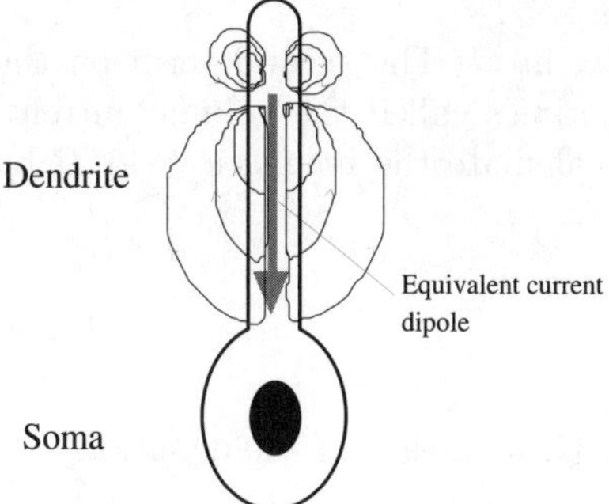

Fig. 3.24. Ionic current by synaptic activity around a neuron (adapted from [18])

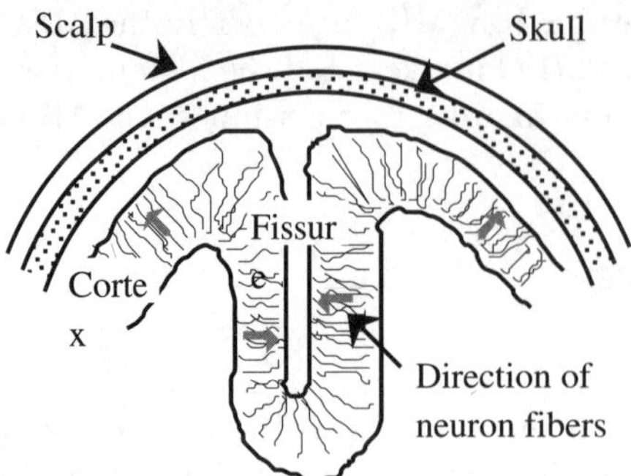

Fig. 3.25. The distribution of the dendrites and axons in the cortex and fissure of the brain

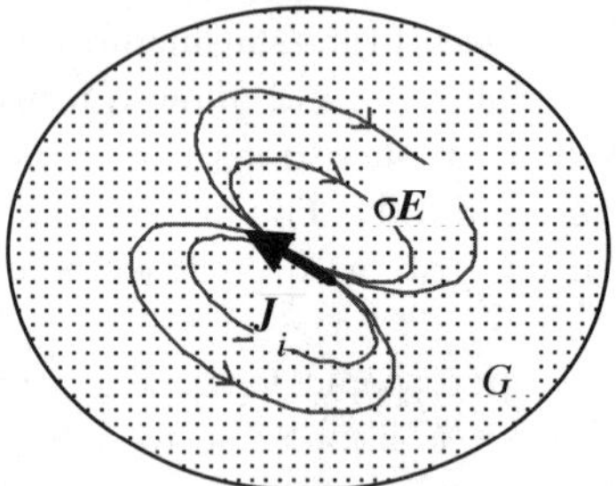

Fig. 3.26. The total current in a volume conductor G

Field Equations in Biomagnetism. We show the basic equations for the calculation of the magnetic field generated by the source current described in the previous section. The following manipulation of equations in this and the next section is derived from reference [21]. Here $\boldsymbol{J}_i$ is the applied current in a volume conductor G which has conductivity σ, as shown in Fig. 3.26. The total current $\boldsymbol{J}$ in G is described as follows:

$$\boldsymbol{J} = \boldsymbol{J}_i + \sigma \boldsymbol{E}\,, \tag{3.59}$$

where $\boldsymbol{E}$ is the electric field caused by $\boldsymbol{J}_i$ in G. The second term on the right-hand side is the ohmic current, sometimes called the volume current. If we obtain the total current $\boldsymbol{J}$, we can calculate the magnetic field $\boldsymbol{B}$ by using the Biot–Savart law:

$$\boldsymbol{B}(r) = \frac{\mu_0}{4\pi} \int_G \boldsymbol{J}(r') \times \frac{\boldsymbol{r} - \boldsymbol{r}'}{|\boldsymbol{r} - \boldsymbol{r}'|^3} \mathrm{d}v'\,. \tag{3.60}$$

Instead of $\boldsymbol{E}$, we use the electric potential V, by means of the equation

$$\boldsymbol{E} = -\nabla V\,, \tag{3.61}$$

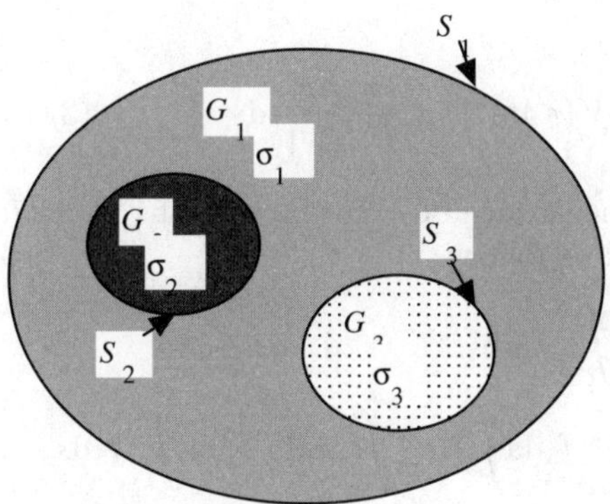

Fig. 3.27. An inhomogeneous conductor

where ∇ means the gradient operator of the scalar potential. Using this equation, (3.59) is transformed into

$$\boldsymbol{J} = J_i - \sigma \nabla V \, . \tag{3.62}$$

If G is divided by surface $S_j, j = 1, , n$, into subregions $G_j, j = 1, , n$, so that $\sigma = \sigma_j$ in each G_j (see Fig. 3.27), (3.60) is transformed into

$$\begin{aligned} \boldsymbol{B}(r) &= \frac{\mu_0}{4\pi} \int_G [\boldsymbol{J}_i(r') - \sigma(\boldsymbol{r}')\nabla V(\boldsymbol{r}')] \times \frac{\boldsymbol{r} - r'}{|\boldsymbol{r} - r'|^3} \mathrm{d}v' \\ &= \boldsymbol{B}_0(\boldsymbol{r}) - \frac{\mu_0}{4\pi} \sum_{j=1}^{n} \sigma_j \int_{G_j} \nabla V(\boldsymbol{r}') \times \frac{\boldsymbol{r} - r'}{|\boldsymbol{r} - r'|^3} \mathrm{d}v' \, , \end{aligned} \tag{3.63}$$

where

$$\boldsymbol{B}_0 = \frac{\mu_0}{4\pi} \int_G J_i(\boldsymbol{r}') \times \frac{\boldsymbol{r} - r'}{|\boldsymbol{r} - r'|^3} \mathrm{d}v' \, . \tag{3.64}$$

$\boldsymbol{B}_0$ is the magnetic field caused by the applied current only. We can derive the identity

$$\nabla V \times \nabla g = \nabla \times (V \nabla g) \, , \tag{3.65}$$

with

$$\nabla g = \frac{\boldsymbol{r} - r'}{|\boldsymbol{r} - r'|^3} \, . \tag{3.66}$$

Using the above identity and Stokes' theorem,

$$\int_G \nabla \times \boldsymbol{A} \, \mathrm{d}v = \int_{\partial G} \boldsymbol{n} \times A \, \mathrm{d}s \, , \tag{3.67}$$

we obtain

$$\int_{G_j} \nabla V(r') \times \frac{r - r'}{|r - r'|^3} \mathrm{d}v' = \int_{\partial G_j} V(r') n(r') \times \frac{r - r'}{|r - r'|^3} \mathrm{d}s\,, \tag{3.68}$$

where $\boldsymbol{n}$ is the outer unit normal of the surface ∂G_j. In the case of Fig. 3.27, the second term in (3.63) is expressed as follows:

$$\begin{aligned} &\sigma_1 \int_{G_1} F_1 \mathrm{d}v + \sigma_2 \int_{G_2} F_2 \mathrm{d}v + \sigma_3 \int_{G3} F_3 \mathrm{d}v \\ &= \sigma_1 \left(\int_{s_1} f_1 \mathrm{d}s - \int_{s_2} f_2 \mathrm{d}s - \int_{s_3} f_3 \mathrm{d}s \right) + \sigma_2 \int_{s_2} f_2 \mathrm{d}s + \sigma_3 \int_{s_3} f_3 \mathrm{d}s \\ &= \sigma_1 \int_{s_1} f_1 \mathrm{d}s + (\sigma_2 - \sigma_1) \int_{s_2} f_2 \mathrm{d}s + (\sigma_3 - \sigma_1) \int_{s_3} f_3 \mathrm{d}s\,, \end{aligned} \tag{3.69}$$

where F and f denote primitive functions in (3.68). From this example, we obtain the general form of the equation follows:

$$\boldsymbol{B}(r) = \boldsymbol{B}_0(\boldsymbol{r}) - \frac{\mu_0}{4\pi} \sum_{j=1}^{n} (\sigma'_j - \sigma''_j) \int_{S_j} V(\boldsymbol{r}') \boldsymbol{n}(r) \times \frac{\boldsymbol{r} - r'}{|\boldsymbol{r} - r'|^3} \mathrm{d}s\,, \tag{3.70}$$

where σ' and σ'' are the conductivities on the inner and outer sides of S_j, respectively. This equation is known as Geselowitz's formula [22]. This equation means that the magnetic field caused by the volume currents is determined by the surface electric potential V. To discover the surface electric potential, we have an equation similar to (3.70) for V:

$$\frac{\sigma'_k + \sigma''_k}{2} V(\boldsymbol{r}) = \sigma_n V_0(r) - \sum_{j=1}^{n} \frac{\sigma'_j - \sigma''_j}{4\pi} \int_{S_j} V(\boldsymbol{r}) \boldsymbol{n}(r) \cdot \frac{\boldsymbol{r} - r'}{|\boldsymbol{r} - r'|^3} \mathrm{d}s\,. \tag{3.71}$$

As a general forward calculation to know the total magnetic field, we first obtain the electric potential by solving (3.71) and then calculate the magnetic field by using (3.70). To solve (3.71) as a practical way, the boundary element method (BEM) is often used. However, it calls for complicated procedures.

The Spherical Conductor Model. In this section we introduce the spherical conductor model, which is often used in MEG. This is one of the special cases where the magnetic field shown in the previous section can be described by simple equations. In this model, the volume conductor G is bounded spherically and symmetrically, as shown in Fig. 3.28. First, we show one of the special features of this model. From (3.70), the radial component of B outside G can be written as follows:

$$B_r(\boldsymbol{r}) = \boldsymbol{B}_0(\boldsymbol{r}) \cdot \boldsymbol{e}_r - \frac{\mu_0}{4\pi} \sum_{j=1}^{n} (\sigma'_j - \sigma''_j) \int_{s_j} V(\boldsymbol{r}) \boldsymbol{n}(r) \times \frac{\boldsymbol{r} - r'}{|\boldsymbol{r} - r'|^3} \cdot \boldsymbol{e}_r \mathrm{d}s\,. \tag{3.72}$$

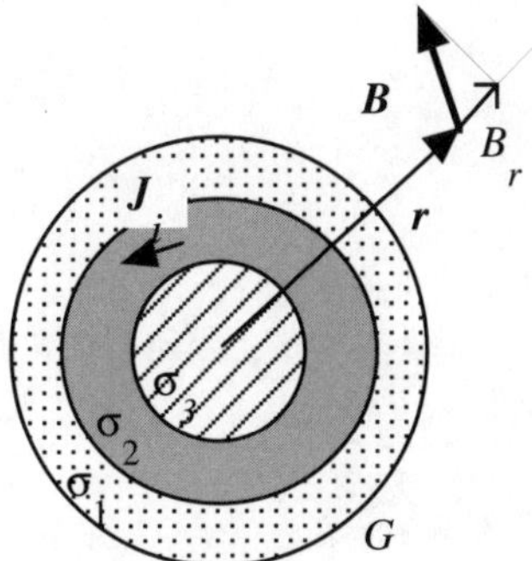

Fig. 3.28. A spherical conductor model

Using $\boldsymbol{n}(\boldsymbol{r}) = \boldsymbol{r}/r, \boldsymbol{e}_r = \boldsymbol{r}/r$,

$$\boldsymbol{n}(r') \times (r - r') \cdot \boldsymbol{e}_r = 0 \tag{3.73}$$

is derived. Thus,

$$B_r(\boldsymbol{r}) = \boldsymbol{B}_{0r}(\boldsymbol{r}) = \frac{\mu_0}{4\pi} \int_G \boldsymbol{J}_i(\boldsymbol{r}') \times \frac{\boldsymbol{r} - r'}{|\boldsymbol{r} - r'|^3} \cdot \boldsymbol{e}_r \mathrm{d}v\,. \tag{3.74}$$

This means that the radial component of the magnetic field is equal to that of the magnetic field caused by the applied current only. When we calculate only the radial component of the magnetic field outside the spherical conductor, we neglect the volume current. The other components of $\boldsymbol{B}$ are affected by the volume currents. To give all the components of $\boldsymbol{B}$, we use a magnetic scalar potential:

$$\boldsymbol{B}(r) = -\mu_0 \nabla U(\boldsymbol{r})\,. \tag{3.75}$$

The scalar potential U is obtained by a line integral of ∇U:

$$\begin{aligned} U(\boldsymbol{r}) &= -\int_0^\infty \nabla U(\boldsymbol{r} + t\boldsymbol{e}_r) \cdot \boldsymbol{e}_r \mathrm{d}t \\ &= \frac{1}{\mu_0} \int_0^\infty B_r(r + te_r)\mathrm{d}t = \frac{1}{\mu_0} \int_0^\infty B_0(r + te_r)\mathrm{d}t \\ &= \frac{1}{4\pi} \boldsymbol{Q} \times (\boldsymbol{r} - r_0) \cdot \boldsymbol{e}_r \int_0^\infty \frac{\mathrm{d}t}{|\boldsymbol{r} + t\boldsymbol{e}_r - \boldsymbol{r}_0|^3}\,. \end{aligned} \tag{3.76}$$

In the above equations, we assumed a current dipole as the applied current. The last integral can be computed; that is,

$$\int_0^\infty \frac{\mathrm{d}t}{|r + te_r - r_0|^3} = \frac{r}{a(ar + r^2 - \boldsymbol{r}_0 \cdot \boldsymbol{r})}\,. \tag{3.77}$$

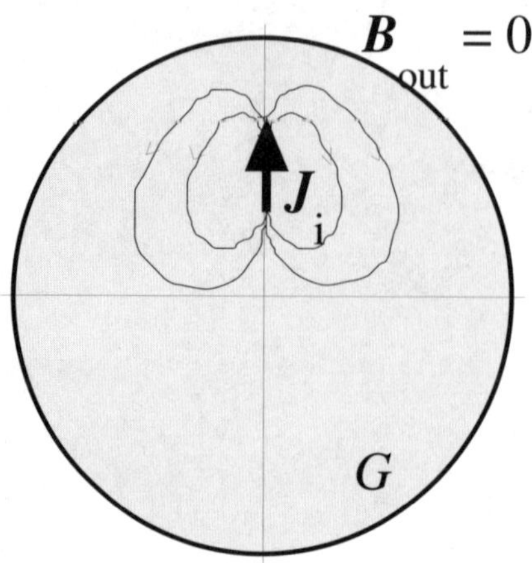

Fig. 3.29. A current dipole oriented parallel to the radius

Therefore,

$$U(\boldsymbol{r}) = -\frac{1}{4\pi}\frac{\boldsymbol{Q} \times r \cdot r_0}{F}, \tag{3.78}$$

where $F = a(ra + r - \boldsymbol{r}_0 \cdot r), \boldsymbol{a} = r - r_0$. As a result, we obtain an equation for the magnetic field outside the spherical conductor G as follows:

$$\boldsymbol{B}(r) = \frac{\mu_0}{4\pi F^2}\left(F\boldsymbol{Q} \times r_0 - \boldsymbol{Q} \times r_0 \cdot \boldsymbol{r}\nabla F\right), \tag{3.79}$$

where,

$$\nabla F = (r^{-1}a^2 + a^{-1}(\boldsymbol{a} \cdot r) + 2a + 2r)\boldsymbol{r} - (a + 2r + a^{-1}(\boldsymbol{a} \cdot r))\boldsymbol{r}_0 . \tag{3.80}$$

From (3.78) or (3.79) we find that if the dipole is parallel to the radial direction, the magnetic field does not appear outside G. Furthermore, the magnetic field outside G does not depend on the value of the conductivity σ.

Examples of Field Patterns. We demonstrate typical field patterns generated by the spherical conductor model (see Fig. 3.30). Figures 3.31, 3.32, and 3.33 show the contour maps of the x, y, and z components of the magnetic field, respectively. For comparison, we show the contour maps due to a current dipole only in Figs. 3.34, 3.35, and 3.36. From these figures, we can show that the tangential components are easily affected by the volume currents in the conductor.

3.3.2 The Inverse Problem

The Equivalent Current Dipole. The MEG inverse problem consists in estimating the neuromagnetic source characteristics from the observed ma-

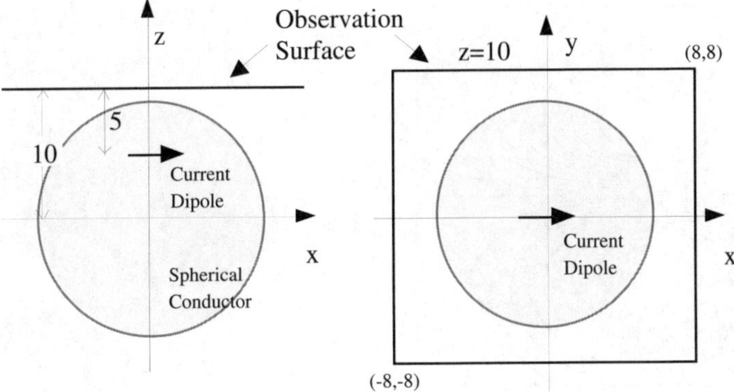

Fig. 3.30. The conditions for the calculation of the field patterns

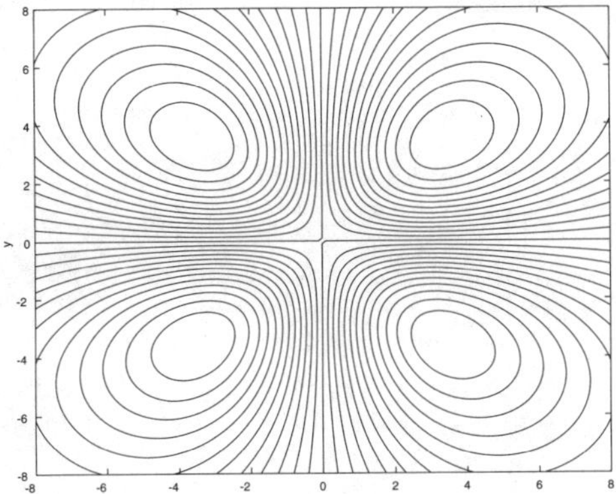

Fig. 3.31. The x component of the magnetic field due to a spherical conductor

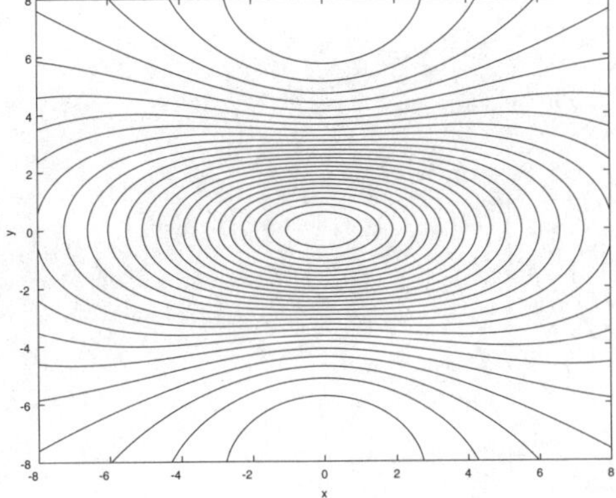

Fig. 3.32. The y component of the magnetic field due to a spherical conductor

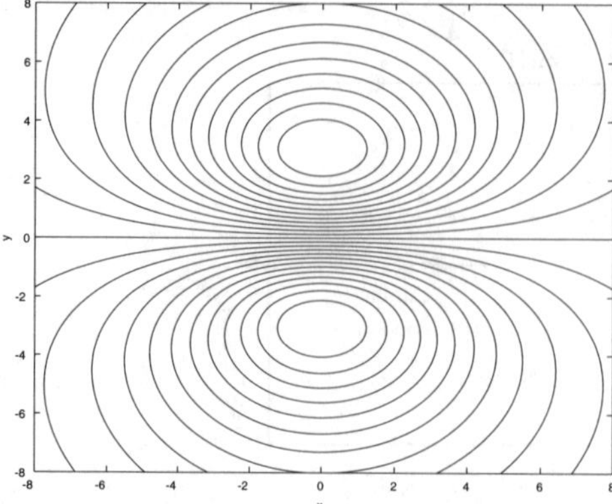

Fig. 3.33. The z component of the magnetic field due to a spherical conductor

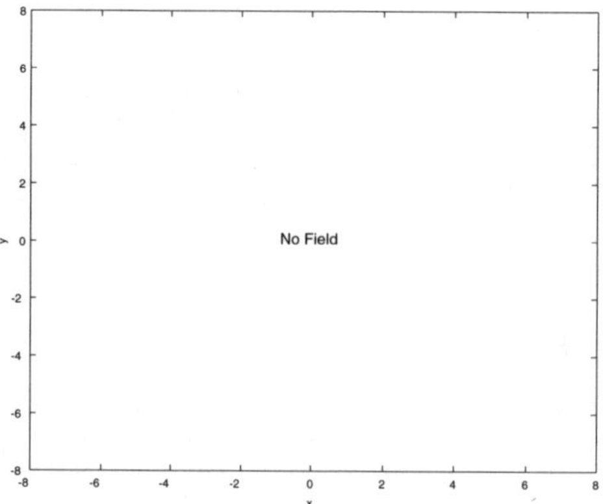

Fig. 3.34. The x component of the magnetic field due to a current dipole only

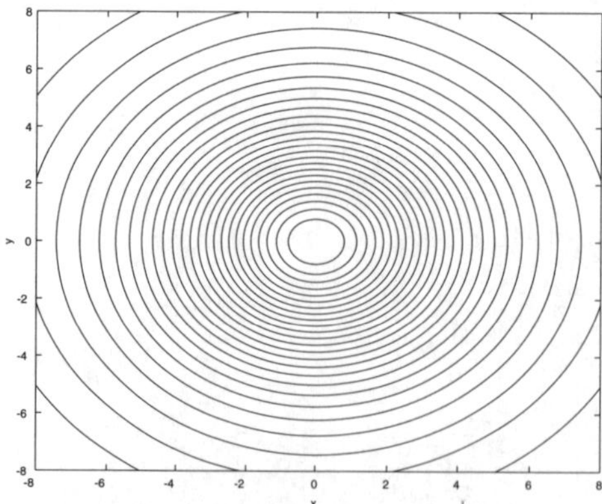

Fig. 3.35. The y component of the magnetic field due to a current dipole only

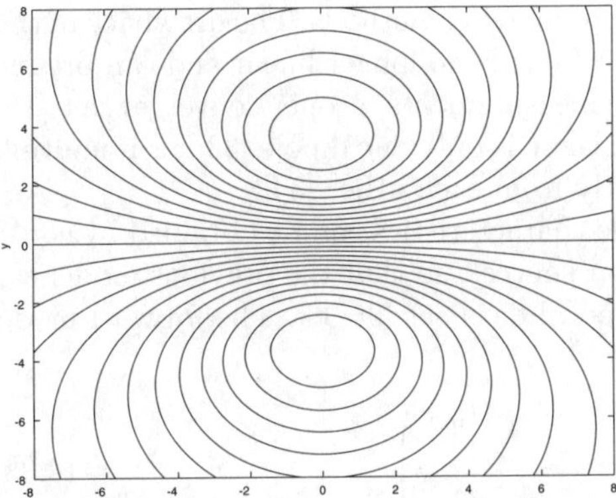

Fig. 3.36. The z component of the magnetic field due to a current dipole only

gnetic measurements. It has been shown [23] that knowledge of the electromagnetic field outside a conductor is not sufficient to uniquely determine the current distribution inside the conductor. This means that there is no unique solution to the inverse problem, so that additional constraints must be added to restrict the number of possible solutions. Basically, one can distinguish two approaches for the resolution of the inverse problem, depending on whether distributed or localized sources are assumed.

Imaging methods [24–29] assume distributed sources and compute an estimate of the full current distribution in the brain. Among the set of all current distributions $\boldsymbol{J}$, an "optimal" solution $\boldsymbol{J}_{\text{opt}}$ is computed by using Tikhonov regularization and finding that

$$\boldsymbol{J}_{\text{opt}} = arg \min_{\boldsymbol{J}} \|\boldsymbol{M} - \boldsymbol{G}\boldsymbol{J}\|^2 + \lambda \boldsymbol{J}^T \left(\boldsymbol{W}\boldsymbol{W}^T\right)^{-1} \boldsymbol{J}\,, \tag{3.81}$$

where $\boldsymbol{M}$ is the observed measurements, $\boldsymbol{G}$ is the lead field matrix, λ is a regularization parameter, and $\boldsymbol{W}$ is a weight matrix that can take several forms, such as the identity matrix or the Laplacian operator, depending on the desired constraints. The drawbacks of imaging methods is that they require manipulation and computation of huge matrices, and that they usually tend to give oversmoothed estimates of the sources.

An alternative method is to assume that the measured electromagnetic field results from the activity of one or a few highly localized neuronal sources of a dipolar nature: this is called the equivalent current dipole model [27,30–32]. In this case, for a single time slice, the inverse problem reduces to finding $\boldsymbol{q}_{\text{opt}}$ such that

$$\boldsymbol{q}_{\text{opt}} = arg \min_{\boldsymbol{q}} \|\boldsymbol{M} - \boldsymbol{G}\boldsymbol{q}\|^2\,. \tag{3.82}$$

Since $\boldsymbol{q}$ is a vector of n dipoles, and each dipole corresponds to six parameters (position, direction, and intensity), the number of parameters to be estimated reduces considerably in comparison to imaging methods. The drawback of this method is that the minimization problem is no longer linear and can present many local minima. For single time measurements, proper convergence to the exact solution is only assured when there is only one dipole [33], or if multiple dipoles can be treated independently from each other [34].

More robust estimates for dipole characteristics can be obtained by considering fixed positions for the dipolar sources, so that the previous model can be extended to the full spatio-temporal data set [35,36]. The forward model is then written as

$$\boldsymbol{M} = \boldsymbol{G} \begin{pmatrix} \boldsymbol{u}_1 & 0 & 0 \\ 0 & \ddots & 0 \\ 0 & 0 & \boldsymbol{u}_n \end{pmatrix} \cdot \begin{pmatrix} q_1(t_1) & \cdots & q_1(t_m) \\ \vdots & \ddots & \vdots \\ q_n(t_1) & \cdots & q_n(t_m) \end{pmatrix}, \tag{3.83}$$

where $\boldsymbol{u}_i$ are unitary orientation vectors and $q_i(t_j)$ is equal to the magnitude of the ith dipole at time j:

$$\forall i \in \{1, \dots, n\},\ \forall j \in \{1, \dots, m\},\ \boldsymbol{q}_i(t_j) = q_i(t_j) \cdot \boldsymbol{u}_i\,. \tag{3.84}$$

Several search algorithms have been proposed for estimating the multi-dipole solution [37–41]. Some of them also make use of prior anatomical or functional information derived from MRI or fMRI data, to restrict the solution space to biologically relevant solutions [42–44]. However, although much research is currently devoted to this subject, solution of the inverse problem for multi-dipole models still presents some unresolved issues. First, the number of dipoles must usually be determined before the inverse problem resolution. Also, most current methods give only a single, presumably best, estimation of the neuromagnetic solution, without any information on other almost equally likely solutions, or on the solution confidence interval. To answer these problems, we have developed two different approaches to the MEG inverse problem: the current element distribution model, and probabilistic resolution of the inverse problem using Markov chain Monte Carlo methods.

A Probabilistic Approach using MCMC Methods. Markov chain Monte Carlo (MCMC) methods are probabilistic tools that have recently been proposed for the resolution of the MEG inverse problem [41,45]. Instead of searching for a single optimal solution to the inverse problem, MCMC methods give an estimate of the full probability distribution of the solutions. The knowledge of this probability distribution gives indications on the number of likely solutions, their characteristics, and their confidence interval. However, simple MCMC methods such as the Metropolis algorithm usually get trapped in local probability maxima, and fail to give robust estimates of the probability distribution. We have proposed the use of two different MCMC schemes to solve this problem [46,47] and to allow for the search for an unknown number of dipoles [48,49]. Here we describe our approach, the two MCMC schemes that we proposed, and results obtained on both simulated and real data.

The first step in using MCMC methods is to define a probability distribution for the inverse problem solutions. Assuming a variable number n of localized dipolar sources and single time measurements, the posterior probability of solutions $(\boldsymbol{q}, \boldsymbol{n})$ knowing the measurements can be computed using Bayes theorem:

$$p(\boldsymbol{q}, n|\boldsymbol{b}_{\text{obs}}) \propto p(\boldsymbol{b}_{\text{obs}}|\boldsymbol{q}, n) \cdot p(\boldsymbol{q}|n) \cdot p(n)\,, \tag{3.85}$$

where $\boldsymbol{q}$ is a vector of n dipoles $q_1, \dots, q_n$, with q_i being defined by its position (x_i, y_i, z_i), its direction (θ_i, φ_i), and its intensity j_i.

The first term is the likelihood of the observed measurements, given the current distribution $\boldsymbol{q}$. We have assumed an m-dimensional Gaussian noise model, with mean 0 and covariance matrix Σ, so that the likelihood is given by

$$p(\boldsymbol{b}_{\text{obs}}|\boldsymbol{q}, n) = \frac{1}{(2\pi)^{m/2}|\Sigma|^{1/2}} \times \exp\left(-\frac{1}{2}(\Delta\boldsymbol{b})^T \Sigma^{-1} (\Delta\boldsymbol{b})\right)\,, \tag{3.86}$$

with $\Delta\boldsymbol{b}$ being equal to $\boldsymbol{b}_{\text{obs}} - \boldsymbol{b}(\boldsymbol{q})$.

For real data, the covariance matrix Σ was estimated from pre-stimuli measurements. The theoretical measurements $\boldsymbol{b}(\boldsymbol{q})$ obtained from the dipole distribution $\boldsymbol{q}$ were computed using the Sarvas formula [50,21].

The second term represents the a priori knowledge about the number and characteristics of the neuromagnetic sources. The source positions were limited to the brain volume by setting a null probability for sources outside the brain. We favored sources in the cerebral cortex by setting a 1 to 100 probability ratio for sources in the cortex. Dipole directions were constrained to be tangential by assuming a normal law of mean 0 and standard deviation $\pi/10$ for the angle between the position and the direction vectors. The source intensity was assumed to follow a constant law in a predefined interval. For variable numbers of sources, a Poisson law was assumed.

To sample the posterior distribution $p(\boldsymbol{q}, n|\boldsymbol{b}_{\text{obs}})$, we used parallel tempering (PT) [51] to prevent being trapped in local modes and to speed up convergence. A number k of Markov chains $C_i = \left(\boldsymbol{q}_i^{(1)}, \dots, \boldsymbol{q}_i^{(n)},\right)$ are constructed, with each realization $\boldsymbol{q}_i^{(j)}$ being a set of n dipoles. The chains C_i are constructed in an iterative process, with each realization $\boldsymbol{q}_i^{(j+1)}$ being determined from the previous state of the chains with a probability distribution $p_i^{(j)}$. The chain C_1 is called the principal chain; it is constructed so that the distribution $p_1^{(n)}$ converges to the posterior distribution:

$$\lim_{n\to\infty} p_1^{(n)} = p(\boldsymbol{q}|\boldsymbol{b}_{\text{obs}})\,. \tag{3.87}$$

Thus, any expectation of a function f with respect to the posterior can be estimated from a delayed series of the realization of C_1:

$$\forall f,\ E[f] = \int_V p\left(\boldsymbol{q}|\boldsymbol{b}_{\text{obs}}\right) \cdot f(\boldsymbol{q}) \cdot \mathrm{d}\boldsymbol{q} \tag{3.88a}$$

$$\simeq \frac{1}{N-M} \sum_{i=M+1}^{i=N} f\left(\boldsymbol{q}_1^{(i)}\right), \tag{3.88b}$$

where the first M iterations of the Markov chain represent the burn-in period (i.e., the convergence process of C_1 from random initial values to the posterior distribution).

Other chains C_i, for $i > 1$, are called auxiliary chains: they prevent the principal chain from getting trapped in a local mode. Each chain C_i is associated with a temperature parameter T_i, such that $T_1 = 1$ and $T_i < T_{i+1}$. An unnormalized probability distribution π_i is then defined for each chain C_i, so that "hot" auxiliary chains are associated with probability distributions that become easier and easier to sample:

$$\pi_i = p\left(\boldsymbol{q}|\boldsymbol{b}_{\text{obs}}\right)^{1/T_i}. \tag{3.89}$$

Transition probabilities $p_i^{(j)}$ are defined according to the Metropolis sampler [52], so that for each chain C_i, the distribution of its elements will converge to the probability distribution π_i. A proposal value $\boldsymbol{q}_i^*$ is drawn at random around $\boldsymbol{q}_i^{(j)}$ and is accepted with probability r:

$$r = min\left(1, \frac{\pi_i(\boldsymbol{q}_i^*)}{\pi_i(\boldsymbol{q}_i^{(j)})}\right). \tag{3.90}$$

If the proposal value is accepted, it becomes the new realization of the chain; otherwise, the previous value is retained and the chain does not move.

Movement between the chains is allowed by PT moves. Two contiguous chains C_i and C_{i+1} are chosen at random and their realizations are swapped with probability

$$r = min\left(1, \frac{\pi_i(\boldsymbol{q}_{i+1}^{(j)}) \cdot \pi_{i+1}(\boldsymbol{q}_i^{(j)})}{\pi_i(\boldsymbol{q}_i^{(j)}) \cdot \pi_{i+1}(\boldsymbol{q}_{i+1}^{(j)})}\right). \tag{3.91}$$

For a variable number of dipoles, we added reversible jump (RJ) moves [53] to the previous sampler. The parameters to be estimated now take the form $(k, \boldsymbol{q}_k)$, where k is the unknown number of dipoles, and $\boldsymbol{q}_k$ is the set of these k dipoles. The move from $(i, \boldsymbol{q}_i)$ to the proposal value $(j, \boldsymbol{q}_j)$ is realized by drawing a random vector $\boldsymbol{u}_i$ independently of $\boldsymbol{q}_i$, and setting $\boldsymbol{q}_j$ to some deterministic function f_i of $\boldsymbol{q}_i$ and $\boldsymbol{u}_i$. This move and the reverse move must satisfy

$$dim\, \boldsymbol{q}_i + dim\, \boldsymbol{u}_i = dim\, \boldsymbol{q}_j + dim\, \boldsymbol{u}_j. \tag{3.92}$$

The new proposal value $(j, \boldsymbol{q}_j)$ is then accepted with probability

$$p = min\left(1, \frac{\pi(j, \boldsymbol{J}_j)\cdot p(j,i)\cdot p_j(\boldsymbol{u}_j)}{\pi(i, \boldsymbol{J}_i)\cdot p(i,j)\cdot p_i(\boldsymbol{u}_i)} \cdot \left|\frac{\partial(\boldsymbol{J}_j, \boldsymbol{u}_j)}{\partial(\boldsymbol{J}_i, \boldsymbol{u}_i)}\right|\right) , \tag{3.93}$$

where $p(i, j)$ is the probability of proposing a move from an i-dipole solution to a j-dipole solution. We used four different types of reversible jumps. Death/birth moves corresponded to deleting a random dipole from the source distribution, or adding to the sources a new dipole with a random position. The Merge move was done by choosing two different dipoles $\boldsymbol{q}_i$, and $\boldsymbol{q}_j$, deleting one of them and updating the other to $(1-r)\boldsymbol{q}_i + r\boldsymbol{q}_j$, with r uniformly drawn in [0;1]. The reverse move, splitting in two a single dipole, was made by first randomly choosing a new dipole $\boldsymbol{q}_j$ and updating an already existing dipole $\boldsymbol{q}_i$ to the new value $[\boldsymbol{q}_i + (1-r)\cdot \boldsymbol{q}_j]/r$. The probability $p_{\text{up}}(k)$ of choosing a move going from dimension k to $k+1$ (respectively, $p_{\text{down}}(k)$ for going from k to $k-1$) was

$$p_{\text{up}}(k) = c(k) \times \frac{p(k+1)}{p(k)} , \tag{3.94a}$$

$$p_{\text{down}}(k) = c(k) \times \frac{p(k-1)}{p(k)} , \tag{3.94b}$$

where $p(k)$ is the Poisson prior on the number of dipoles, and $c(k)$ is a normalizing term such that $p_{\text{up}}(k) + p_{\text{down}}(k)$ is equal to 1.

The RJ plus PT algorithm has first been applied to simulated data [48,49] and compared to simulated annealing (SA) that has been previously proposed for resolution of the MEG inverse problem [54,55]. Measurements for a two-dipole pattern were simulated with 20 dB SNR white noise and 20 dB SNR neuromagnetic noise originating from 60 random sources located in the brain. An unknown number of sources between 1 and 4 was assumed for reversible jump. Results showed a clear improvement of our RJ/PT algorithm over simulated annealing: on 30 trials, the mean error was 4.5 mm, less than simulated annealing (6.3 mm) even with the added difficulty of an unknown number of sources. These results were even more striking when comparing parallel tempering alone with simulated annealing: 90% of the trials made with PT found the true solution to within 5 mm, but only half of SA runs achieved this result. Moreover, the mean error in PT (3.0 mm) was reduced to almost half of the SA error value.

Moreover, the PT/RJ sampler was able to correctly estimate the source number and found a 65% probability of a two-dipole source configuration, in accordance with the model used. The algorithm found also a 24% and 11% probability for three- and four-dipole source configurations, thus showing a slight bias toward a higher number of sources. More information about this simulation study can be found in [49].

A second simulation was conducted to study the influence of noise on the parallel tempering algorithm, without RJ moves [46]. Measurements for two

Table 3.2. Algorithm comparison: percentage of success and mean error

Algorithm	Number of trials	Number of dipoles	Correct position (%) to within 5 mm	10 mm	15 mm	Mean error (mm)
PT	30	2	90	100	100	3.0
PT/RJ	30	1–4	73	92	98	4.5
SA	30	2	53	83	92	6.3
Metropolis	10	2	35	55	70	14.5

Table 3.3. Influence of noise level on PT and SA

Total SNR (dB)	SA valid runs (%)	PT valid runs (%)	SA mean error (mm)	PT mean error (mm)
15.8	100	90	8.7	7.0
10.1	70	90	9.5	9.4
7.8	45	95	10.6	11.5
4.1	0	15	–	17.7

dipoles were simulated with both white noise (20 dB SNR) and neuromagnetic noise (5, 10, 15, and 20 dB SNR). Sources were restricted to the brain volume, and constrained to the cortex (approximated by the outer 5 mm of the brain) by a 1 to 1000 probability ratio. Solutions explaining at least 95% of the total variance were considered as valid. Although SA and PT gave equivalent results for a low noise level (16 dB SNR), SA results degraded quickly when the noise level increased (45% of valid runs for 8 dB SNR). On the contrary, PT was little affected by the noise level, and it was only for the very high noise level (4.1 dB SNR) that the PT performance suddenly degraded.

We also applied the PT algorithm to real MEG data obtained in different conditions (auditory evoked fields, somatosensory evoked fields [47,56]) and with signal-to-noise ratios ranging from 5.4 dB to 20 dB. For AEF, magnetic measurements were recorded by 148 axial magnetometers (baseline = 50 mm) and averaged on 100 runs. Measurements were analyzed at a latency of 92 ms. The noise standard deviation was equal to 15 fT, for a SNR of 20 dB. The prior probability for the source positions was the same as in the previous study. The results clearly showed two dipoles located in the left and right primary auditory cortices, and that explained 95.7% of the measurement variance (Fig. 3.37). In the second experiment, SEF data were obtained after electrical stimulation of the right index, below the movement

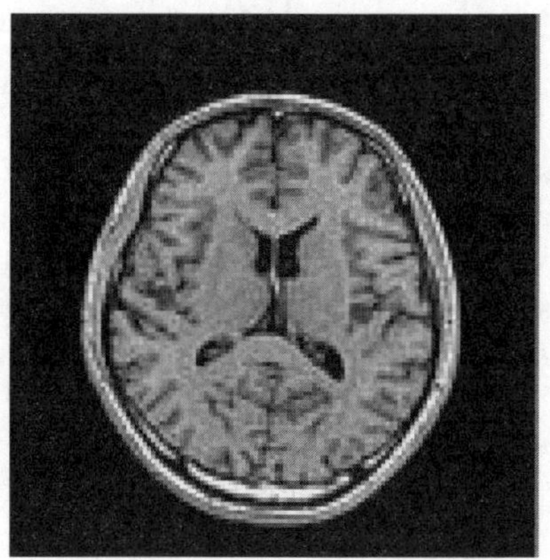

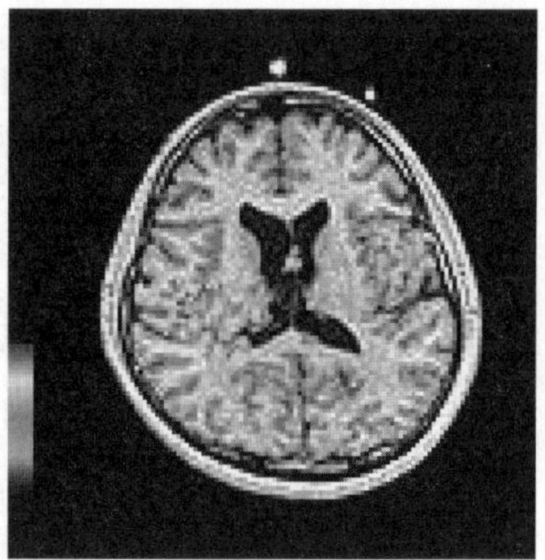

Fig. 3.37. Results of dipole estimation using the PT algorithm: AEF (*left*) and SEF (*right*). The increased area of the dipole locations for SEF reflects the lower SNR. The intensity is color coded for the SEF image from blue (lowest intensity) to red

threshold. Measurements were recorded for five subjects with an 80-channel magnetometer system, and were averaged from 70 runs. The noise standard deviation ranged from 17 to 27 fT, for a 5 to 15 dB SNR. Depending on the subjects, the source locations showed different patterns, that corresponded to the contralateral, and sometimes the ipsilateral, first and second somatosensory cortices (Fig. 3.37). These first results obtained on real data, even with a low noise ratio, are consistent with previous studies. Other practical applications of the MCMC method are planned, both for cognitive sciences and clinical applications.

Current Element Distribution. The conventional procedures for analyzing the magnetic field source in biomagnetism are to estimate the location, the direction, and the intensity of an equivalent current dipole which is assumed to be representative of biological activity. Iso-magnetic field mapping is mostly used to determine the location of the dipole, by finding the two extrema of the magnetic field on the surface being observed. The dipole is estimated directly from the extrema or calculated by some mathematical procedure such as least square fitting of a dipole model.

To estimate a single dipole, the mathematical operation yields a unique solution in the solution space, which consists of six coordinate corresponding with the dipole parameters. A dipole is described by parameters of three components of position, two of direction, and one of intensity. If we have magnetic field signals without any noise, these unknown parameters are estimated with at least six independent observations. In this case, the model is evaluated by the value of the residue between the observed data and magnetic field generated by model. If we expand the model and adopt a more complex system [31,57,58], such as the current multiple expansion (CME), multi source model (MSM), or the distributed current model, there is no longer a unique solution. We encounter a kind of combinatorial optimization problem and the mathematical approach should introduce a novel concept of evaluation. If we assume that the MSM acts like multiple current dipoles – even just two or three current dipoles – we should observe 12 or 18 signals that should be strongly independent of each other and should be free from noise. Furthermore, in the course of parameter estimation, there will be many local minima. Simulated annealing [59] is one of the ways of escaping from local minima. However, there is still no concrete strategy for a model that has large numbers of signal sources.

To consider a more general aspect of the inverse problem, we must first think of a model of a system (such as the cardiac or central nervous system in biomagnetism). There is no practical idea that teaches us to adopt a particular model. The CME reduces the residue, such as the $\mathcal{L}_2$ norm [59], between the measured data and the magnetic field generated by the model. We still have no rational idea of how to select a model, even if the locations of the estimated poles happen to be an incident with the excited portion of the biological system.

Model. In this study, we tried to establish an approach to describe magnetic sources due to a distribution of current elements that converge to a specific source distribution. We believe that every observable magnetic field signal consists of a summation of magnetic fields, each of which is generated by an independent current element. The model of the signal source consists of a large number of current elements. Each element is described as a current dipole. Each element has the same intensity. The position and direction of each element are not specified. With this model, the inverse problem is that, by using an algorithm mentioned below, the distribution of the elements is expected to converge to a signal source current pattern.

Algorithm. Assume N current elements $\boldsymbol{I}_{i,i=1,\ldots,N}$. The position of $\boldsymbol{I}_i$ is $\boldsymbol{r}_i$, and the direction of $\boldsymbol{I}_i$ is $\boldsymbol{n}_i$.

Step 1: To distribute the elements randomly into the volume that is subject to a biological function being active. Then, to calculate the magnetic field $\boldsymbol{B}_{\text{model}}$ generated by distributed current elements $(\boldsymbol{I}_1, \ldots, \boldsymbol{I}_N)$ at measurement positions $(\boldsymbol{R}_j, \ldots, \boldsymbol{R}_M)$:

$$\boldsymbol{B}_{j,\text{model}} = \sum_{j=1}^{M} \boldsymbol{B}_i \,, \tag{3.95}$$

where $\boldsymbol{B}_i = B(r_i, n_i)$ is the magnetic field generated by current element $\boldsymbol{I}_i$ at position $\boldsymbol{r}_i$ and direction $\boldsymbol{n}_i$.

Step 2: To calculate an evaluation value L of model fitting by norm between model field $\boldsymbol{B}_{j,\text{model}}$ and measured magnetic field $\boldsymbol{B}_{j,\text{measured}}$ at measurement position $\boldsymbol{R}_j$:

$$L = \sum_{j}^{M} \left|\boldsymbol{B}_{j,\text{model}} - \boldsymbol{B}_{j,\text{measured}}\right|^2 , \tag{3.96}$$

where the norm is the $\mathcal{L}_2$ criterion of two vectors.

Step 3: To select an arbitrary current element. The element moves to the vicinity of its original position in the solution space $\boldsymbol{S}_i$ to reduce L. The element motion is limited by the algorithm parameters. There are N solution spaces, each of which belongs to an independent current element $\boldsymbol{I}_i$:

$$\boldsymbol{S}_i = \boldsymbol{S}_i(\boldsymbol{r}_i, \boldsymbol{n}_i) . \tag{3.97}$$

The solution space is specified by the position coordinate and the direction vector of each current element.

Step 4: In the case of L being reduced, the element has a new position. In the case of L being not reduced, the element stays at its original position. The algorithm operation returns to Step 3 and the algorithm selects the next arbitrary element.

The mathematical charactersistics of the algorithm are not clarified. It might be quite difficult to analyze the characteristics by means of an analytical method, rather than by empirical or heuristic means. The major purpose of this study is to verify that the algorithm can calculate the distribution of current elements that describes the magnetic source in good agreement, within a practical cpu time.

Simulation 1. As the first step of mathematical clarification of the algorithm, the following simulation has been carried out. The circular current was assumed to be a magnetic source located in a cube. Magnetic field vectors at several positions on the surface of a cube generated by the circular current were adopted as the measured magnetic field vectors $\boldsymbol{B}_{\text{measured}}$. The magnetic field vectors at measurement positions generated by elements are calculated.

In the initial stage of the simulation, we should determine all of the parameters that specify the signal source, the model, and the algorithm. The parameters of the signal source that are not essential but are practical for this method are the size of the cube, the diameter of the circle current, the position and the direction of the circle and the current intensity. The intensity of the element, the number of elements, and the distribution are model parameters. The algorithm is specified by the rule of element motion and the termination rule of iteration.

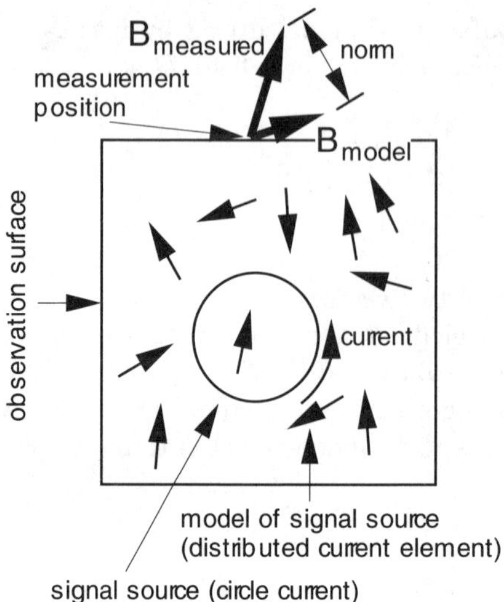

Fig. 3.38. The circular current which is the source of the magnetic field in a cube. The model is evaluated by a value of summation of the norm between model field $\boldsymbol{B}_{\mathrm{model}}$ and measured field $\boldsymbol{B}_{\mathrm{measured}}$

In the steps of the algorithm, an arbitrarily selected element moves within its vicinity in the solution space $\boldsymbol{S}_i$, where a five-dimensional space is specified by the position and direction of the element $\boldsymbol{I}_i$. Figure 3.38 shows the signal source (the circular current in the cube). Figure 3.39 shows the successive results of current elements converging from random distribution to the signal source patterns. 3600 iterations were carried out.

Discussion 1. The result is quite well posed to the signal source, even though it is not certain whether the algorithm may be effective for an arbitrary ill-posed situation. In a practical application, we should introduce many heuristics and a priori knowledge about the measurement, in order to restrict the solution space and to avoid the local minimum solution.

Returning to the specification of this simulation, number and intensity of elements are determined arbitrarily. However, it is necessary to select concrete values for the parameters. It is proposed that, before the algorithm is run, the intensity of an equivalent current dipole is estimated and, from histological and physiological knowledge, the number of elements is determined. Then, the current intensity of the element can be determined from the energy value of the equivalent dipole divided by the number. As the iteration number increases, the element becomes a pair of elements. The pair of elements, that face each other, produce a very small magnetic field and make a small contribution to the norm. It is a kind of local minimum.

(a)

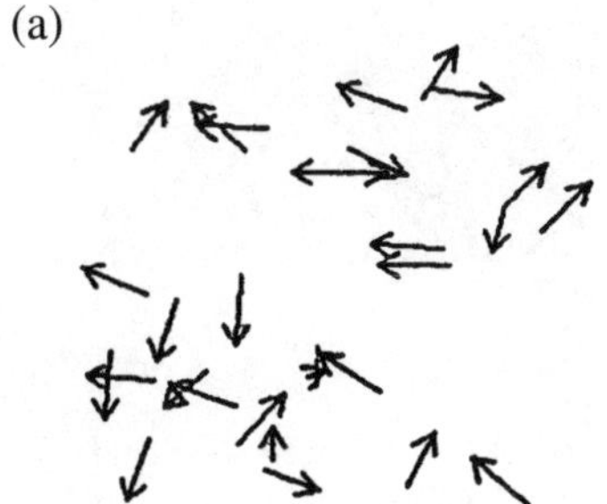

(b)

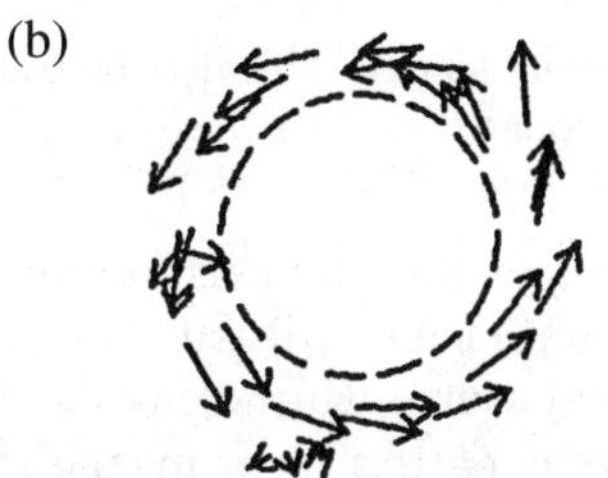

(c)

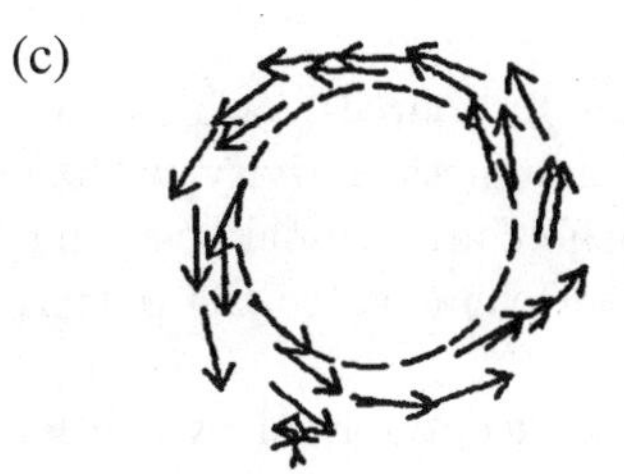

Fig. 3.39. a The initial distribution of current elements in the algorithm: 28 current elements were distributed at random as a starting situation for the inverse problem. **b** The distribution of the current elements after 2400 iterations. They seem to going to converge to the pattern of magnetic source. **c** The distribution of the current elements after 3600 iterations. They have converged to the pattern of a magnetic source, which is a solution to this inverse problem

There might be many pairs when the number of elements is large. The appearance of the pair of elements is a sign that the iteration of the algorithm should be terminated. This algorithm is quite heuristic. The algorithm was applied to an inverse problem for a circular current signal source. We could not gain a perspective on the possibility of applying the algorithm to a general inverse problem. However, we would like to insist that this element approach could figure out the three-dimensional pattern of an electric current that is hidden in a volume without any strict parametric model such as CME. If we introduce a strategy such as CME, we have to measure more and more data as the number of unknown parameters increases.

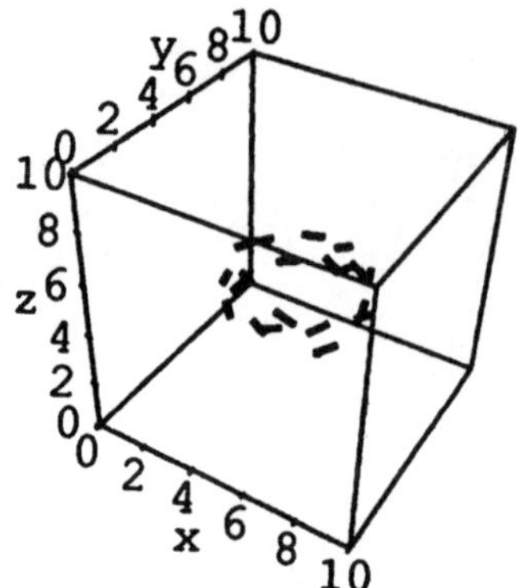

Fig. 3.40. The result of estimating a single circular current source. The distribution of 20 current elements was obtained after 10 000 iterations

Improvement of the Algorithm. As discussed in the previous section, the algorithm of magnetic source imaging by current element distribution is capable of describing the pattern of a specific signal source which consists of circular currents. The algorithm describes a current pattern as a magnetic source without the assumption of a parametric source model The estimated solutions for a single circular current source and a double circular current source are shown in Figs. 3.40 and 3.41, respectively. These solutions were obtained by computer simulations. The circular current sources are described by the distribution of current elements.

To apply this algorithm to other types of magnetic sources, such as a volume current source, which is more suitable for biomagnetic activity, we have introduced a stabilized condition in addition to the conventional cost function. As a result of this, the convergence of current elements to the pattern of the magnetic source is improved.

The major part of the optimization strategy, except for an evaluation criterion, is the same as discussed previously. The proposed evaluation criterion is composed of the least-square norm and the restrictive condition: the so-called stabilized condition [60]. The stabilized condition is intended to evaluate more complicated distributions of current elements. The least-square norm of this optimization procedure is (3.96). To improve this algorithm, the stabilized condition is combined with the evaluation function as follows:

$$L + w \times Op(g)\,, \tag{3.98}$$

where $Op(g)$ is a stabilized condition, g is a set of parameters to be estimated, and w is the weight with which the stabilized condition contributes to the evaluation function.

The Stabilized Condition. Generally, the result of optimization of a multiple current element distribution shows a complicated pattern, even when

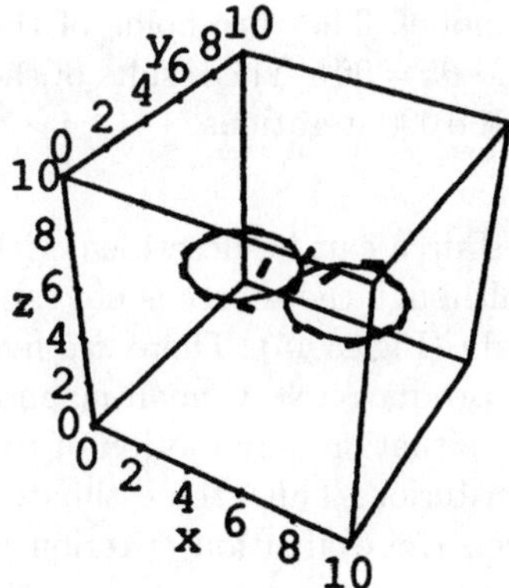

Fig. 3.41. The result of estimating a pair of circular current sources with the current flowing in opposite directions. The distribution of 20 current elements was obtained after 30 000 iterations

the distribution is a smoothly curved line or circle. Two current elements are frequently located closely together and face each other. In this state, the value of the evaluation criterion does not decrease any further. This is one kind of the local minimum. To avoid this problem, a stabilized condition as shown in Fig. 3.42 is introduced. The condition is that the current flow has to be continuous and smooth (Fig. 3.42a). Any two current elements, whose positions are close to each other, must not be in opposite directions. If the distance d between two elements is less than the threshold dt, the angle θ between them must be less than θt (Fig. 3.42b). If there is a pair of current elements that does not satisfy this condition (Fig. 3.42c), one of the pair moves away, and its direction is changed. The motion is controlled by random numbers.

Simulation 2. The mathematical characteristics of the algorithm have not been fully clarified. It might be quite difficult to analyze the characteristics using an analytic method. Computer simulations were carried out to confirm that the algorithm is capable of finding that distribution of elements which best describes the source. The algorithm was applied to a volume current source consisting of 45 elements, as shown in Fig. 3.43. The initial state of the estimation strategy is shown in Fig. 3.44, where 20 current elements, the model of the magnetic source, are randomly distributed in a subject region. The results of the estimation using the least-square norm of (3.96) and the improved evaluation criterion of (3.98) are shown in Figs. 3.45 and 3.46, respectively. Two hundred and sixteen ($= 6 \times 6 \times 6$) measurement points are placed on the whole surface of a cube. The measured magnetic field $\boldsymbol{B}_{\text{measured}}$ is calculated as the summation of the magnetic field generated by each element of the volume current source using the Biot–Savart law. $\boldsymbol{B}_{\text{model}}$

was generated by 20 current elements of the model. The thresholds of the stabilized condition were as follows: $dt = 0.5$ and $\theta t = 90°$. The results of the estimation were obtained after approximately 100 000 iterations.

Discussion 2. The assumed volume current source can be described with the proposed algorithm quite well (Fig. 3.43), although the source is not estimated correctly with the least-square norm only (Fig. 3.45). There are five pairs of elements facing each other. Such pairs produce a very small magnetic field as if no current element existed. In this situation, any motion of the element increases the value of the evaluation criterion. Thus, the estimated distribution is not improved in this case. When the evaluation criterion is the least square norm only, it frequently happens that two current elements are close together and face each other. The proposed algorithm, however, is capable of describing a volume current source by introducing current continuity as a stabilizing factor. The assumption of current continuity shown in Fig. 3.42 appears to be reasonable, even when this algorithm is applied to biomagnetic measurements. However, to apply this algorithm to complex magnetic source distributions, it is necessary to introduce suitable stabilized conditions which depend on the characteristics of the subject.

3.3.3 Visualization

Field Contour Map. A contour map is the illustration of the variation of $b = f(G)$, which is valuable on a surface G, by contours as a result of connection of the points that have the same value of b.

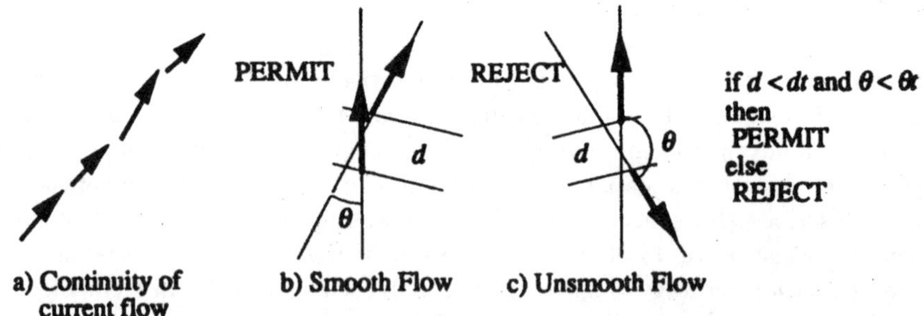

Fig. 3.42. The stabilized condition for optimizing the distribution of current elements. **a** The situation in which the current flow is continuous and smooth. **b** The definition of smooth current flow. The distance d between two elements is less than the threshold dt, and the angle θ between them is less than the threshold θt. In this case, the pair is allowed to stay in its present position. **c** The definition of a nonsmooth current flow. The angle θ is larger than θt. In this case, the pair is rejected

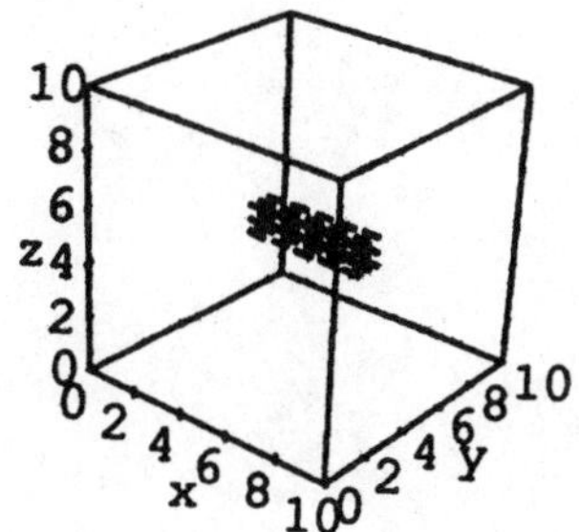

Fig. 3.43. The volume current source used in the computer simulation. The current is composed of 45 elements

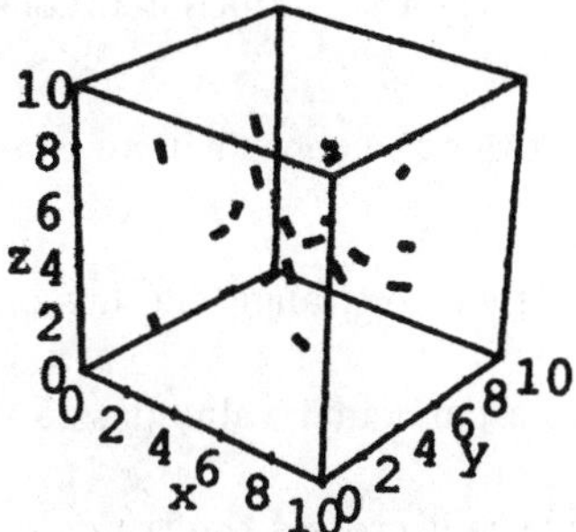

Fig. 3.44. The initial state of the estimation strategy: 20 elements are distributed at random in the subject region

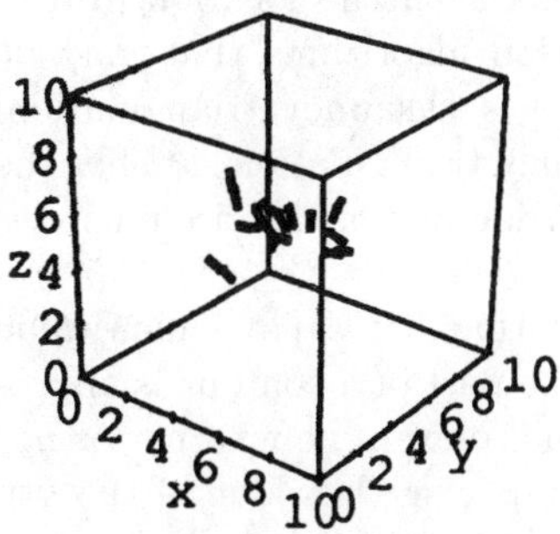

Fig. 3.45. The result of estimation using only the least-square norm without the stabilized condition. The distribution of 20 current elements was obtained after 100 000 iterations. There are five pairs of elements in opposite directions. The estimated distribution was not improved any more in this case

Drawing a field contour map is one of the most general ways to express the distribution of the magnetic field. In this section, a simple way to draw a field contour map from a set of the data measured by a multiple SQUID

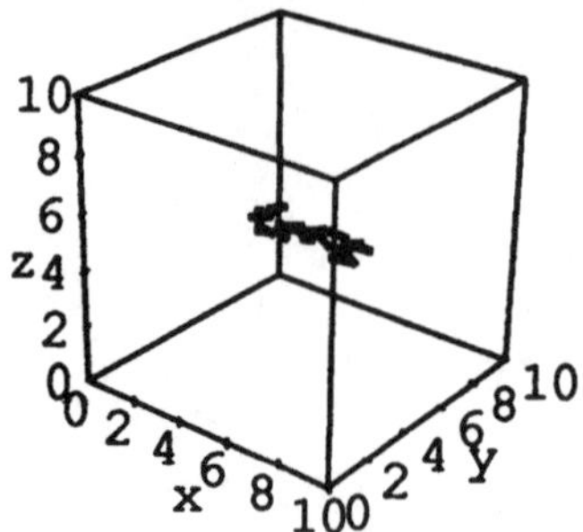

Fig. 3.46. The result of optimization with the proposed stabilized method. The current elements converged to the pattern of the magnetic source quite well. The estimated distribution of 20 current elements was obtained after 100 000 iterations

sensor array is described. The basic procedure to draw the contour map is as follow:

(1) Each observation site is regarded as a grid. Generate triangle meshes on the observation surface.
(2) Determine whether or not a contour that has a particular value passes a triangle.
(3) If the contour passes a triangle, determine where it crosses the triangle.

Part of the contour is drawn at steps (2) and (3).

Triangulation, as the first step in drawing a contour map, is used to create a set of nonoverlapping triangular bounded cells from the location grid of the sensors. There are a number of triangulation algorithms that may be proposed. One of the more popular algorithms is Delauney triangulation. The triangular network that results from Delauney triangulation satisfies the criterion that states that no vertex should lie inside any of the circumcircles of the triangles in the network.

At the next step, whether or not a contour that has a particular value crosses a triangle is examined. In this section, the part of a contour is approximated as a segment for simplicity. It is possible to examine whether or not a segment of a contour crosses a triangle by comparing the value of the kth contour, b_k, with three values at each vertex of the triangle $\mathrm{A_1A_2A_3}$, b_{A_1}, b_{A_2}, and b_{A_3}. If

$$(b_k - b_{\max})(b_k - b_{\min}) \leq 0\,, \tag{3.99}$$

where b_{max} is the maximum of b_{A_1}, b_{A_2}, and b_{A_3}, and $b_{\min}$ is their minimum is satisfied, the contour crosses $\triangle\mathrm{A_1A_2A_3}$. Otherwise, it is outside the triangle.

When the contour k goes on $\triangle\mathrm{A_1A_2A_3}$ and $b_{\mathrm{A}_n} \leq b_k < b_{\mathrm{A}_m}$ is satisfied, an edge of the segment of the contour is at the point that separates A_m and A_n in the ratio of $(b_{\mathrm{A}_m} - b_k) : (b_k - b_{\mathrm{A}_n})$.

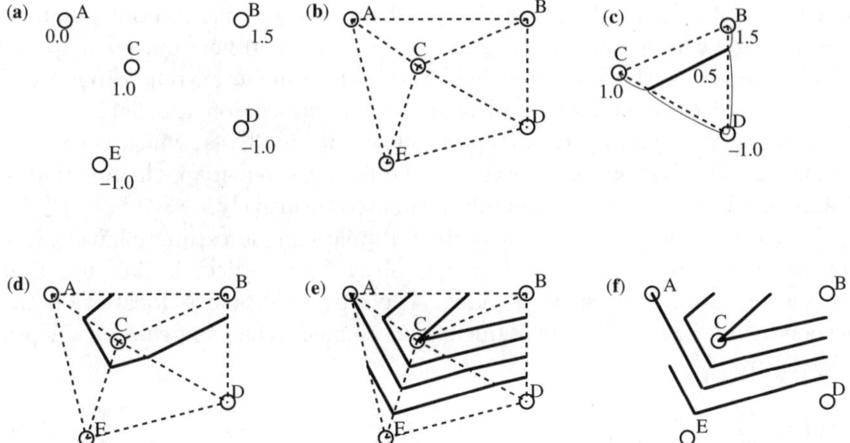

Fig. 3.47. An example of drawing a contour map. **a** A set of data at five observation points was given. **b** Triangular meshes were produced. **c** A segment of a contour was determined on one of the triangular meshes. **d** A contour was formed after examining all the triangles. **e** The same procedure was repeated, with different values of the contours. **f** The whole contour was obtained

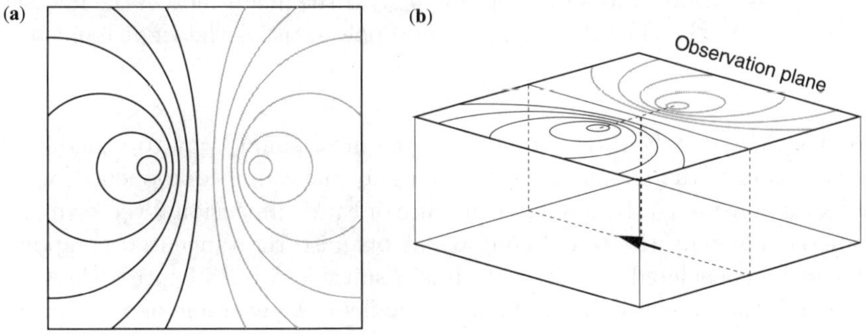

Fig. 3.48. A field contour map with an ideal single current dipole. **a** An example of a contour map with a dipole-like pattern. Black lines represent an upward field. Gray lines represent a downward field. **b** The estimation of a magnetic source. The arrow represents the estimated current dipole

After examining the contour for all the triangles, a whole connected contour is represented. Furthermore, a whole contour map is obtained by repeating the same procedure with different values for the contours.

In Fig. 3.47, an example of making a contour map from a set of data at five observation points is shown. The contour map in the figure is very angular because only a few observation points are used for simplicity. More observation points located at narrower intervals can provide a smoother contour map. Otherwise, the interpolation of the value at the interior points of each triangle can be applied in order to obtain a smoother contour map.

If a source of the magnetic field is an ideal single current dipole, two extrema, called an inhalant and an exhalant, appear on the field contour map. When this symmetric pattern as shown in Fig. 3.48a, which is called a dipole-like distribution, is observed on the field contour map, the location of the magnetic source can be roughly estimated intuitively.

In the case of Fig. 3.48b, a current dipole can be estimated using the approximation that the surface of the observation object is flat and that the sensors are arranged on a plane. The current dipole is located on the perpendicular bisector to the segment that connects the two peaks at a depth d, given by

$$d = \frac{L}{\sqrt{2}}, \tag{3.100}$$

where L is the distance between the two peaks. The magnitude of the dipole q is

$$q = \frac{6\sqrt{3}\pi d^2}{\mu_0} b_{\mathrm{max}}, \tag{3.101}$$

where μ_0 is permeability of the air and b_{max} is the magnitude of the magnetic field at the peaks. The direction of the dipole satisfies the right-hand screw rule.

Vector Map. A field contour map is the most simple way to visualize the distribution of the magnetic field, when only the radial component from the object is considered. However, in the case of three-dimensional vector magnetometry, not only the radial component but also the tangential component should be considered. A vector map is a suitable way to express the distribution of the tangential component of the field. An example of a vector map is shown in Fig. 3.49.

Extraction of Head and Brain Volume from MRI. Because of the difficulty of the MEG inverse problem, it is important to use as much information as possible to restrict the search only to the solutions that are biologically consistent [42–44]. One way to achieve this goal is to use magnetic resonance images (MRI) to obtain information about the head and brain volume. For example, this MRI information could be used to discard solutions located outside the brain, to favor certain parts of the brain, or to force the dipole orientations to follow certain constraints. Also, the 3D representation of the head and of the brain could also be used as input for realistic modeling of the forward fields. And, of course, this MRI information can also be used to construct three-dimensional representations of the head to better visualize the pattern of the neuromagnetic sources and their evolution with time.

Manual selection of the brain and head parts on the MRI images is precluded, since it would require several hours of human work for each subject. It is therefore necessary to develop an automatic algorithm to analyze these images, with minimum intervention from the user. We present here a method that we have devised to extract the brain and head volume [61,62] and that we currently use to set prior knowledge, and to display the solutions.

The initial data are T1-weighted MRI images. Each data set contains from 122 to 125 axial slices of 256×256 voxels. In-plane resolution is 0.9375×0.9375 mm, and out-of-plane resolution is 1.5 mm. Since MRI suffers from numerous artifacts, a direct extraction of 3D objects from only gray level information is not possible. Therefore, our algorithm proceeds in two steps. The first step is the automatic extraction of estimates of the brain and head objects from gray level information. In the second step, these estimates are filtered and errors are corrected. For the first step, the gray level histogram is computed for all of the MRI volume, and is smoothed by using the Silverman bump-hunting method [63]. For that method, the histogram is considered as a probability density function and is estimated by

$$p_h(x) = \frac{1}{h \cdot N} \sum_{i,j,k}^{N} f\left(\frac{x - g(i,j,k)}{h}\right), \tag{3.102}$$

where N is the number of voxels, f is a Gaussian function N(0,1), $g(i,j,k)$ is the gray level of voxel (i,j,k), and h is a parameter controlling the degree of smoothing. A large value of h will produce a unimodal histogram, while smaller values of h will give histograms with more and more modes. Our algorithm begins with a large value of h and makes a dichotomic search for the smallest value of h that will produce a distribution with three modes

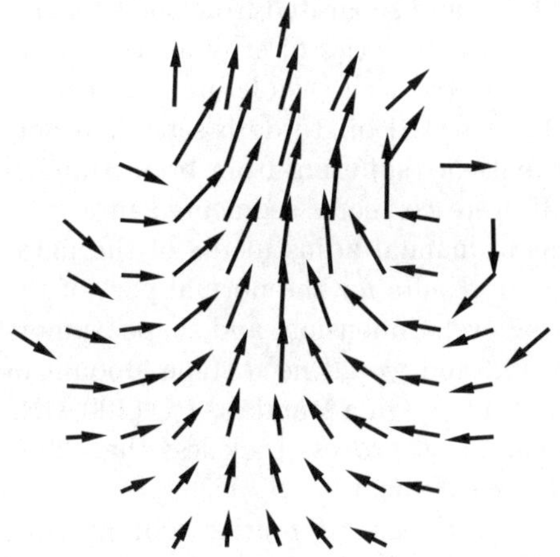

Fig. 3.49. An example of a vector map. The tangential component of the field is projected onto a plane

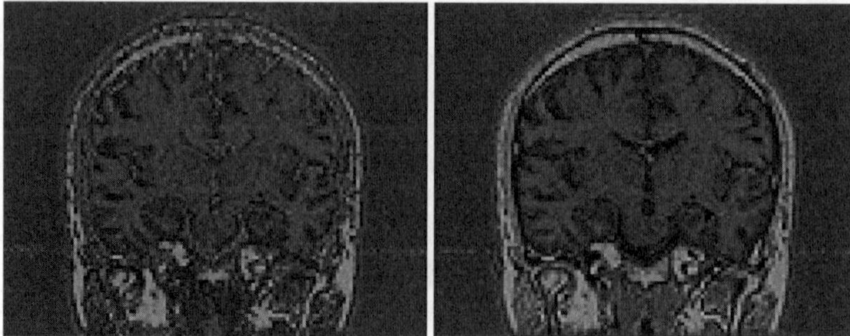

Fig. 3.50. Segmention of MR images. Histogram segmentation (*left*) gives an approximate segmentation between brain, head, and background, but many details are still erroneously resolved. Morphological filtering (*right*) corrects them and gives an accurate segmentation of the head and of the brain

(background, head, and brain). The smoothed histogram is then automatically thresholded into background, head, and brain by searching for points with a derivative smaller than a given threshold, and corresponding to the beginning or end of the peaks.

Since the MRI data suffer from various sources of noise, and so that there is no unambiguous correspondence between gray level and anatomical structures, a simple segmentation based on gray values does not give adequate results, so it must be refined by appropriate post-filtering. We make extensive use of two functions of mathematical morphology [64]: erosion (which corresponds to removing the outer voxels of an object), and the opposite operation, which is called dilation. The combination of these two operations with logical operations on binary images is sufficient to correctly extract the head and brain objects from the segmentation results previously obtained. For this, the whole head object is first filled and separated from the background. The brain is then correctly extracted by a series of erosions to remove spurious brain areas, and dilations to reconstruct the whole brain object.

This segmentation algorithm has been tested on 15 data sets from normal subjects, and four data sets from patients suffering from brain tumors. All the data sets from normal patients were correctly segmented in a fully automated way. For the four other cases, manual adjustments of the parameters were necessary to obtain adequate results for the normal part of the brain. Due to an extremely variable gray level, dimension, and shape, tumors were generally not satisfactorily extracted and would need some amount of manual contouring to be accurately segmented. On a standard PC (400 MHz Pentium II processor), the whole segmentation process took less than 20 s, thus allowing its use for routine MEG measurements.

The segmentation information was directly used for prior computation. For 3D visualization, the next step is the computation of the boundaries of the brain and of the head. This is done with the marching cube algorithm

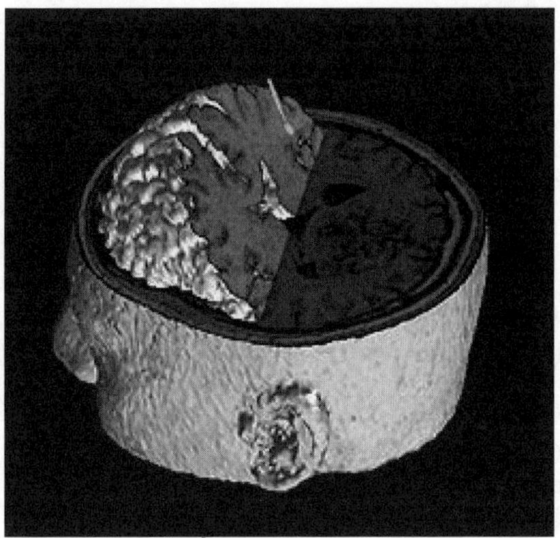

Fig. 3.51. The display of MEG sources (AEF) on the reconstructed 3D head and brain

[65], modified according to [66] to resolve the ambiguities of the initial algorithm. Since the marching cube algorithm usually creates a very large number of triangles, the user can parameterize the precision of the boundary computation by adjusting the size of the marching cubes. This provides an easy way to obtain different level of detail, allowing the user to adjust the tradeoff between realism and speed of rendering. The obtained triangulation is then used to display a three-dimensional representation of the head and of the brain, along with the neuronal source estimations (Fig. 3.51).

3.4 Biomagnetic Measurement

3.4.1 Magnetoencephalography

In this section, we describe a specific MEG system developed at Kanazawa Institute of Technology, and also some examples of MEG measurements.

Magnetoencephalogram Systems Developed at KIT. We have developed new systems for magnetoencephalography (MEG). The pickup coil is a coaxial type first-order gradiometer with a 50 mm baseline. The magnetic field resolution of the system is about 4 fT/rtHz or 0.8 (fT/cm)/rtHz in the white noise region. The unique feature of the system is the gantry-free horizontal dewar, which is fabricated through what we call a "ship-in-a-bottle approach". Less than 10-liter/day liquid helium consumption for the 100-liter capacity is realized. One of the merits of the horizontal dewar is that a small room is sufficient for installation because of the low height (0.89 m) of the dewar. Another merit is that the patient can be measured in a lying position, which is more relaxing compared to the conventional vertical type.

System Configuration. The array of SQUID sensors is installed in the liquid helium dewar, which keeps the sensors at around 4.2 K. The dewar is placed inside the magnetically shielded room (MSR) in order to eliminate the environmental noise. The sensors are operated with flux-locked loop (FLL) electronics, which are located outside the MSR. The magnetic signals sensed by SQUID are transformed to voltage signals and processed through analog signal processors, and are acquired and stored by a data acquisition (DAQ) system.

The data are collected by the analyzing software, which operates on Windows NT on the host computer. Data analysis such as equivalent current dipole (ECD) estimation is done on this console. Operators can easily compile reports with the analyzed results using common word processing, desktop publishing software, spreadsheet software, or graphing according to their preferences. The operation of the SQUID sensor, such as the lock or unlock mode, is also controlled through this console. The post-processing software and the acquired data are transportable to other PCs so that the users can do further analysis using their own desktop or laptop Windows PCs. The stimulation apparatus for the subjects is also provided and can be controlled through the console. Another peripheral instrument is the automatic liquid refilling system, which reduces the operator's burden. The subjects lie down on a bed in a relaxed position to have their MEG responses to various stimuli measured. Figure 3.52 shows an overview of the system. We describe the components of the systems in the following sections.

Sensor and Electronics. A coaxial type first-order gradiometer is adopted for the SQUID sensor. The sensing coil and reference coil have a diameter of 15.5 mm and the baseline between the two coils is 50 mm. The magnetic signal is transferred to a double-washer SQUID placed about 75 mm from the sensing coil. There is small possibility of flux trapping caused by magnetic materials that some subjects might bring into the tail of the dewar carelessly, such as eyeglasses.

We have employed an external feedback coil configuration [67]. The feedback current to compensate the input magnetic signal is fed to the sensing loop, the reference, the coil, and the input coil, so that the field to be measured suffers no disturbance. Consequently, the cross-talk between adjacent sensors is reduced. To measure the cross-talk, two gradiometers are placed in parallel and adjacent to each other at a separation of 23 mm. The level of cross-talk is measured by monitoring one of the SQUID outputs when a sinusoidal flux of known amplitude, 2.0 Φ_0, is applied to another SQUID, both being in FLL mode [68]. The result is excellent, −93.3 dB, thanks to the external feedback coil configuration.

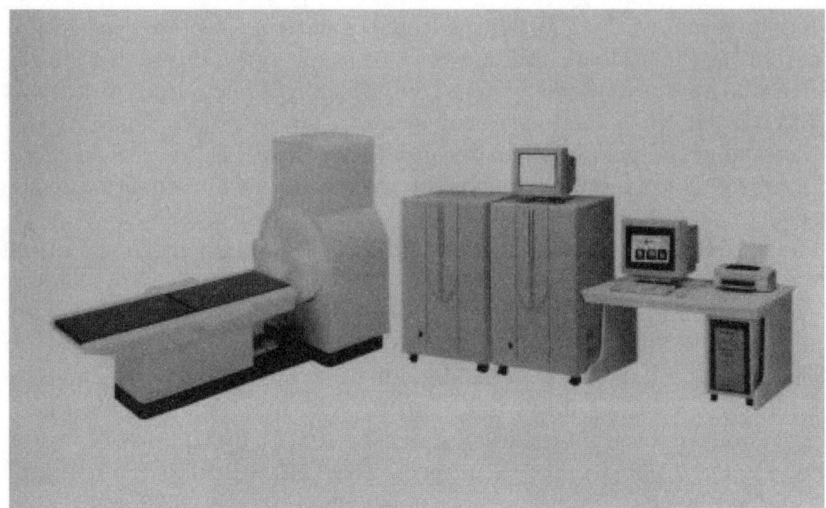

Fig. 3.52. An overview of the MEG system

Fig. 3.53. A first-order gradiometer with 50 mm baseline and 15.5 mm diameter coils

The SQUID is driven by the direct offset integration technique (DOIT) [69], which is useful for cost and complexity reduction, especially MEG systems with over 100 channels.

Sensitivity. The term sensitivity is often used as a power spectrum of the noise of the sensor or system. Here, we use the term as follows: the field sensitivity [nT/Φ_0] is the magnetic field at the pickup coil necessary to couple one flux quantum to the SQUID; the feedback current sensitivity [$\mu A/\Phi_0$] is the current at the feedback coil necessary to couple one flux quantum to the SQUID; and the magnetometer sensitivity [nT/V] is the output voltage at the FLL circuit when the pickup coil senses a unit magnetic field. We can also determine the field gradient sensitivity [$(nT/cm)/\Phi_0$] as the magnetic field gradient at the pickup coil necessary to couple one flux quantum to the SQUID.

The feedback current sensitivity is 49.5 $\mu A/\Phi_0$. We adopt 20 kΩ as the feedback resistor. This means 49.5 $\mu A \times 20$ kΩ = 0.99 V is the output voltage of the FLL circuit, while the feedback current balances one flux quantum in the SQUID. The field sensitivity of the sensor is designed to be 0.68 nT/Φ_0. In the sense of field-gradient sensitivity, it corresponds to $(0.68\ nT/\Phi_0)/(5\ cm) = 0.136\ (nT/cm)/\Phi_0$. The magnetometer sensitivity and gradiometer sensitivity are calculated by $(0.68\ nT/\Phi_0)/(0.99\ V/\Phi_0) =$ 0.69 nT/V and 0.138 (nT/cm)/V, respectively. These calculated values must be confirmed by the calibration process described in Sect. 3.2.2 and the results are 0.59 nT/V and 0.118 (nT/cm)/V. The differences of about 17% are due to field sensitivity and the feedback resistor. We adopt the calibrated value for MEG measurement.

Noise and Flux Trapping. It is important and practical to represent the noise performance for the overall system, not just for the SQUID sensor. SQUID sensors have low intrinsic noise; for example, 2.0 fT/rtHz in the white noise region and 4.1 fT/rtHz at 1 Hz. However, additional noise sources which we must include are the electronic noise of FLL circuits, Johnson noise from the radiation shield foil in the vacuum chamber of the dewar, and electromagnetic interference from the DAQ system/host PC/the stimulators. The measured noise, shown in Fig. 3.54, is 2.3 fT/rtHz in the white noise region, above 50 Hz, and 7.0 fT/rtHz at 1 Hz with the overall system, which includes the noise due to the SQUID, FLL, dewar, radiation shield, cables, MSR, and an analog signal processor with a low-pass filter set to 2 kHz cutoff.

It is possible to assume that a signal is applied only to the sensing coil and not to the reference coil. In such a case, if a signal of 2.7 fT/rtHz is applied to the sensing coil, we can interpret this signal as being detectable with SNR = 1. Therefore, it is convenient to describe the noise performance in terms of "field noise [fT/rtHz]". However, in the actual MEG measurement, the signal is detected by the sensing coil, and some decayed signal at the reference coil is subtracted. ECD estimation is carried out by taking this subtraction into account. Therefore, a notation of "field-gradient noise" is sometimes useful for the expression of noise performance. The gradiometer baseline in this case is 5 cm, which leads to another notation of 2.3/5 = 0.46 (fT/cm)/rtHz in the white noise region and 1.4 (fT/cm)/rtHz at 1 Hz with overall system.

Incidentally, the range of white noise observed over all channels is spread around the average value of 4.0 fT/rtHz, which corresponds to 0.8 (fT/cm)/rtHz, with a standard deviation of 1.3 fT/rtHz.

The sensors are usually cooled down by liquid helium inside an MSR. Otherwise, magnetic flux is easily trapped in the SQUID chip, where it reduces the critical current and consequently prevents the sensor from operating properly. We have observed this phenomenon just once in 1151 accumulated trials. In this case, the SQUID washer with an area of 0.01 mm^2 is exposed to a magnetic field of about 0.2 μT inside the MSR, and is supposed to trap (0.2 μT × 0.01 mm^2 = 2 fWb = 1.0 Φ_0, should this rare case happen. For a magnetic field less than that, it is impossible for the SQUID to trap the flux. On the other hand, we have observed that almost all trials lead to flux trapping when cooled down in the Earth's magnetic field of about 50 μT, which corresponds to 250 Φ_0. Therefore, it is crucial to cool down the sensors inside the MSR.

The Dewar. The unique feature of the system is the gantry-free horizontal dewar, which is fabricated through what we call a "ship-in-a-bottle approach". The size of the sensor complex is larger than the diameter of the neck of the dewar bottle. The larger the sensor complex is, the wider the head surface is covered. On the other hand, the smaller the neck of the dewar is, the lower is the boil-off rate of the liquid helium. Therefore, we assemble the large sensor complex inside the dewar just as hobbyists make ship models in an empty bottle, using tools which go through the neck of the bottle, instead of inserting the sensor complex after the assembly. With this technology, less than 10 liter/day for the 100 liter capacity is realized. There is a 20 mm distance between the sensing coil of the gradiometer and the

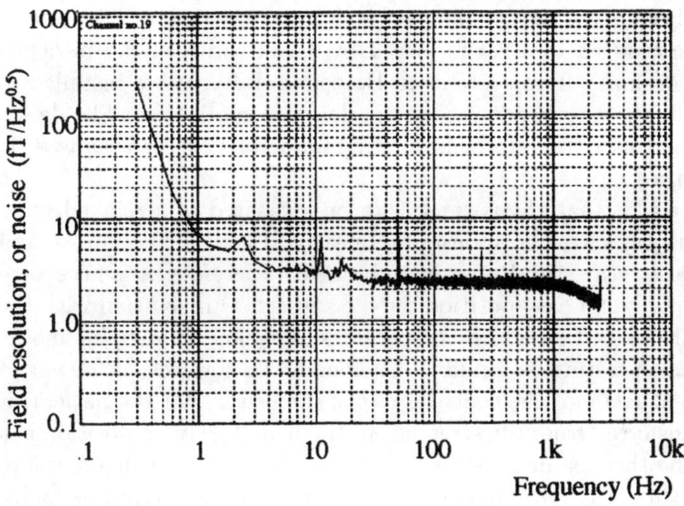

Fig. 3.54. The noise spectrum for the overall system, which includes the SQUID, FLL, dewar, radiation shield, cables, and MSR: 2.3 fT/rtHz in the white noise region, 7.0 fT/rtHz at 1 Hz. The cutoff of the low-pass filter is set to 2 kHz

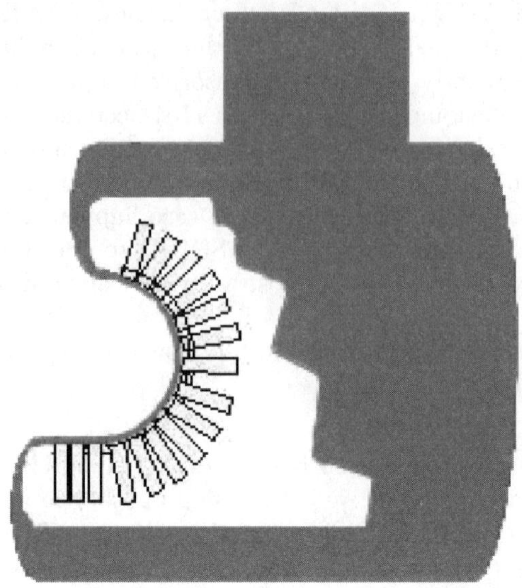

Fig. 3.55. A cut-away diagram of the horizontal dewar made using a "ship-in-a-bottle" technique

surface of the dewar where the subject's/patient's head touches. Figure 3.55 shows a cut-away diagram of this dewar. The helmet-shaped manifold is the sensor-holder, with many sockets for the gradiometers.

One of the merits of the horizontal dewar is that a small room is sufficient for installation because of the low height (0.89 m) of the dewar. The size of the MSR can be reduced to a 2.4 m height by 2.2 m width by 3.5 m depth (outer shell dimensions). Two systems in Boston and Tokyo are installed in this size of MSR. The next merit is that the dewar is gantry free. The dewar is placed on a rigid base, therefore eliminating numerous concerns caused by moving mechanisms.

The third merit is that the patients can be measured in a relaxed lying position. The subject's MEG responses are measured for 10–20 minutes, and they are required to stay still during measurement. Therefore, it is necessary to provide a relaxing and easy position for the subject. Our solution with the horizontal dewar is a bed on which the patient can recline in a supine mode, with his or her head resting on a pillow. The bed is slid towards the dewar so that the head is placed into the concave part of the dewar. For the patient, it is easy to be transferred from the stretcher to the bed. The lying position is a good match with other scanners, such as MRI and X-ray CT, which measure patients in the same way. We made a vertical type of dewar with the "ship-in-a-bottle approach" in order to compare the vibration of the subject's head during the MEG measurement. To confirm the improvement, the vibration of the head was measured with a normal subject and the amplitudes were compared between the vertical dewar and the horizontal one.

Figure 3.56 shows the time variation of the magnetic field in 5 s caused by small magnetic object, a 2 mm square part of a subway ticket, attached to the forehead of the subject. These are not the noise signals that appear in an ordinary MEG measurement, but they are a good indication of the relative value of the vibration of the head position of the subjects. The subject was asked to stay still during the measurements in a vertical-type dewar case and a horizontal dewar case. The vibration was about 65 pT peak-to-peak for the horizontal case, but 320 pT peak-to-peak for the vertical case. The improvement factor is about 5. The factor for a longer time span, such as 20 minutes, would be larger. The result for a seriously ill patient is imagined to be much larger, because it is hard to maintain a straight and rigid pose in a vertical system. Recently, swinging type dewars which allow patients to lie in a supine mode have been described [70–72], and those would show a similar improvement.

It is practical for the operator to refer to the helium consumption rate in terms of the necessary amount including loss during transfer, rather than in terms of the boiling-off rate at the time in-between the refills. If the cycle for refill is frequent, the overhead loss is not negligible. The necessary amount of liquid helium to maintain this system is 100 liter/week or 5200 liter/year, and the refill cycle is one refill per week.

User Friendliness. As described in the previous sections, the systems have some merit as far as "user friendliness" is concerned. We now summarize the issue from this point of view, adding some other functions which have not been referred to before.

Operator Friendliness. The liquid helium refill operation with the LHe storage dewar can be done outside the MSR with the door closed. The transfer tube is installed in the dewar with one port sticking out of the MSR. The operator connects the extension tube to be inserted in the 100 liter helium container. There is no need to open the door of the MSR or to jolt the dewar to be connected to the LHe container outside the MSR.

An automatic helium gas pressurizing system makes it easier to transfer liquid from the container to the dewar. A heater inserted in the container heats up the liquid helium and the vaporized gas forces the liquid helium to transfer from 100 liter container to the dewar. The level gauge monitors the liquid helium in the dewar and automatically stops the heating up when the dewar is filled. There is no need for an extra helium gas cylinder or an operator's rubber balloon. The whole process, including setting-up, is accomplished in about 45 min.

Parameter tuning for the DOIT type FLL circuit is done automatically. It takes only 30 s and saves 3 hours compared to manual tuning. Flux-locked

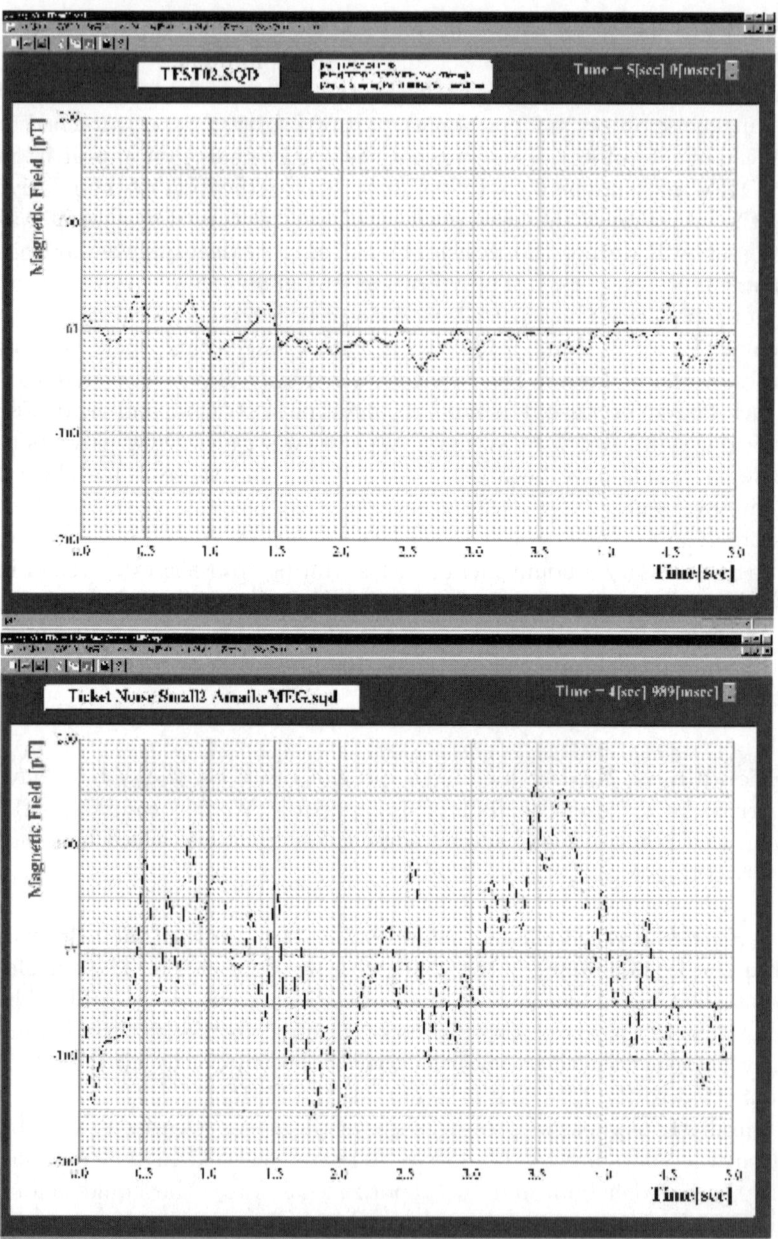

Fig. 3.56. The time variation of the magnetic field measured in the horizontal dewar (*upper*) and the vertical dewar (*lower*) for 5 s caused by a small magnetic object attached to the forehead of the subject

mode operation, the data acquisition operation, stimulus control, and data reports can be done on a single PC, allowing the operator to remain in focused on a single monitor.

Owner Friendliness. A small MSR is sufficient for installation because the dewar is compact: it is 0.89 m in height and needs no gantry. Prospective owners of the system need not be concerned about available space.

The necessary liquid He amount to maintain the system is 100 liter/week or 5200 liter/year, and the refill cycle is 1 refill/week for this system.

An upgrade plan is easy to realize because the components of the system are based on a 16-channel unit. It is easy to add on another 16 channels or more. It is also easy to install additional sensors.

Subject/Patient Friendliness. Conventional MEG dewars are of a vertical type and their height is more than 1 m. The subjects/patients must sit beneath the dewar and put their heads into the concave-shaped tail: they must sit with their necks straight and stiff during measurement. Especially for patients in a serious condition, it is a hardship to be measured beneath such a vertical dewar. The dewars are settled on a gantry in order to adjust the alignment of the angle of the head and the tail of the dewar. The gantry structure causes some vibration problems. This might give the subject an uncomfortable feeling under the heavy weight.

When using this horizontal dewar, the subject can lie down during the MEG measurement. The reclining type of dewar is gantry free and the subjects/patients feel less weight. The swinging type dewars of commercial companies [70–72] have similar merits in relaxing subjects in a supine mode, but those dewars still have gantry structures.

An Example of MEG Measurement. Figure 3.57 shows an example of the MEG response and the estimated equivalent current dipole of a normal subject, a right-handed 35-year old male, with electrical stimulation to his right-hand thumb. Measurement is done with a filtering band of 3–500 Hz, a sampling rate of 2 kHz and 400× synchronized averaging. The time chart is recorded with an origin time at the moment of stimulation. The signals measured with 160 channel sensors are superimposed on the same chart, in the upper left of Fig. 3.57. A clear response at 20.5 ms after the stimulus is observed with a maximum amplitude of about 110 fT. The large signal at 0 ms is an artifact due to the stray current from the electrodes attached to the subject for stimulation. The artifact does not cause deterioration of the 20.5 ms response because the latency is separated far enough. The contour map of the isomagnetic field at 20.5 ms around the head is also displayed in the upper right in reference to the position of the SQUID sensors. The increment for the contour is 10 fT/step. The dark lines are positive, which means that the flux goes from the inside to the outside of the head. The gray lines are negative values. L and R represent the left and right hemispheres, respectively. The local minimum and part of the local maximum are observed on the left hemisphere. From these spatial data and the calibration data for markers on the head, equivalent current dipoles are estimated using a spherical model and displayed as a white dot superimposed on the MRI image (lower three diagrams). The white line stretching from the white dot shows the direction and the amplitude of the estimated ECD. The position of the ECD is on the posterior part of the central sulcus of the brain. The result is in good agreement with anatomical knowledge. The intensity of the ECD is also calculated to be 9.75 nAm. The reproducibility of an equivalent current dipole model for measurements on actual patients/subjects is also good and the spread is in a box of 1–2 mm, thanks to the horizontal dewar, which eliminates subject vibrations.

Clinical Applications. With MEG, neural activity may be imaged with a time resolution of 1 ms or better and a spatial resolution of a few to several mm. MEG is basically noninvasive measurement, so it gives profound benefits not only to the researcher but also to the people to be investigated. In clinical application, MEG is used to study psychiatric conditions such as epilepsy, to carry out presurgical localization of brain function, and to evaluate the level of recovery of brain function. In scientific research, MEG is used to study human cognitive functions or so-called higher brain functions.

In this section, the actual clinical applications of MEG are described. They are neurosurgery, ophthalmology, otorhinolaryngology, neurology, the examination of epilepsy, and psychiatric conditions.

Presurgical Localization of Brain Function. A 31-year-old patient with left frontal lobe astrocytoma was examined before surgical operation in order to identify the primary somatosensory area (Fig. 3.58). This information helps a medical doctor to avoid damage to the important parts of the brain. Before surgery, electrical stimulation to the median nerve in the right-hand joint was applied to the patient and the position of the primary somatosensory area in the left brain was identified with MEG.

Postoperative examination was done to confirm that the important part had survived.

Electrical stimulation was applied to the median nerve in the right hand of the patient. The signals on the affected side of the brain were more or less weak, but a somatosensory evoked field was observed 20.5 ms after stimulation, as shown in Fig. 3.59, which proves that the important part has survived and the sensory has remained.

Objective Examination. MEG sometimes can examine defects in brain function that cannot be detected by other modalities. A 56-year-old female was examined, who complained about the absence of the right half of the visual field after injury in a traffic accident. MR images failed to identify any

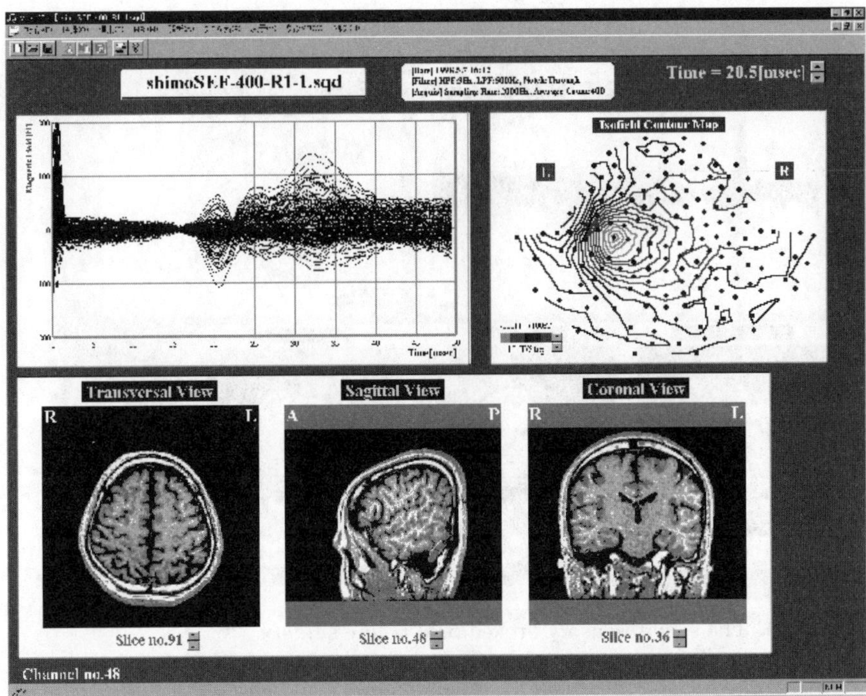

Fig. 3.57. An example of the MEG response and estimated equivalent current dipole of a normal subject, a right-handed 35-year-old male, with electrical stimulation on his right-hand thumb. The time chart (*upper left*) shows the signals measured with 160 channel sensors. The iso-field contour map is displayed in the upper right in reference to the positions of the sensors (*dots*). The ECD is displayed as a white dot superimposed on the MRI image (*lower three diagrams*). The graphs in the lower left, lower center, and lower right are transversal, sagittal, and coronal views. The *white lines* stretching from the *white dots* show the direction and the amplitude of the estimated ECD

responsible lesion. Electroencephalography failed to find homonymous hemianopsia, because the right and left visual areas are close to each other. The whole visual field of the right eye was stimulated. The right occipital region responded. MEG measurement shows that the left occipital region did not respond, as the patient claims (Fig. 3.60).

MEG is also useful when medical doctors would like to diagnose children who are too young to have objective communication. A five-year-old female with left occipital brain neoplasms was examined with MEG. Upon measuring visual evoked fields resulting from flash stimulation to the left eye, the right and left occipital regions responded. Since the subject was too young to fix her eyes, visual field examination was impossible. Visual evoked potential measurement can hardly help to find hemianopsia, because the right and left visual areas are close to each other. Magnetoencephalograms allo-

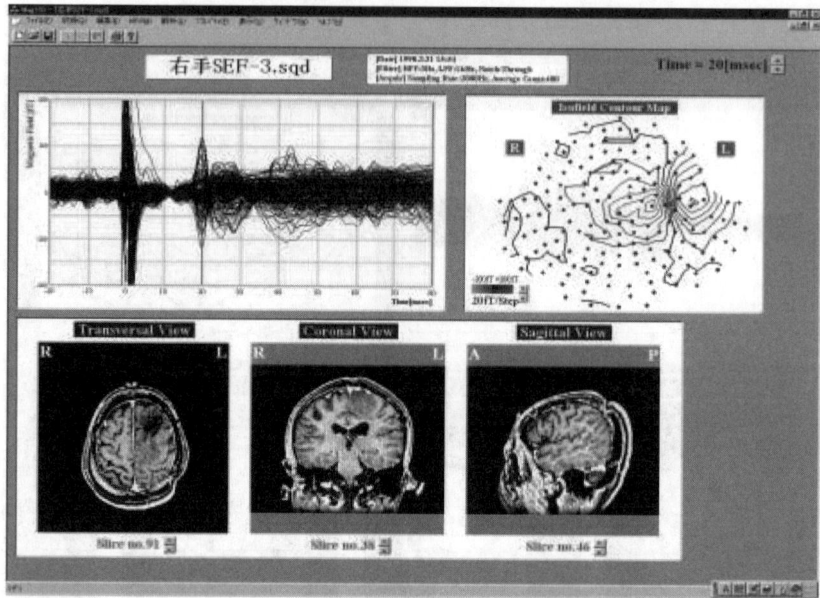

Fig. 3.58. The somatosensory evoked field before surgery

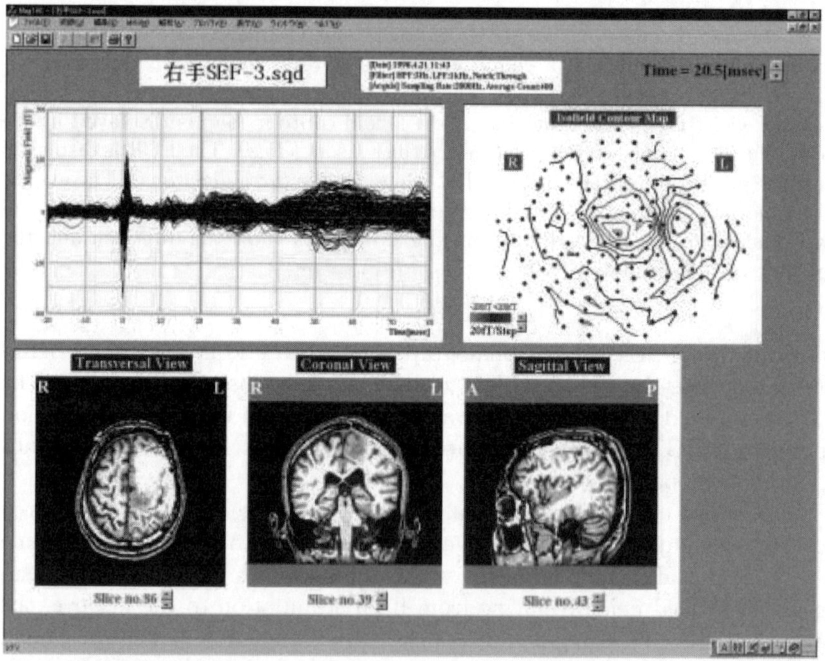

Fig. 3.59. The somatosensory evoked field after surgery

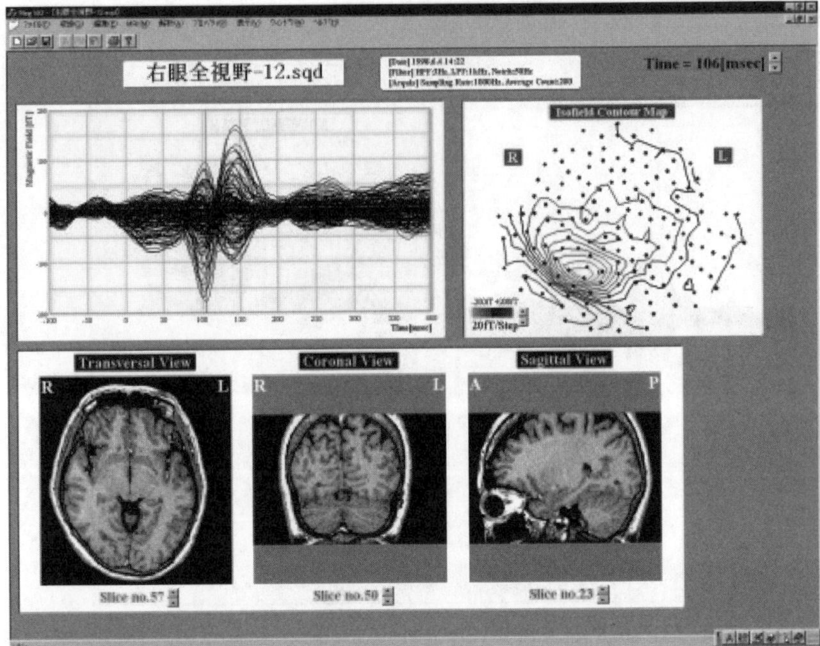

Fig. 3.60. Homonymous hemianopsia found by MEG

wed objective diagnosis of functional normality of the subject's visual sense (Fig. 3.61).

Epilepsy Examination. In some cases of focal epilepsy, medical doctors introduce electrical cortical measurement, which is invasive examination, to localize the foci before surgery. Noninvasive localization of foci has recently been attempted with MEG. Figure 3.62 is an example of a 23-year-old male with Lennox–Gastaut syndrome. A two-dipole pattern was observed on the iso-contour map at the point of diffuse slow spike and wave discharges in the EEG, and the equivalent current dipoles were estimated to originate in the region of a cortical dysplasia lesion.

The results of the ECD estimation are as follows. All of the ECDs were superimposed on to the area of the cortical dysplasia. The difference in the gray levels of the rounded marker in Fig. 3.63 indicates the time sequence of estimation. Cortical dysplasia bilaterally in the parietotemporal area was found on MRI.

Measurement of Event-Related Field. In addition to clinical applications, the MEG system has been introduced into the field of brain and cognitive science. One of the most frequently used methods with MEG in cognitive science is the measurement of an event-related field. The MEG signals under

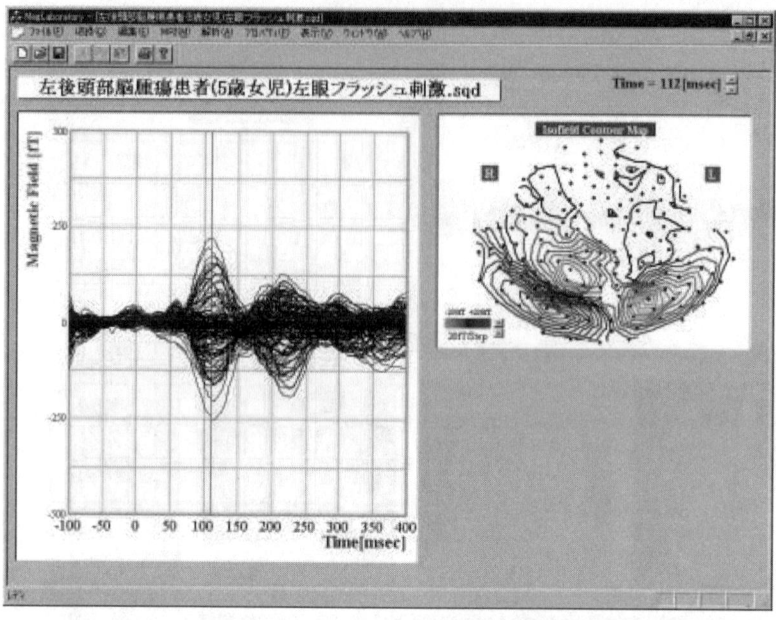

Fig. 3.61. Objective diagnosis of a young child's visual sense with MEG

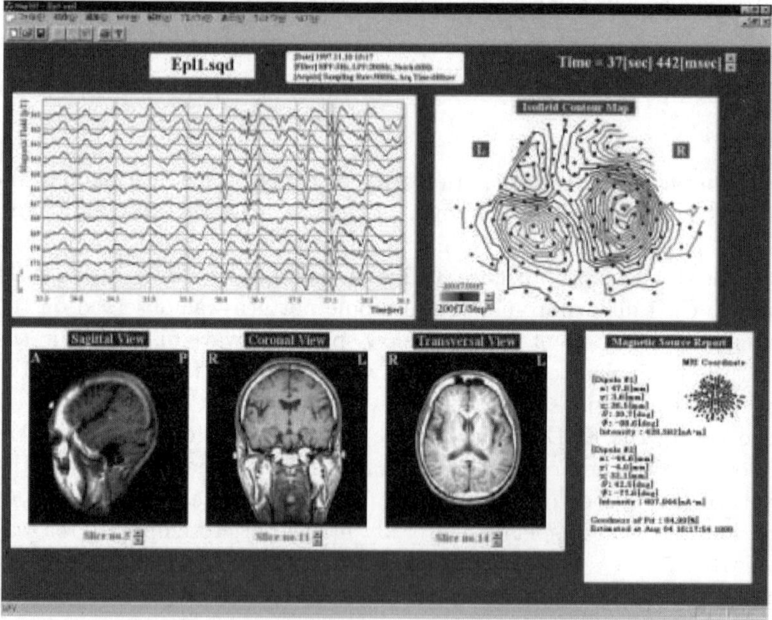

Fig. 3.62. Noninvasive localization of foci of epilepsy with MEG

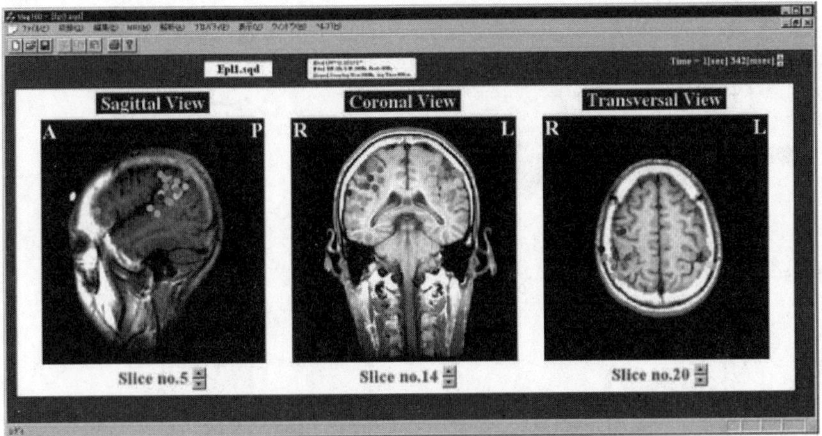

Fig. 3.63. A time sequence of the estimation of foci

the auditory oddball paradigm [73] were measured. Subjects were presented with two types of auditory stimuli: a frequent tone (1000 Hz) and an infrequent tone (2000 Hz). The stimulus sequence consisted of 80% frequent tone and 20% infrequent tone. The duration of both tones was 100 ms. The interval between stimuli was 2 s. The tones were delivered binaurally through a plastic tube. Subjects were tasked to count mentally the number of infrequent stimuli. The length of the recorded MEG signals was about 300 s and, consequently, they contained at least 30 epochs for the infrequent stimulus. The time series of the responses of one of the subjects to both stimuli is shown in Fig. 3.64a. Figure 3.64b shows the isofield contour map at 300 ms of the latency. In the response to the infrequent tone, some broad peaks in addition to the N100 component were observed.

Summary. We have developed new systems for magnetoencephalography (MEG), with the number of sensors ranging from 96 to 160. The gantry-free horizontal dewar is realized with a liquid helium consumption rate less than 10 liter/day. A small room is sufficient for installation because of the low height (0.89 m) of the dewar. The patient can be measured in a lying position, which is more relaxing compared to the conventional vertical type. The vibration of the subject's head during the MEG measurement is reduced by a factor of 5 compared to that of a vertical dewar supported by a gantry. The magnetic field resolution of the overall system is about 4 fT/rtHz or 0.8 (fT/cm)/rtHz in the white noise region. The systems have sufficient noise performance and are friendly to owners, operators, subjects, and patients.

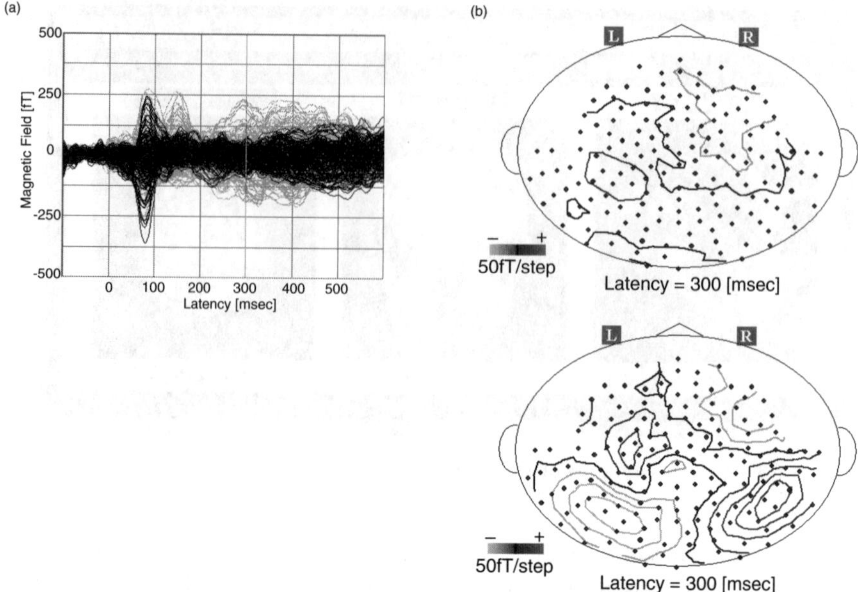

Fig. 3.64. **a** The comparison between the waveforms of the responses to the frequent tone (*black lines*) and the infrequent tone (*gray lines*). Both are the averaged data of 30 epochs. The MEG signals from all 150 sensors are overlapped. **b** The isofield contour maps at 300 ms of the latency. The left and right maps correspond to the responses to the frequent and infrequent tones, respectively

3.4.2 Other Biomagnetic Measurements

In this section, we consider the biomagnetometer system for the measurement of spinal cord evoked fields.

As shown in the preceding section, one of the most successful applications of SQUIDs is the magnetoencephalography (MEG) system [74,75]. On the other hand, there is a strong demand for the measurement of the magnetic field from the spinal cord and peripheral nerves, that enables orthopaedic surgeons to observe the neuronal activity noninvasively. The biomagnetic recording of signal propagation in the peripheral nerves and spinal cord is expected to be a highly effective tool in diagnostic study of the conduction blocks, such as carpal tunnel syndrome [76], and in the analysis of spinal cord function, due to the fact that magnetic recording can provide information on the propagation of neuronal signals with a much higher spatial resolution than the conventional electrical potential measurement with electrodes attached to the skin.

Several measurements of the biomagnetic fields from peripheral nerves have already been reported [77,78]. We have developed a biomagnetometer system especially designed to measure the spinal cord evoked magnetic field (SCEF), taking clinical applications into consideration.

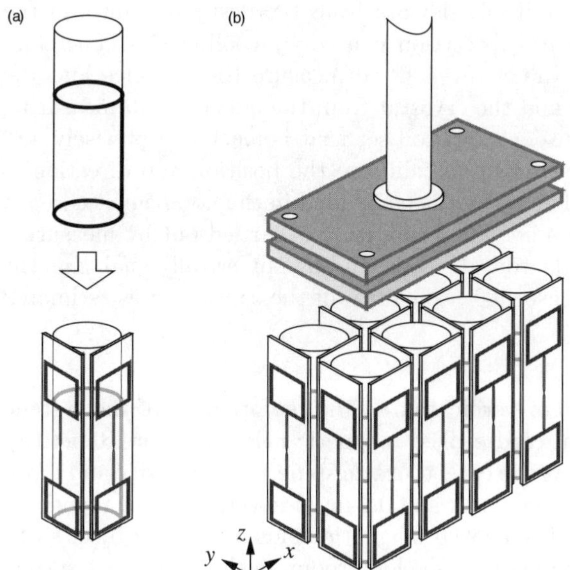

Fig. 3.65. **a** The configuration of the composite gradiometer. **b** The configuration of the probe of the system and its x–y–z coordination

Instrumentation. We used composite SQUID gradiometers in the system to observe not only the radial components of the biomagnetic fields from the subjects but also the tangential components. Each combination sensor is composed of one axial-type gradiometer and two planar-type gradiometers, located perpendicular to each other as shown in Fig. 3.65a; therefore the three-dimensional components of the magnetic field vectors can be observed. The pickup of axial type is a round coil with a diameter of 15.5 mm. The planar-type pickup is a square coil with a 14 mm side length. Both gradiometers have a baseline length of 50 mm.

Eight composite gradiometers arranged in a 2×4 matrix are attached to a flat plane as shown in Fig. 3.65b.

The SQUIDs are driven by the direct offset integration technique (DOIT) [69], which is useful for the reduction of both cost and complexity, especially in a multi-channel magnetometer system.

The array of composite gradiometers is installed in a glass fiber reinforced plastic (GFRP) cryostat. The cryostat has a capacity of 28 liters of liquid helium and can maintain the superconductivity of the sensors for about 84 hours without refilling. The thickness of the bottom of the cryostat is less than 7 mm.

It is necessary to identify the exact values of the location and direction of the sensors relative to the subjects for precise measurement of the spatial distribution of the field.

The sensor positions inevitably deviate from the designed values in the cryostat because of structural distortion caused by cooling [79]. Therefore we cannot know the real values, even if we measure the distance and the angle between the subject and the cryostat from the outside with measuring scales. In this regard, we use a multi-coil set that is machined precisely and fed with a critical electric current to calibrate the position and direction of the sensors after cooling. The coil set is embedded in the nonmagnetic x–y–z stage on which the subjects lie. The calibration is carried out by measuring the reference magnetic fields from the coils, and by numerically analyzing the signals detected by each sensor [80]. The error of the calibration is estimated to within ± 0.3 mm [74].

SCEF Measurements. For verification of the performance of the system, we recorded cervical spinal cord evoked magnetic fields in a cat. Repetitive electric stimuli were given at the cat's thoracic spine. The intensity and duration of the electric stimuli were 3 mA and 0.2 ms, respectively. The repetition rate of stimuli was 10 Hz. The evoked magnetic fields over the subject's cervix were measured in a magnetically shielded room, in the coordinate system shown in Fig. 3.66. The signals from the 24 gradiometers were filtered with a band-pass filter of 500 Hz–2 kHz and were recorded digitally at a sampling rate of 40 kHz. To improve the signal-to-noise ratio, 1000 responses were averaged. Eight positions were scanned, and consequently the three-dimensional components of the fields at 64 sites (= 8 x–y–z-gradiometers $\times$ 8 positions) were acquired.

Figure 3.67 shows the result of the measurement. We can see that the biphasic time domain waveforms and the peaks of the evoked fields were traveling on the cat's spine. The propagation velocity estimated from Fig. 3.67a was 90–100 m/s. The two poles of the distribution of the x components shown in Fig. 3.67c were located on the points between poles in Fig. 3.67b. This indicates that there were two current dipoles as sources of the fields. These results are in good agreement with the earlier physiological study [19].

Summary. We developed a 24-channel SQUID vector magnetometer system for SCEF measurements and successfully observed the spatial distribution and propagation of SCEF in a cat with this system. It was demonstrated that the hardware of the system had enough performance to detect the magnetic signals from spinal cords.

3.5 Other Applications of Magnetic Source Imaging

In this section, a field observation system is described as one of the other applications of magnetic source imaging.

3.5.1 Field Observation

We have developed a compact low-Tc three-axis superconducting quantum interference device magnetometers (SQUID magnetometer) for geophysical applications at low frequency from dc up to 1 kHz, such as for the search for magnetized metals buried under the ground, or for moving magnetized objects, and for measurements of geomagnetic fields. The SQUIDs used here are relaxation oscillation SQUIDs (ROSs), which have a larger voltage output compared to that of standard the SQUIDs used for the magnetoencephalogram (MEG) system, so that greater dynamic performance is obtained for stable operation in the field.

Currently, the major application of low-Tc superconducting quantum interference devices (SQUIDs) is for magnetoencephalograms (MEG) to detect extremely small magnetic fields in the order of ~100 fT generated in a human brain [81]. We have developed 160-channel systems for human brain research and clinical diagnosis, which can help brain researchers and medical doctors to investigate brain functions without harming patients [82]. In that case, the SQUIDs are operated in a magnetically shielded room (MSR) to prevent environmental noise.

While being widely applied for MEG systems, SQUIDs have also been expected to function as promising sensors for geophysical applications. In particular, the unrivaled sensitivity in the range at extremely low frequency (ELF), lower than ~1 kHz, enables one to obtain more detailed properties of the Earth deep below the surface. In the 1980s, Clarke et al. proposed remote-reference magnetotellurics (MTs) by using SQUID magnetometers, and indicated their availability for geological survey [83–85]. Besides this use in MT methods, SQUIDs have a big potential in geophysics. Recently, some geophysicists have been observing electromagnetic phenomena related to seismic and volcanic activity [86,87].

It is expected that SQUID magnetometers will yield new results, not obtainable with conventional sensors, not only in geophysics but also in astro-

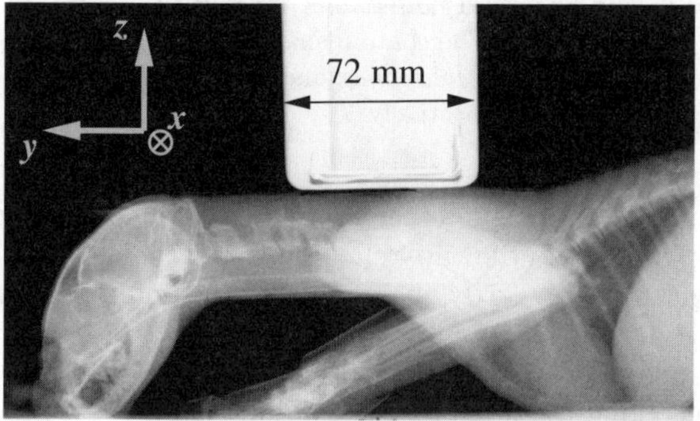

Fig. 3.66. A lateral X-ray image of the subject and the tail of the cryostat

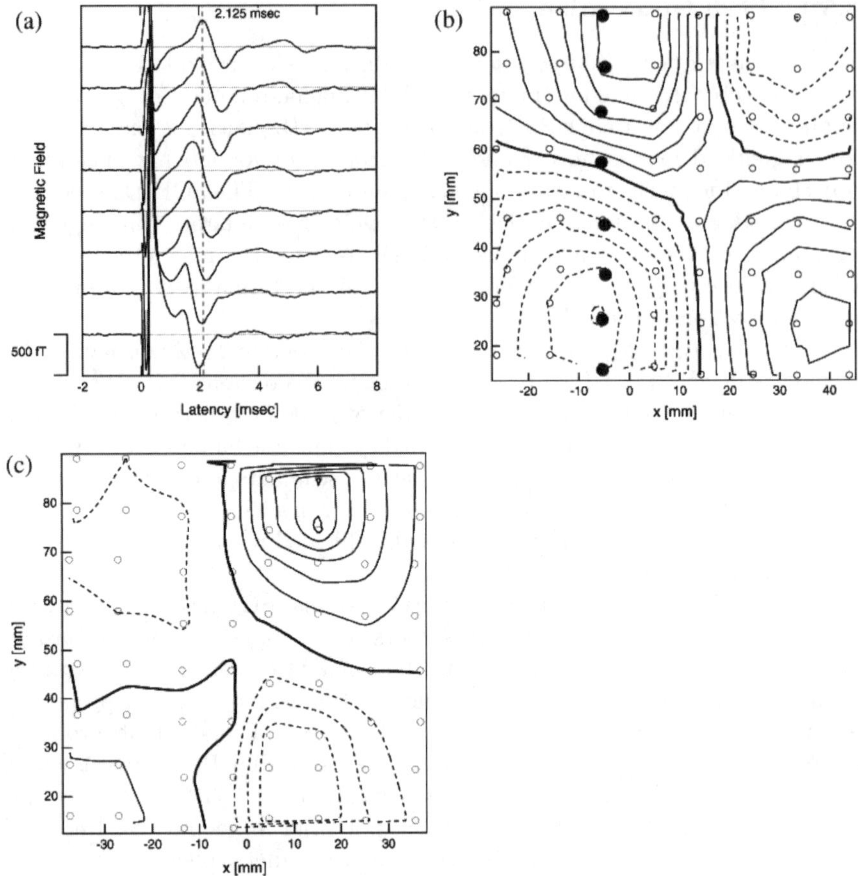

Fig. 3.67. The results of the measurement of SCEFs. **a** The time dependence of the z components of the magnetic fields measured at the sites represented by the filled round markers in (**b**). The lower rows correspond to the sites closer to the stimulation point. **b** The contour map of z components of the magnetic fields at 2.125 ms of latency. *Bold lines*, *solid lines*, and *dotted lines* represent the zero, upwards, and downwards fields, respectively. **c** The contour map of the x components of the fields at 2.125 ms of latency. *Bold lines*, *solid lines*, and *dotted lines* represent the zero, rightwards, and leftwards fields, respectively

nomy. At present, however, SQUID magnetometers are not popular among researchers involved in geophysics, and they are reluctant to use them. This is because SQUIDs are thought to be unwieldy, mainly due to the liquid helium supply and the complicated driving electronics.

We have been developing a SQUID magnetometer system aimed at long-term operation in the field, taking the reliability and stability of the system, easy operation, and lower consumption of liquid helium into consideration.

Instrumentation

The SQUID Magnetometer and the Flux-Locked Loop (FLL). Considering geophysical applications, not only high sensitivity but also a large dynamic range and a high slewing rate are necessary in order for a SQUID to work stably in the field, where SQUIDs are exposed to large artificial magnetic noises and the Earth's field. A relaxation oscillation SQUID (ROS) is adopted as a magnetometer because it has a 2–3 times larger output voltage and typically a more than 10 times larger magnetic field-to-voltage transfer function, compared to those of a standard dc SQUID used for MEG working inside a magnetically shielded room (MSR) [88,89].

A ROS magnetometer is integrated on a silicon substrate by a thin-film technology, using niobium as a superconductor. It has more than 100 µV of the voltage modulation and 5–10 $\mu V/\Phi_0$ of the transfer function. Due to the large output voltage and the large transfer function, the output signal is read directly with a preamplifier, without any transformer matching, which is helpful in designing an uncomplicated FLL circuit. In addition, the noise attributed to the preamplifier can be neglected, which means that the $1/f$ noise of a preamplifier can be reduced, and the closed loop gain can be decreased to obtain a wide frequency bandwidth.

The chip is mounted on a PCB (1 cm × 1 cm) and is bonded to the terminals with aluminum wires. A small resistor is placed adjacent to the chip, to heat up if the SQUID is trapped by magnetic noise while cooling down. The chip and the resistor are covered with epoxy glue for protection. Three PCB chips are fixed perpendicular to each other on a GFRP block. In order to eliminate radio frequency (rf) noise, which degrades the properties of a SQUID, a small *RC* filter of about 3 MHz is connected to the SQUID. To reduce helium consumption, stainless steel shielded wires are used between the *RC* filters and a connector attached to the top of a probe at room temperature.

A compact electronics, with a size of 30 cm × 20 cm × 80 cm, are developed for easy carrying. Sixteen sensors are driven at one time. An FLL is realized with a simple direct readout technique. High- and low-pass filters, band elimination filters, and amplifiers of various magnifications are prepared for each channel.

In designing an FLL for a magnetically unshielded SQUID, the maximum feedback field and the slewing rate, which means the allowed detectable field and the frequency response, are important parameters. They have been determined by optimizing the preamplifier gain, the time constant of an integrator, and the feedback resistor consisting of FLL, taking the SQUID parameters into consideration. So far, the frequency response extends from dc to 100 kHz under optimum conditions. The output gain in the magnetic field is set to be 24 nT/V.

The Properties of a Magnetometer. Figure 3.68 shows the magnetic field resolution of a magnetometer measured with FLL operation, being magnetically shielded with a superconductive tube. Resolutions of 10 fT/rtHz (10 µγ/rtHz) at 1 kHz and 20 fT/rtHz (20 µγ/rtHz) at 1 Hz are achieved. These values are 10^3 smaller compared with those of a commercially available fluxgate magnetometer. These properties at lower frequency are confirmed by measuring the dc fluctuation, which is mainly due to the thermal fluctuation of the preamplifier. A fluctuation lower than 1 pT_{pp}/h is achieved.

The maximum feedback field and the slewing rate are shown in Fig. 3.69. More than 100 µT/s ($1 \times 10^5 \gamma$/rtHz) is realized at 10 kHz, which is expected to be large enough for geophysical measurements carried out in rural areas. At a frequency lower than 4 Hz, the maximum feedback field is limited by the voltage of the power supply, corresponding to 360 nT (360γ). This value is large enough compared to the diurnal variation in the Earth's field. A dynamic range of 10^7 in a 1 Hz bandwidth and 10^3 in a 1 kHz bandwidth are obtained.

Liquid Helium Cryostat. A probe with three SQUID magnetometers is inserted into a liquid helium cryostat, and connected to driving electronics. Two types of cryostat are prepared. One is made of aluminum and is 0.8 m in height and 0.26 m in diameter, for low-frequency measurements. It contains 10 liters of liquid helium, which keeps the SQUIDs in a superconductive condition for a week. This cryostat functions as a rf shield because of the eddy current effect. For measurements of high-frequency signals, we have also developed a GFRP cryostat with a height of 1 m and a diameter of 0.3 m which has a capacity of 35 liters of liquid helium, and a liquid helium consumption of about 1 liter per day. It enables us to use SQUID magnetometers for a month without a further supply of liquid helium.

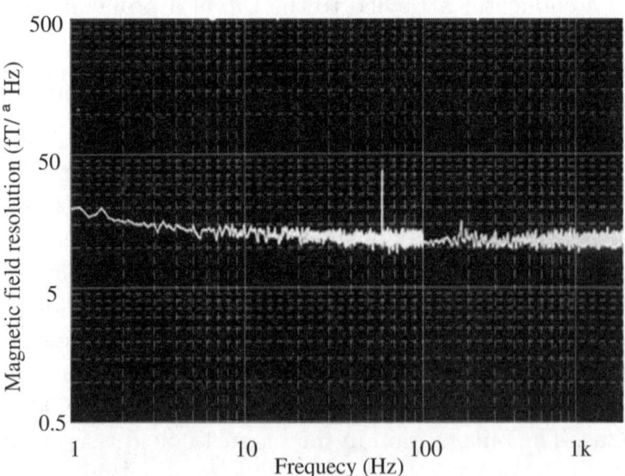

Fig. 3.68. The magnetic field resolution of a magnetometer

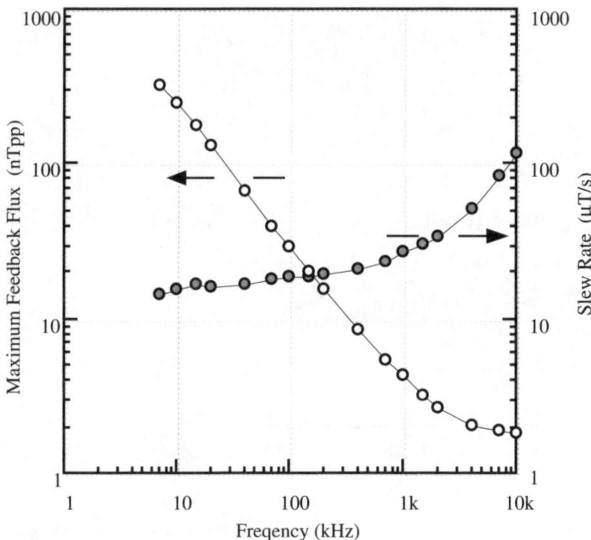

Fig. 3.69. The maximum feedback field and the slewing rate of a magnetometer operated with FLL

Results and Discussion. We operated the three SQUID magnetometers in the aluminium cryostat in our laboratory building and measured the variation in the Earth's field for one night. Since some other electrical equipment was working and people were going into and out of the room during the day time, opening and closing an iron door, the measurements were made in the middle of the night. Signals were filtered at 10 Hz by low-pass filters, and 60 Hz noise from the power line was eliminated by band elimination filters. The measured data recorded with a time recorder are shown in Fig. 3.70. The horizontal axis represents Japanese Standard Time (JST) and the vertical axis is the variation in the magnetic field, corresponding to 12 nT (12γ) per division. H, D, and Z denote the direction of each component of the field in terms of geophysical notation, where H and D are the directions from north to south and east to west, respectively, and Z is the direction perpendicular to the surface of the Earth. It is indicated that the system is working stably. No lock-off and no flux jump can be seen, where a flux jump per flux quanta (Φ_0) corresponds to 1/5 division. We have also confirmed that the SQUID magnetometers work stably in the GFRP cryostat without a rf shield.

There are many unexpected noises with various amplitudes and frequencies in the field, even in rural areas. It is clear that a wider frequency bandwidth and a larger slewing rate are strongly required so that the feedback loop of an FLL does not result in flux jumps and lock-off. However, it is impossible to predict how large a noise a SQUID magnetometer would be exposed to. For stable operation, the system should be carefully shielded from the rf noise system, including the SQUIDs. An aluminum cryostat as used

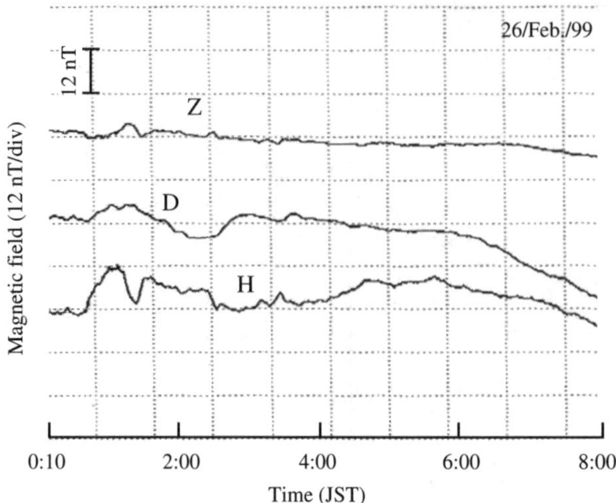

Fig. 3.70. An example of time variation in the Earth's field measured with SQUID magnetometers

here is available for that purpose. However, the cutoff frequency is in the order of around 10 Hz, because a metal cryostat with a capacity of several tens of liters of liquid helium is normally ∼10 mm thick. If a higher-frequency signal is focused on, a nonmetal cryostat, such as one made of GFRP, should be used, and it is desirable that the cryostat should be shielded with metal mesh or foil, ensuring that the thickness is equal to the skin depth for the maximum frequency of interest. In the case where a magnetometer itself is shielded with metal, the thermal noise from the metal should be considered.

For confirmation of the longer-term stability of the system, we are planning to bury the three-axis SQUID magnetometer developed here underground and attempt to measure the Earth's field for a month.

References

1. A. Barone: *Superconducting Quantum Interference Devices* (World Scientific, 1992)
2. A. Barone and G. Paterno: *Physics and Applications of the Josephson Effect* (John Wiley & Sons, 1982)
3. J.C. Gallop: *SQUIDs, the Josephson Effects and Superconducting Electronics* (Adam Hilger, 1991)
4. J. Clarke: Proc. IEEE **77**, 1208 (1989)
5. M.B. Ketchen: IEEE Trans. Mag. **17**, 387 (1981)
6. C.D. Tesche, J. Clarke: J. Low. Temp. Phys. **29**, 301 (1977)
7. C.T. Rogers, P.A. Buhrman: IEEE Trans. Mag. **29**, 453 (1983)
8. F. Wellstood, C. Urbina, J. Clarke: Appl. Phys. Lett. **50**, 772 (1987)

9. V. Foglietti, W.J. Gallagher, M.B. Ketchen, A.W. Kleinsasser, R.H. Koch, S.I. Raider, R.L. Sandstorm: Appl. Phys. Lett. **40**, 1343 (1986)
10. K. Goto, N. Fujimaki, T. Imamura, S. Hasuo: SCE91-22 **43**, (1991)
11. J.W. McWane, J.E. Neighbor, R.S. Newbower: Rev. Sci. Instrum. **37**, 1602 (1966)
12. D. Drung, R. Cantor, M. Peters, H.J. Scheer, H. Koch: Appl. Phys. Lett. **57**, 406 (1990)
13. T. Ryhanen, R. Cantor, D. Drung, H. Koch: Appl. Phys. Lett. **59**, 228 (1991)
14. Y. Takada, K. Kazami, G. Uehara, H. Kado: Proceedings of SPIE'94 **2160**, 195 (1994)
15. Y. Takada, G. Uehara, N. Matsuda, H. Kado: Jpn. J. Appl. Phys. **33**, L1595 (1994)
16. G.L. Romani, S.J. Williamson, L. Kaufman: Rev. Sci. Instrum. **53**, 1815 (1982)
17. G. Kajiwara, K. Harakawa, H. Ogata, H. Kado: IEEE Trans. Mag. **32**, 2582 (1996)
18. S.J. Williamson, L. Kaufman: IEEE Trans. Mag. **MAG-19**, 835–844 (1983)
19. J.P. Wikswo: "Biomagnetic sources and their models", in: *Advances in Biomagnetism*, ed. by S. J. Williamson, M. Hoke, G. Stroink, and M. Kotani, pp. 1–18 (Plenum Press, New York, 1990)
20. J.P. Wikswo: IEEE Trans. Appl. Supercond. **5**, 74 (1995)
21. J. Sarvas: Phys. Med. Biol. **32**, 11 (1987)
22. D.B. Geselowitz: IEEE Trans. Mag. **MAG-6**, 346 (1970)
23. H. von Helmholtz: Ann. Phys. Chem. **89**, 211–233 and 353–377 (1853)
24. A.A. Ioannides, J.P.R. Bolton, C.J.S. Clarke: Inverse Probl. **6**, 523 (1990)
25. J.Z. Wang, S.J. Williamson, L. Kaufman: IEEE Trans. Biomed. Eng. **39**, 665 (1992)
26. A.M. Dale, M.I. Sereno: J. Cognitive Neurosci. **5**:2, 162 (1993)
27. M. Hamalainen, R. Hari, R.J. Ilmoniemi, J. Knuutila, O.V. Lounasmaa: Rev. Mod. Phys. **65**, 413 (1993)
28. R.D. Pascual-Marqui, C.M. Michel, D. Lehmann: Int. J. Psychophysiol. **18**, 49 (1994)
29. M.S. Hamalainen, R.J. Ilmoniemi: Med. Biol. Eng. Comput. **32**, 35 (1994)
30. D. Brenner, J. Lipton, L. Kaufman, S.J. Williamson: Science **199**, 81 (1978)
31. S.J. Williamson, L. Kaufman: J. Magn. Mat. **22**, 129 (1981)
32. M. Scherg: "Fundamentals of dipole source potential analysis", in: *Advances in Audiology – Auditory Evoked Magnetic Fields and Electric Potentials*, ed. by F. Grandori, M. Hoke, and G.L. Romani, pp. 40–69 (Karger, Basel, 1990)
33. M.S. Hamalainen: Brain Topogr. **7**, 283 (1995)
34. R. Hari, K. Reinikainen, E. Kaukorenta, M. Hamalainen, R. Ilmoniemi, A. Pentinnen, J. Salminen, D. Teszner: Electroencephalogr. Clin. Neurophysiol. **57**, 254 (1984)
35. M. Scherg, R. Hari, M.S. Hamalainen: "Frequency-specific sources of the auditory N19-P30-P50 response detected by a multiple source analysis of evoked magnetic fields and potentials", in: *Advances in Biomagnetism*, ed. by S.J. Williamson, M. Hoke, G. Stroink, and M. Kotani, pp. 97–100 (Plenum, New York, 1989)
36. M. Scherg, D. von Cramon: Elec. Clin. Neurol. **62**, 32 (1985)
37. J.C. Mosher, P.S. Lewis, R.M. Leahy: IEEE Trans. Biomed. Eng. **39**, 541 (1992)
38. S. Baillet, L. Garnero: IEEE Trans. Biomed. Eng. **44**, 374 (1997)

39. J.C. Mosher, R.M. Leahy: IEEE Trans. Biomed. Eng. **45**, 1342 (1998)
40. K. Uutela, M. Hamalainen, R. Salmelin: IEEE Trans. Biomed. Eng. **45**, 716 (1998)
41. D.M. Schmidt, J.S. George, C.C. Wood: Hum. Brain Mapp. **7**, 195 (1999)
42. C.C. Wood, J.S. George, P.S. Lewis, D.M. Ranken, L. Heller: Society for Neuroscience, Abstracts **16**, 1241 (1990)
43. M. Scherg, P. Berg: Brain Topogr. **4**, 143 (1991)
44. J.S. George, C.J. Aine, J.C. Mosher, D.M. Schmidt, D.M. Ranken, H.A. Schlitt, C.C. Wood, J.D. Lewine, J.A. Sanders, J.W. Belliveau: J. Clin. Neurophysiol. **12**, 406 (1995)
45. D.M. Schmidt, J.S. George, D.M. Ranken, C.C. Wood: "Spatio-temporal Bayesian inference for MEG/EEG", in: *Biomag 2000*, 12th International Conference on Biomagnetism at Helsinki, Finland, August 13–17, 2000, p. 7
46. C. Bertrand, M. Ohmi, R. Suzuki, Y. Haruta, M. Ochiai, H. Kado: "Resolution of the MEG inverse problem by Markov Chain Monte Carlo methods: Algorithm comparison and 3-dimensional visualization tools", in: *Biomag 2000*, 12th International Conference on Biomagnetism at Helsinki, Finland, August 13–17, 2000, p. 171
47. C. Bertrand, Y. Hamada, H. Kado: "MRI prior computation and Parallel Tempering algorithm for a probabilistic resolution of the MEG/EEG inverse problem", Brain Topgr. **4**, no. 1, 2001
48. C. Bertrand, M. Ohmi, R. Suzuki, H. Kado: "Resolution of the MEG inverse problem by Markov Chain Monte Carlo methods: Parallel tempering and reversible jump algorithms", in: *Proceedings of the Second Annual Meeting of the Japan Human Brain Mapping Society*, Tokyo, Japan, 2000, p. 109.
49. C. Bertrand, M. Ohmi, R. Suzuki, H. Kado: "A probabilistic solution to the MEG inverse problem via MCMC methods: The reversible jump and parallel tempering algorithms", to appear in the IEEE Trans. Biomed. Eng.
50. J.C. Mosher, R.M. Leahy, P.S. Lewis: IEEE Trans. Biomed. Eng. **46**, 245 (1999)
51. C.J. Geyer: "Markov chain Monte Carlo maximum likelihood", in: *Computing Science and Statistics: Proceedings of the 23rd Symposium on the Interface, Fairfax: Interface Foundation*, ed. by E. M. Keramigas (1991), pp. 156–163
52. N. Metropolis, A.W. Rosenbluth, M.N. Rosenbluth, A.H. Teller, E. Teller: J. Chem. Phys. **21**, 1087 (1953)
53. P.J. Green: Biometrika **82**, 711 (1995)
54. J. Gerson, V.A. Vardenas, G. Fein: Electroenceph. Clin. Neurophysiol. **92**, 161 (1994)
55. H. Haneishi, N. Ohyama, K. Sekihara, T. Honda: IEEE Trans. Biomed. Eng. **41**, 1004 (1994)
56. C. Bertrand, Y. Hamada, H. Kado: "Probabilistic current mapping of somatosensory fields", in *ISBET2001, 12th World Congress of the International Society for Brain Electromagnetic Topography*, pp. 1–4
57. T. Katila, P. Karp:"Magnetocardiology: Morphology and multipole presentations", in: *Biomagnetism: An Interdisciplinary Approach*, ed. by S.J. Williamson, G.L. Onmani, L. Kaufman, and I. Modena (Plenum Press, New York and London, 1983)
58. S. Kirkpatrick, C.D. Gelatt Jr. M.P. Vecchi: "Optimization by simulated annealing", Science **220**, 671 (1983)

59. R.S. Gonnelli, M. Sicuro: "Use of the current multipole model for the cardiac source localization in normal subjects", in: *Biomagnetism '87*, pp. 314–317 (1987)
60. A. Tarantola: *Inverse Problem Theory* (Elsevier, New York, 1987)
61. C. Bertrand, H. Kado, Y. Adachi: "A 3D visualization software for biomedical data and images: An application to magnetic resonance image and magnetoencephalography", in: *Proceedings of the Third International Conference on Human and Computer*, University of Aizu, Tsuruga, Japan, September 6–9, 2000, pp. 127–132.
62. C. Bertrand, H. Kado, Y. Adachi: J. Three Dimension. Images, **14**-4, 94 (2000)
63. B. Silverman: *Density Estimation for Statistics and Data Analysis* (Chapman and Hall, London, 1993)
64. J. Serra: *Image Analysis and Mathematical Morphology* (Academic Press, London, 1982)
65. W.E. Lorensen, H.E. Cline: Comput. Graph. **21**, 163 (1987)
66. C. Montani, R. Scateni, R. Scopigno: Visual Comp. **10**, 353 (1994)
67. H.J.M. ter Brake, F.H. Fleuren, J.A. Ulfman, J. Flokstra: Cryogenics **26**, 667 (1986)
68. G. Uehara, N. Matsuda, K. Kazami, Y. Takada, H. Kado: "Wafer scale integration of drung type SQUIDs", in: *4th International Superconductive Electronics Conference Proceedings*, Boulder, CO, 1993, pp. 184–185
69. D. Drung, R. Cantor, M. Peters, T. Ryhanen, H. Koch: IEEE Trans. Magn. **27**, 3001 (1991)
70. J. Vrba et al.: "151-channel hole-cortex MEG system for seated or supine positions", in: *Proceedings of 11th International Conference on Biomagnetism*, Sendai, Japan, 1998, P-I-30, p. 60
71. D.S. Buchanan, R.T. Johnson, K.C. Squire: "Performance of whole head biomagnetic sensor using magnetometer signal coils", in: *Proceedings of 11th International Conference on Biomagnetism*, Sendai, Japan, 1998, P-I-5, p. 57
72. K. Sata et al.: "A helmet-shaped MEG measurement system cooled by a GM/JT cryocooler", in: *Proceedings of 11th International Conference on Biomagnetism*, Sendai, Japan, 1998, P-I-21, p. 65
73. J. Polich, S.E. Eischen, G.E. Collins: Electroencephalogr. Clin. Neurophys. **92**, 253 (1994)
74. H. Kado, M. Higuchi, M. Shimogawara, Y. Haruta, Y. Adachi, J. Kawai, H. Ogata, G. Uehara: IEEE Trans. Appl. Supercond. **9**, 4057 (1999)
75. J. Vrba: in: *NATO ASI Series: E Applied Sciences* **365**, ed. by H. Weinstock, pp. 61–138 (Kluwer Academic Publishers, Dordrecht, 2000)
76. On the web site of the American Academy of Orthopaedic Surgeons (http://orthoinfo.aaos.org/)
77. I. Hashimoto, T. Mashiko, T. Mizuta, T. Imada, K. Iwase, and H. Okazaki: Electroencephalogr. Clin. Neurophysiol. **93**, 259 (1994)
78. B.M. Mackert, G. Curio, M. Burghoff, P. Marx: Electroencephalogr. Clin. Neurophysiol. **104**, 322 (1997)
79. M. Kawakatsu, K. Kobayashi, A. Fujimoto, H. Ishibashi, T. Ishikura, Y. Uchikawa, M. Kotani: J. Jpn. Biomag. Bioelect. Soc. **12**, 3 (1999)
80. M. Higuchi, K. Chinone, N. Ishikawa, H. Kado, N. Kasai, M. Nakanishi, M. Koyanagi, Y. Ishibashi: in: *Advances in Biomagnetism*, pp. 701–704 (Plenum Press, New York, 1989)

81. J.P. Wikswo: IEEE Trans. Appl. Supercond. **5**, 74 (1995)
82. M. Higuchi, M. Shimogawara, Y. Haruta, G. Uehara, J. Kawai, H. Ogata, H. Kado: Appl. Supercond. **5**, Nos 7–12, 441 (1998)
83. J. Clarke, T.D. Gamble, W.M. Goubau, R.H. Koch, R.F. Miracky: Geophys. Prospect. **31**, 149 (1983)
84. E.A. Nichols, H.F. Morrison, J. Clarke: J. Geophys. Res. **93**, 13743 (1988)
85. W.M. Goubau: "Geophysical applications of SQUIDs", in: *Proceedings of SQUID80*, pp. 603–613 (Walter de Gruyter & Co., Berlin New York)
86. P. Varotsos, K. Alexopoulos: Tectonophys. **110**, 73 (1984)
87. Y. Tanaka: J. Volcanol. Geotherm. Res. **56**, 319 (1993)
88. D.J. Adelerhof, J. Kawai, G. Uehara, H. Kado: Appl. Phys. Lett. **65**, 2606 (1994)
89. J. Kawai: *Relaxation Oscillation SQUIDs (ROSs) Based on Nb/AlOx/Nb Josephson Tunnel Junctions*, Thesis (1995)

4 Bioanalyses Using Electrochemical and Electrophysiological Methods

E. Tamiya, K. Mabuchi, K. Yokoyama, Y. Murakami, M. Kobayashi, M. Suzuki, H. Suzuki, T. Suzuki, and M. Kunimoto

4.1 Introduction

Biosensors are devices that combine the advantages of the specificity and sensitivity of biological systems with the rapid and quantitative transducing of the response of electrochemical or other instruments [1–3]. Biosensors generally require three elements, as shown in Fig. 4.1.

A molecular recognition element is composed of biological materials that react selectively with the specified substrate. Transducers convert the related information from the bio-catalytic reaction into electrical signals and an element for recording these electrical signals. The sensing elements used in biosensors can be generally classified into several groups, such as proteins, organelles, cells, and tissues. On the other hand, the conversion elements used in biosensors employ most of the methods used in the field of physical and chemical analysis. Spectrophotometry, amperometry, potentiometry, thermometry, and fluorometry are widely adopted for this purpose.

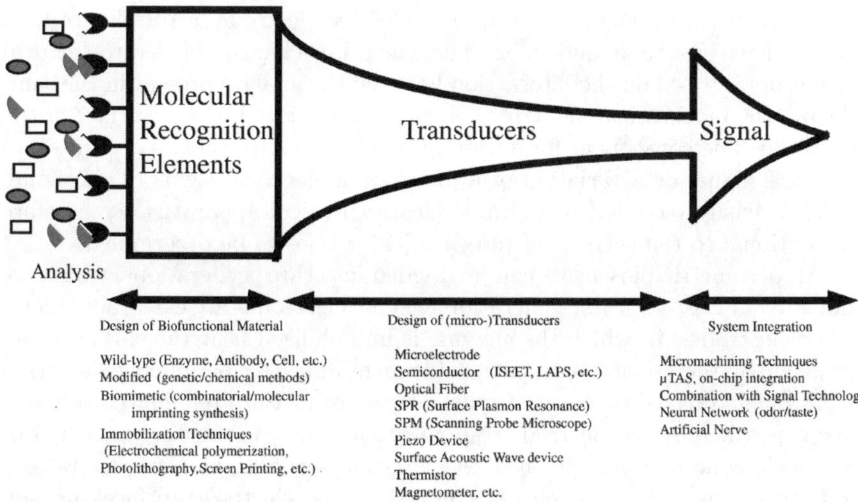

Fig. 4.1. The design and development of biosensors

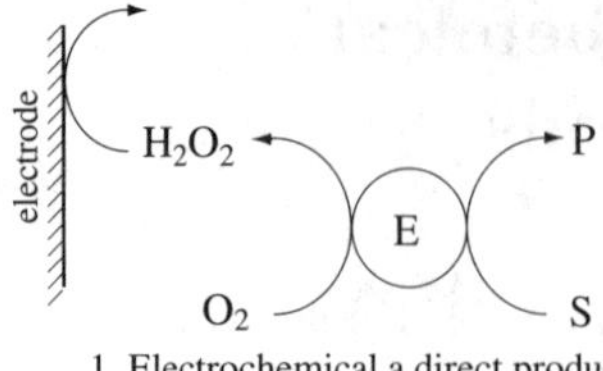

1. Electrochemical a direct product

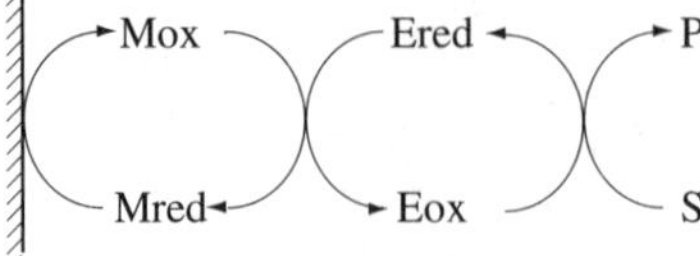

2. Mediated electron transfer

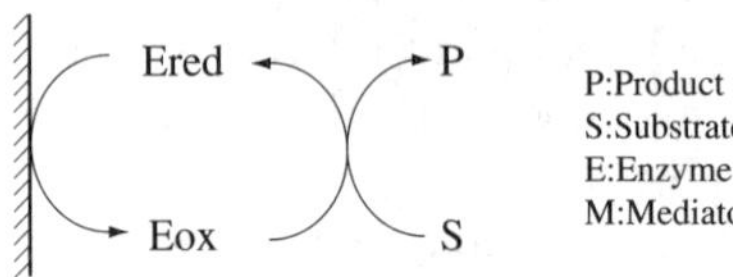

3. Direct electron transfer

Fig. 4.2. The development of amperometric enzyme sensors

Electrochemical methods were first used as transducers of biosensors. The use of biosensors begin in 1962, when Clark and Lyons combined an oxygen probe with glucose oxidase to determine glucose levels.

Electrochemical biosensors are formed by coupling a biochemical mediator, such as an enzyme, a microorganism, and an antibody, to a transducer of the electrical signal, such as an ion-selective electrode, a pH electrode, a gas-sensitive electrode, and so on. The general mechanism of electrochemical biosensors is based on the interaction between the analyte present in the sample and the biochemical mediator, which is immobilized on the surface of the electrode. The consequent formation of an electroactive species generates an electrical signal or a variation of a preexisting electrical signal. This signal, which is easily recorded by suitable electrochemical apparatus, is therefore proportional to the activity of the chemical species to be determined.

Amperometric biosensors can be divided into three generations, which are indicated in Fig. 4.2. First-generation enzyme electrodes are essentially membrane electrodes, in which the enzyme is immobilized near the surface of an electrode by the use of a semi-permeable membrane. Amperometric detection of enzyme-catalyzed turnover of substrate to product is often limited to redox enzymes, whereby the electrode responds to a change in the oxygen or hydrogen oxide concentration. In second-generation sensors, the electron transfer activity of the redox enzymes is coupled to the electrode surface through a low molecular weight electron transfer mediator. This arrangement lessens interference from competing species in the analyte solution by requiring a less

1. Electroactive hybridization indicator

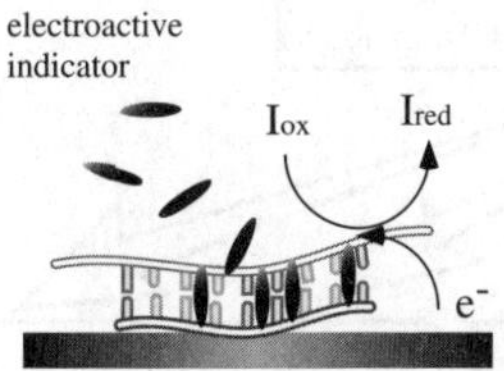

2. Redox enzyme label

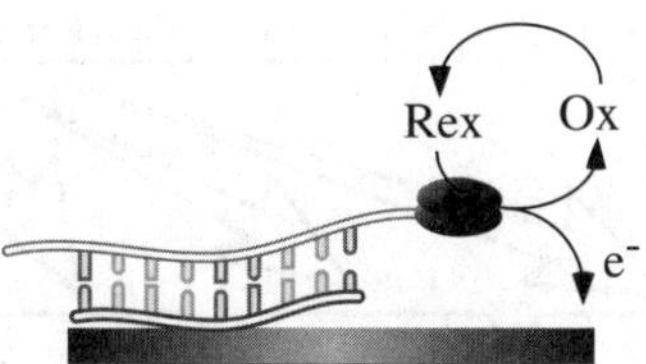

3. Guanine (G) oxidation

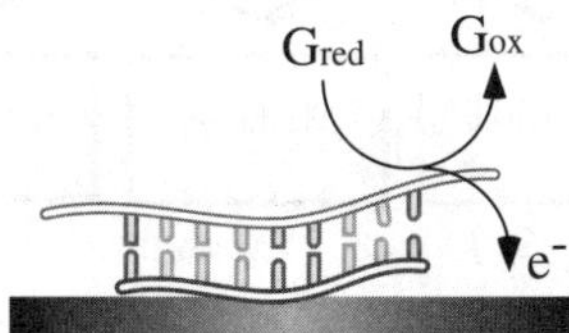

Fig. 4.3. Various kinds of electrochemical DNA detection

extreme potential than is normally used in a first-generation sensor. In the third-generation enzyme electrode there is direct electron exchange between the immobilized enzyme and the electrode surface. The third-generation enzyme electrode can form the basis for a complete amperometric biosensor, which does not require the addition of external reagents.

Electrochemical sensors also can detect DNA and antibodies labeled with redox species such as oxidoreductase and mediators. Figure 4.3 indicates the classification of electrochemical detection of specific DNA and antigens.

Electrochemical cell sensors are mainly classified into two types on the basis of the respiration activity and production of electroactive metabolites (Fig. 4.4). In the former type, the respiration activity can be directly monitored by an oxygen electrode. One of the successful cell sensors consisting of immobilized yeasts and an oxygen probe have been developed and commercialized for five-day BOD (biochemical oxygen demand), which is one of the most important and widely used tests in the measurement of organic pollution. In this type of sensor, several kinds of metabolites, such as carbon dioxide, ammonia, pH, and organic acids, can be determined by the corresponding electrodes and semiconductor devices such as the ISFET and the SPV.

Microfabrication technology can be used to integrate microelectrodes and semiconductor sensors such as the ISFET and the SPV [4]. This fabrication method has become an advantageous approach in the development of the miniaturization of the total analysis system, which involves several functional units for detection, separation, reaction, sample injection, and so on. It is also suitable for in vivo measurement, microanalysis, and disposable uses. The

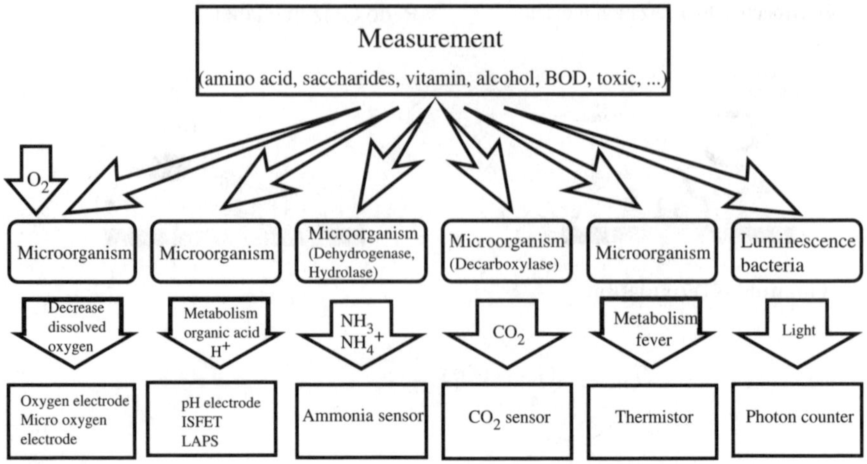

Fig. 4.4. The construction of various cell sensors

integration of the enzyme and the antibody biosensor was developed for the simultaneous analysis of multiple samples of multiple analytes. A patterned array of DNA can be realized for smart gene sensors.

The first half of this chapter (Sects. 4.2–4.6) is focused on the principles and application of electrochemical biosensors. Redox enzyme activity and DNA can be electrochemically monitored by linking with a suitable mediator and intercalator, respectively. Electrochemical active metabolite was estimated by SPV devices and applied to BOD sensor. Microfabrication technology gives advantageous examples of electrochemical biosensors. A GLOD-immobilized microelectrode can directly approach the living neural network and monitor neurotransmitter activity. Advances in integration of chemical and biosensor systems are also discussed.

In the latter part of this chapter (Sects. 4.7–4.9), we explain: 1) neural interfaces, including the development of the devices, methods and techniques, and their application for controlling external devices or for evoking artificial sensations; and 2) trials to measure humoral factors of the living body in vivo and on-line, and their applications to the control of external devices using humoral information from the living body.

We are currently developing several kinds of artificial organs, such as internal organs (artificial heart, etc.), sensory organs (artificial eye, artificial somato-sensory system for the skin, etc.) and locomotive organs (artificial limbs, etc.), and we are currently focusing our interests on the development of techniques to control these artificial organs using some information from the living body (e.g., neural signals, concentrations of the humoral factors, etc.). In the case of artificial sensory organs, we are developing techniques to evoke artificial sensations by inputting signals to the nervous system of the living body in accordance with the output of the artificial sensory system.

In order to achieve these studies, it is necessary to develop a biochemical sensor that allows in vivo and on-line measurement of humoral factors with both a very high sensitivity and specificity, or a multi-channel neural interface that is able both to record signals from each nerve fiber (or nerve cell) individually and also to stimulate them individually.

Recent striking advances in micro-electromechanical systems, tissue engineering, biotechnology, biomaterials, and electrochemical techniques are making it possible to realize these sensors or neural interfaces, and the development of these devices will surely mean that the next generation of artificial organs or bionic human systems will make notable progress.

4.2 The Electrochemical DNA Chip Sensor

4.2.1 Introduction

The detection of sequence-specific DNA is of great significance in biomedical, environmental, and food analyses. Recent progress in fabrication on chip, the DNA microarray, using photolithography or arrayer methods, has opened up a new window to nucleic acid analysis applications. This includes sequence analysis, genotyping, and gene expression monitoring. Electrochemical detection has many advantages, such as reducing the size of the total detection system, compared to a current fluorescence DNA chip system. Therefore, portability and point-of-care testing will be realized. Sensor chips can be fabricated by a mass production system based on lithography and microarray technology, and it is possible to an produce inexpensive sensor. By combination with mTAS, which is a micro total analysis system based on micromachined technology, more intelligent chip design can be possible for automated flow analyses. Table 4.1 indicates examples of previously reported electrochemical DNA sensors, especially focused on the intercalator type DNA sensor. Metal complexes such as $[Co(bpy)_3]^{3+}$, $[Co(Phen)_3]^{3+}$, and $[Os(bpy)_3]^{2+}$ function as cationic redox mediators, because DNA polymer has much negatively charged phosphate and interact easily with cationic compounds. On the other hand, the antibiotic Daunomycin or the dye Hoechst 33258 have different kinds of affinity; for example, Hoechst is considered to bind with the minor groove of duplexed DNA. Ishiimori first reported a DNA sensor using Hoechst dye [5]. The ferrocenyl naphtalene diimide developed by Takenaka was a different type of intercalator that was a unique and specific bind with duplex DNA [6]. In this study, we used Hoechst intercalator and designed and prepared an integrated DNA chip sensor.

In this section, we report the electrochemical detection of DNA and PCR by microfabricated electrodes and microfluidic chips, respectively.

4.2.2 The Multiplexed Electrochemical DNA Sensor

The principle and preparation of our electrochemical DNA sensor is shown in Fig. 4.5.

Table 4.1. Reported electrochemical DNA sensors

	Year	Indicator	Sensitivity		DNA immob.	Potential/V Curren t/μ A	Electro de	Ch.	Target D
			DNA conc.	DNA quantity					
Single electrode	1993	$[Co(bpy)_3]^{3+}$ $[Co(phen)_3]^{3+}$	8.6mM	—	Adsorption	2 .5 μ A	GC (0.5mm)	1	20bp-ol
	1994	$Co(bpy)_3(ClO_4)_3$ $Co(phen)_3ClO_4)_3$ $Os(bpy)_3Cl_2$	0.19nM	4.8fmol		—	Pt (0.5mm)	1	4.0kbp poly(dT)
	1994	Hoechst 33258	1.8fM	—	-SH	550 mV, 192nA	Au (0.3mm)	1	4.2kbp-pV plasm
	1996	$[Co(phen)_3]^{3+}$	4nM	200fmol	Adsorption	500mV 8 μA	Carb on (1.6mm)	1	21bp-ol (HIV-1 U5
	1996	Ferrocenyl oligo-deoxynucleotides	0.5 μM	—	—	450 mV (CV)	Pt (1.6mm)	1	25bp-ol
	1998	Ruthenium hexaammine	—	—	SAM	—	Au (1.3 mm^2)	1	20bp-ol
	1999	DM (daunomycin) MB(meth ylene blue)	—	50pmol	-SH linker	—	Au (0.2 mm^2)	1	15bp-ol
	1999	DM	0.6 μM	—	Avidin-bioti n	—	Graphite (7 mm^2)	1	12bp-ol
	2000	Ferrocenyl-	—	10zmol	-SH linker	460-570mV (CV, DPV)	Au (2 mm^2)	1	20bp-dA2
Multi-electrode	—	ferrocen deriv.	—	—	-SH	—	Au	14	—
	—	Ferrocenyl-naphtalene diimide	—	—	—	—	Au	100	—
	—	Hoechst 33258, etc.	—	—	-SH	550 mV	Au	32	28bp HIV, H

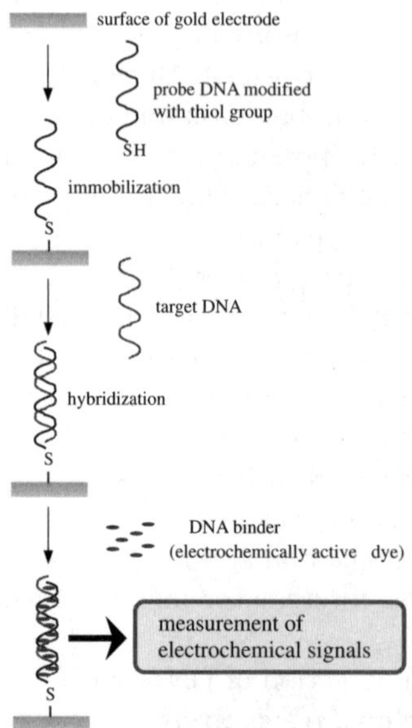

Fig. 4.5. The principle of a DNA chip using an electrochemical reaction on the gold electrode

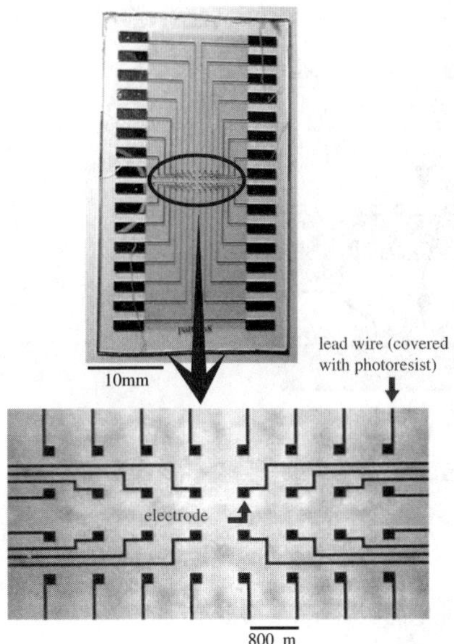

Fig. 4.6. The microelectrode array (electrode area 200 × 200 µm^2)

First, we fabricated a microelectrode array and then the DNA probes were synthesized and had thiol group placed at the 5'-phosphate ends for covalent binding with the surface of each gold electrode. By using a microarrayer, each DNA probe was selectively arrayed on each electrode. The fixed probes were complementarily bound with targeted 20 mer oligonucleotides chosen from the conservative region of HIV and HCV genes, as shown later. After a hybridization process with target DNA, an electrochemically active DNA intercalator specifically binds with the duplexed DNA. Then electrochemical signals were obtained from the intercalators fixed with DNA on the electrode. The electrochemical signal provides a correlation with the amount of DNA duplex formation.

The 32 individually addressable gold electrodes were arranged on a glass plate (Fig. 4.6). Each electrode measured 200×200 µm^2. Each microelectrode was connected to an external potentiostat by an insulated gold track. The fabrication procedure of the microelectrodes is shown in Fig. 4.7. At first, a 200 nm gold layer was deposited over a 20 nm chromium adhesion layer on a glass chip by vacuum evaporation. Next, the chip was spin-coated with photoresist and was irradiated with UV light. Each metal layer was etched to form electrodes, lead wires, and their connections. The lead wires were photolithographically covered with photoresist for insulation. Only the electrode region was open to the solution.

Figure 4.8 shows a cyclic voltammogram of intercalator, Hoechst 33258, which consists of piperazine and imidazole derivative. This has been reported to be a DNA minor groove binder and an electrochemically active dye. This

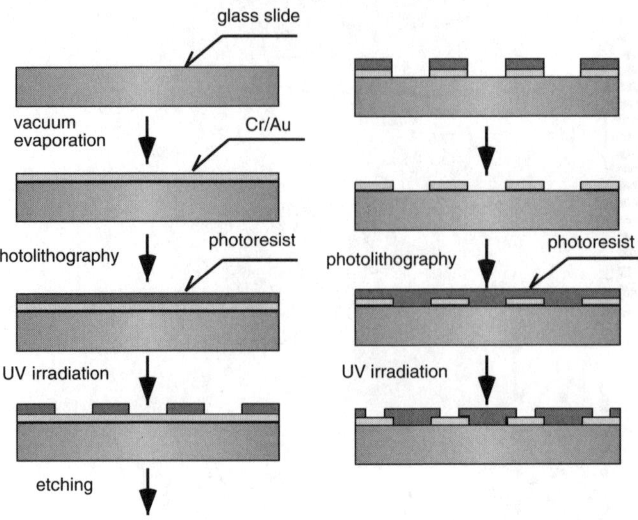

Fig. 4.7. The fabrication procedure of microelectrodes on a chip

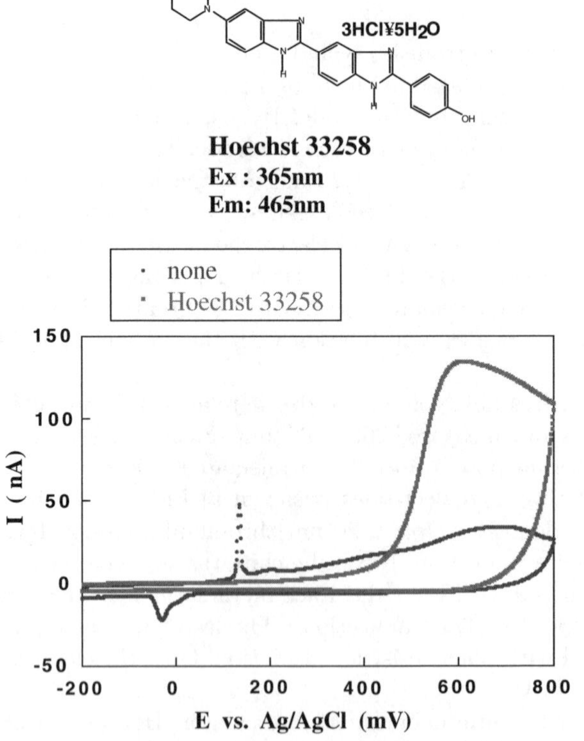

Fig. 4.8. The cyclic voltammogram of 100 µM Hoechst 33258 in 0.2 M phosphate buffer at 100 mV/s. Electrode area 200 × 200 µm^2

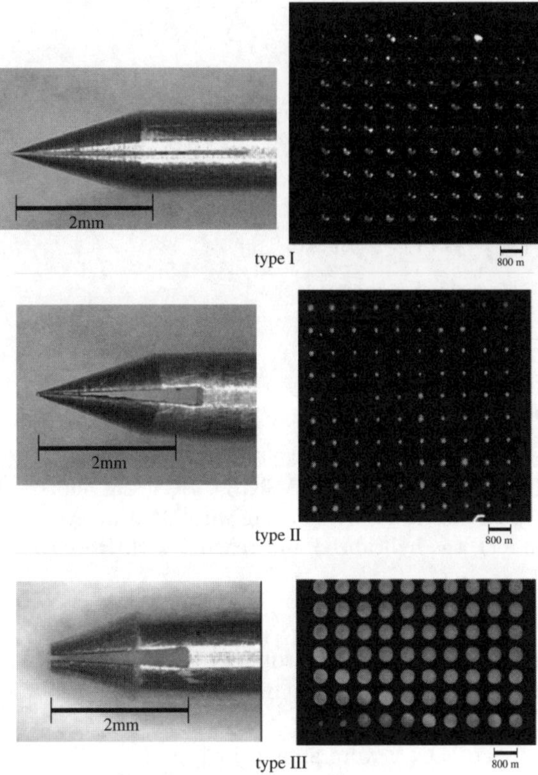

Fig. 4.9. Three types of printing tips

compound is concentrated at the electrode surface through association with formed hybrid, and is irreversibly oxidized on the bare gold electrodes at about 550 mV. Ishimori has reported sequence-specific gene detection using a gold electrode modified with probe DNA and Hoechst dye.

Each probe DNA was spotted on the microelectrodes using a microarrayer. The printing tips can be moved two-dimensionally and used for loading of probe DNA onto each electrode. We prepared three types of printing tip, as shown in Fig. 4.9. Each tip had a different size of groove and then different spotted areas were obtained. In this study, the type III tip was used because the spotted area was sufficient to cover the whole electrode.

We targeted HIV-1 and Hepatitis C virus DNA. Target sequences were selected from sequence-conserved regions. The target DNA or control DNA was hybridized at 37°C for 1 h and allowed to react with 100 nM intercalator dye for 10 min. After washing, the electrode electrochemical signals derived from the dye were measured by linear sweep voltammetry or differential pulse voltammetry. Four kinds of probe DNA were immobilized onto each electrode. The anodic currents were compared between the four types

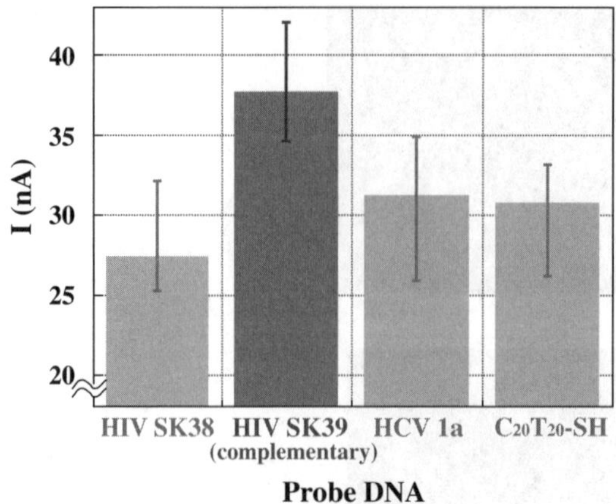

Fig. 4.10. Anodic peak current values of Hoechst 33258 on different probe-modified electrodes. 100 nM probe DNA (HIV SK 38, HIV SK 39, HCV 1a, and $C_{20}T_{20}$-SH) was immobilized on the gold electrodes by spotting with DNA arrayer. 10 nM target DNA (HIV SK 39 target) was hybridized and reacted with Hoechst 33258 on the chip

of electrode. When we used the HIV SK 39 probe as the target DNA, the electrode modified with complementary DNA against HIV SK 39 probe gave the biggest anodic current (Fig. 4.10).

Calibration curves for target DNA were shown in Fig. 4.11. The anodic current increased with the concentration of the target DNA from 0.01 to 100 nM. The control DNA also gave anodic current because Hoechst dye was not completely specific to duplex DNA. If we have target DNA over 1 nM, we can discriminate target DNA. When nontarget DNA was less than 0.01 nM, the sensitivity improved up to 0.1 nM, because the background level was reduced. Each electrode may react with a sample solution of approximately 1 nl volume. On this assumption, approximately 10^5 molecules of target DNA can be determined by each electrode. Figure 4.12 shows the results of detection of a single base pair mismatch. When target or one base mismatched DNA was used at 100 nM, they were detected using differential pulse voltammometry. However, it was difficult to detect SNPs less than 10 nM. A PCR-amplified sample is also acceptable in this case.

4.2.3 The Microfluidic PCR Chamber and the Electrochemical Detector

The determination of a specific DNA sequence presented in biological samples such as blood, serum, tissues, and body fluid is of great significance in the biomedical field. The results from such analyses can be used to detect infelicitous diseases of bacterial and viral origin. Therefore, there has been

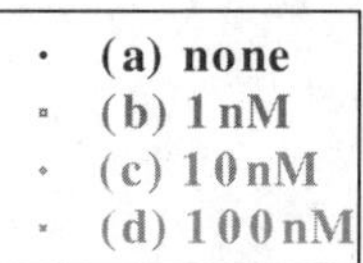

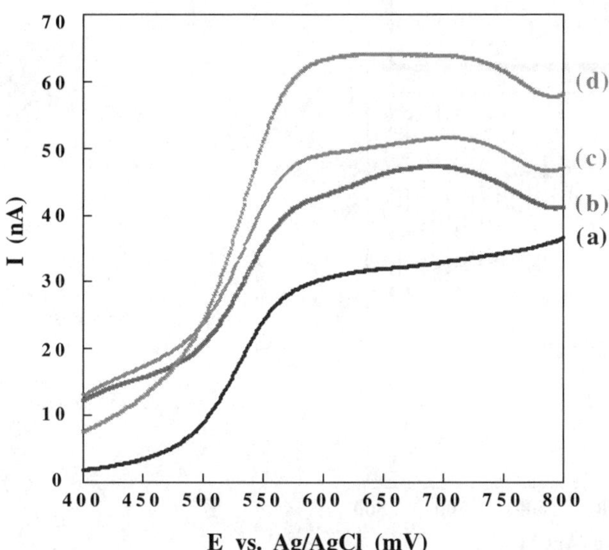

Fig. 4.11. Linear sweep voltammograms of Hoechst 33258 on gold electrodes at different concentrations of DNA probes: **a** none; **b** 1 nM; **c** 10 nM; **d** 100 nM reacted with 100 nM target DNA at 100 mV/s. Electrode area $200 \times 200\ \mu m^2$

considerable interest in developing reliable methods for single nucleotide polymorphism analysis [5].

When we need to lower concentration of sample DNA, PCR amplification of target DNA is effective. Recently, many methods for amplification and detection of DNA, such as 5' nuclease assay [7] and electrophoretic analysis [8] on a microchip, have become available. These methods are based on the phenomenon of fluorescence resonance energy transfer [9] (FRET), or use intercalating dye for fluorescence detection, and there are few reports of DNA amplification and electrochemical detection on a microchip. Therefore, this section introduces a microfluidic PCR chamber and an electrochemical detector.

Microchip Fabrication. We designed integrated chips for electrochemical PCR and gene detection. The structure of our chip is shown in Fig. 4.13. This chips couples a PCR chamber with a chamber for electrochemical gene detection, and provides on-chip DNA amplification and detection. The chip was fabricated using standard micromachining techniques. The chip design

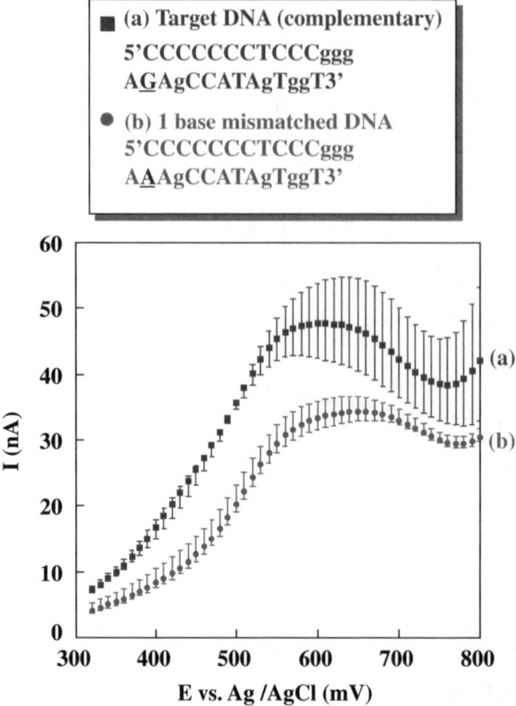

Fig. 4.12. Differential pulse voltammograms of Hoechst 33258 on HCV 1a probe-modified electrode reacted with 100 nM target DNA (complementary) or one base mismatched DNA at 100 mV/s. Electrode area $200 \times 200\ \mu m^2$

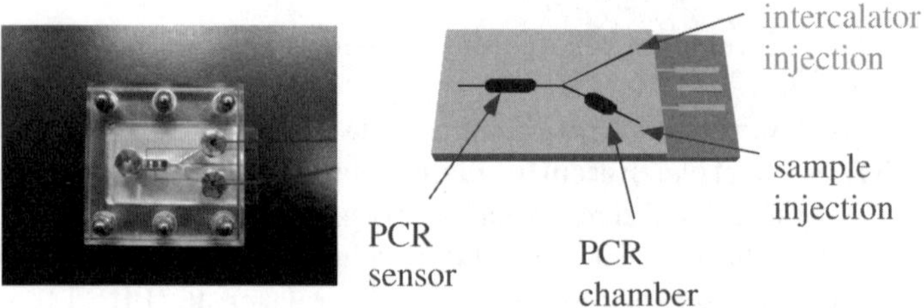

Fig. 4.13. The structure of the PCR chip

was transferred onto the glass substrates using a photoresist, a photomask, and UV exposure, and the chip was coated in a photoresist. The chamber plate was also transferred onto the substrate using same techniques, and etched into the substrate in dilute HF/HNO_3. Then, the chip was covered with the chamber plate. This structure is simple and versatile, enabling easy accommodation of different chip designs. This kind of integration is not difficult using micromachined technology, and it gives us a sensitive and selective DNA chip sensor.

The Detection of DNA Amplification Using an Electrochemical Method. In this subsection, the principles and advantages of DNA biosensing using electrochemical methods are illustrated by examples. Figure 4.14 shows the principle of electrochemical detection. Hoechst 33258 is a DNA minor groove binder and an electrochemically active dye [10]. Before PCR, Hoechst 33258 hardly binds to DNA, because the concentrations of target DNA are low. On the other hand, After PCR the DNA concentration is high, and therefore this dye will almost bind to DNA.

Figure 4.15 shows linear sweep voltammograms of Hoechst 33258 on DNA. By using an electrochemical method after PCR, DNA amplification was detected. The target DNA is HBV, its length is 1 kbp, and the amplified area was a selected high preserve sequence. The electrochemical measurements were carried out using an electrochemical analyzer (BAS-50W). The electrochemical signals of 50 µM Hoechst 33258 were measured. The anodic current of Hoechst 33258 after PCR is smaller than that before PCR. The reason for this may be that a lot of Hoechst 33258 is bound to amplified DNA and so the electrochemical response is decreased compared to that before PCR. In fact, this result means that electrochemical response of intercalated Hoechst

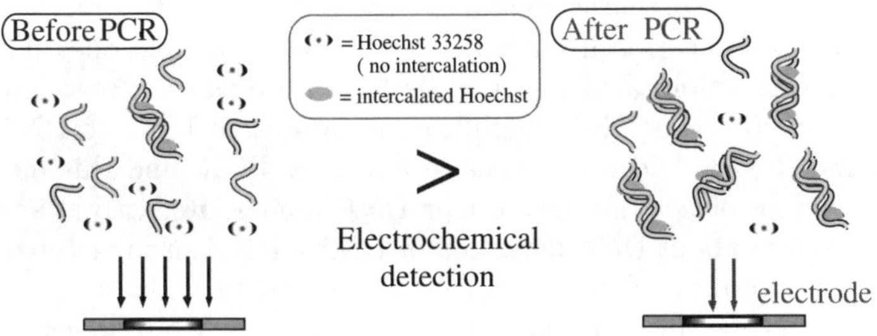

Fig. 4.14. The method of electrochemical gene detection

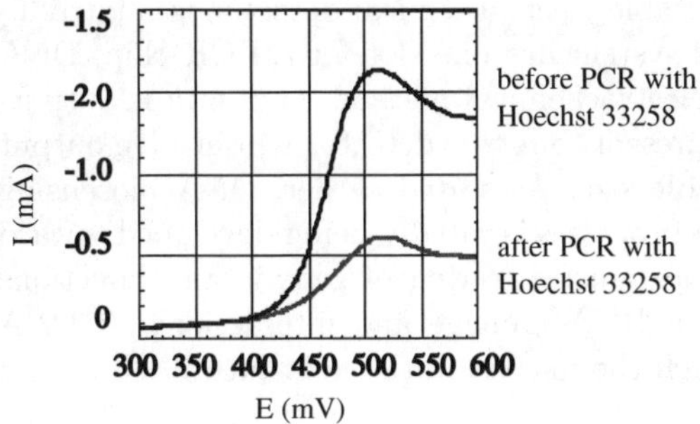

Fig. 4.15. Linear sweep voltammograms of Hoechst 33258 on HBV

Fig. 4.16. The portable electrochemical detector

33258 is smaller than that with no intercalation. As a result, this method can detect target DNA. A great advantage of this method is that it does not need to immobilize the DNA probe on the electrode, like current DNA chips, and so this system enables easy preparation and detection. However, this method has some weak points. The Hoechst 33258 used in this principle does not undergo a reduction reaction, although it does have an oxidation reaction. Furthermore, this dye protects DNA amplification, because it binds to DNA on PCR. Therefore the dye did not proper to real time PCR, and this method requires screening of new intercalator or DNA binder. Recently, there have been many reports about DNA detection or DNA analysis using electrochemical [11] and fluorescence methods [12]. It is difficult to know which of these methods is better, because each has some advantages for DNA detection and analysis. However, the system for DNA detection using an electrochemical method can be compact, and this is one of the main advantages of the method. Therefore, finally, a portable electrochemical detector was introduced, as shown in Fig. 4.16. This system has one slot for a PCR chip, DNA amplification is detected by an electrochemical method, and data is outputted on a notebook computer at present, but this detector will involve output system, and will be more portable size. As stated earlier, DNA biosensing based on nucleic acid recognition processes is rapidly being developed toward the goal of rapid, simple, and inexpensive testing of genetic and infectious diseases, and for the detection of DNA damage and interactions. A DNA biosensor is thus expected to reach the market in the near future.

4.3 Enzyme-Based Electrochemical Biosensors

Biosensors are analytic devices that are capable of providing either qualitative or quantitative results. Biosensors combine the excellent selectivity of biology with the processing power of modern microelectronics and optoelectronics to offer powerful new analytic tools with major applications in medicine, environmental diagnostics, and the food and processing industries. This review covers enzyme-based electrochemical biosensors, particularly amperometric enzyme sensors for glucose monitoring.

4.3.1 Introduction

The measurement of the blood glucose level is very important for the diagnosis of diabetes mellitus, and many glucose sensors have been developed [13]. Most glucose sensors are enzyme-based amperometric biosensors. Three general strategies are used for the electrochemical sensing of glucose, most of which use immobilized glucose oxidase (GOx), an enzyme that catalyzes the oxidation of glucose to gluconic acid with the production of hydrogen peroxide. The first detection scheme measures oxygen consumption; the second measures the hydrogen peroxide produced by the enzyme reaction; and a third uses a diffusable or immobilized mediator to transfer the electrons from GOx to the electrode.

It is easy to construct glucose sensors based on the principle of measuring the consumption of oxygen or the production of hydrogen peroxide (Fig. 4.17). However, they are disadvantageous in that the signal depends on the dissolved oxygen concentration. Blood glucose concentration is comparatively high, and it is around 5 mM in the hungry healthy human. It reaches 20 mM for diabetic patients, and measurement of the accurate glucose concentration by the sensor on the basis of this principle is difficult, since the oxygen concentration is insufficient. Furthermore, a high potential is necessary to reduce oxygen or to oxidize hydrogen peroxide. For example, in the case of a metal electrode, when 700 mV is applied to a silver/silver chloride

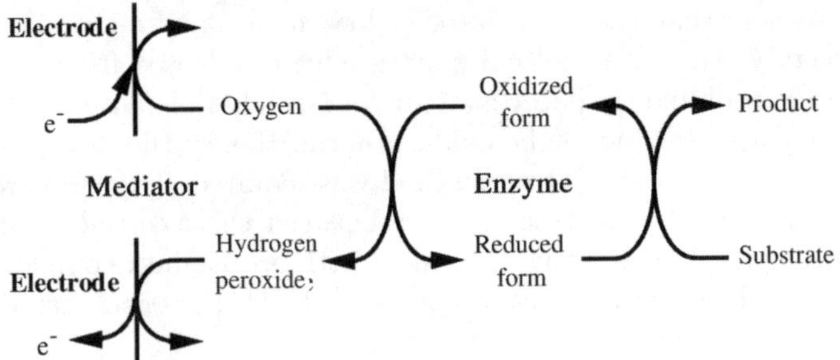

Fig. 4.17. The rinciple of a conventional electrochemical biosensor

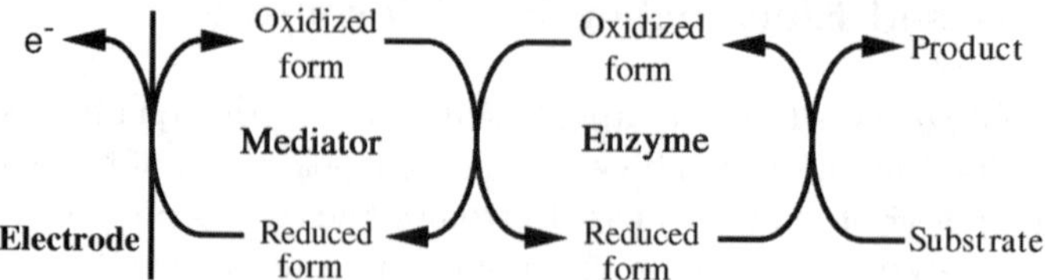

Fig. 4.18. The principle of a biosensor based on electron transfer mediator

electrode, the reducing compounds, such as ascorbic acid and uric acid, in the blood sample will be oxidized. Hence, these compounds usually influence the sensor output when hydrogen peroxide is measured. That is why a method that uses a nonnatural electron acceptor instead of oxygen attracts attention (Fig. 4.18). This acceptor is called a mediator, since it mediates the electron transfer reaction between the electrode and the enzyme, and electrochemical biosensors using mediator are hardly affected by the dissolved oxygen concentration. Hence, a sensor can be constructed to measure highly concentrated glucose. Ferrocene derivatives [14], benzoquinone [15–17], and a phenazine derivative [18] have been used for mediators.

Cass et al. [14] have shown that the ferrocene/ferricinium ion couple acts as an effective mediator between reduced glucose oxidase and a graphite electrode. The incorporation of 1,1-dimethylferrocene into a graphite electrode upon which glucose oxidase has been immobilized provides a glucose sensor that has many of the features required for the analysis of clinical glucose samples. Its performance depends on a number of features: the rapid rate of electron transfer between the reduced enzyme and the ferriciniun ion; the good electrochemical properties of ferrocene; and the low solubility of the ferrocene, which results in the effective confinement of the mediator to the electrode surface. The greater solubility of the ferricinium ion allows diffusion between the immobilized enzyme and the electrode surface.

4.3.2 Biosensors Based on Redox Active Polymers

Although the properties of biosensors using a mediator seem to be excellent, defects in the biosensors that use a mediator of low molecular weight have been indicated recently. Mediator molecules are gradually released from the electrode, even if the insoluble mediator is used. An insoluble mediator such as ferrocene will be water-soluble in the oxidized form. Hence, the development of a novel mediator immobilization method was required. Therefore, a polymer that can mediate the electron transfer between the electrode and GOx has been developed. Redox active polymers with an osmium complex of pyridine and imidazole group have been reported [19–25]. Ferrocene-based redox active polymers have also been developed [26–33].

Gregg and co-workers have reported biosensor applications of cross-linked redox gels containing GOx based on polyvinylpyridine (PVP) with complexed $[Os(bpy)_2Cl]^+$ (bpy = 2,2'-bipyridine) [19]. Free pyridine groups were quaternized with either the bromoacetic or the bromopropionic acid ester of *N*-hydroxysuccinimide to promote cross-linking with lysines on the enzyme surface. However, competition between the desired aminolysis of the *N*-hydroxysuccinimide groups and their spontaneous hydrolysis caused unsatisfactory variation from film to film. Subsequently, they developed an improved redox polymer, PVP containing complexed $[Os(bpy)_2Cl]^+$ and partially quaternized with bromoethylamine, for the immobilization of oxidoreductases [20]. The electrochemical characteristics of this redox polymer in the absence of the enzyme and the steady-state electrochemical response of the GOx immobilized electrode were examined. Elmgren and co-workers have evaluated the details of the electrochemical characterization for PVP containing complexed $[Os(bpy)_2Cl]^+$ [21]. Ohara and co-workers have reported novel osmium-based redox hydrogel enzyme electrodes. These were formed by cross-linking poly(1-vinylimidazole) (PVI) complexed with $[Os(bpy)_2Cl]^+$ [22], [Os(4,4'-dimethyl-bpy)$_2$Cl]$^+$ [23,24], or [Os(4,4'-dimethoxy-bpy)$_2$Cl]$^+$ [25] and polyethyleneglycol diglycidylether in the presence of GOx or lactate oxidase.

Tatsuma and co-workers have reported an enzyme electrode using GOx and poly(*N*-isopropylacrylamide-co-vinylferrocene) (PAF) [26]. Since PAF was soluble in water at $< 20°C$ and insoluble at $> 28°C$, the electrode was loaded with PAF and GOx at $< 20°C$ and employed at $> 28°C$. However, leakage and swelling of this redox polymer occurred at room temperature.

Calvo and co-workers have investigated the electrical communication between a novel ferrocene-containing acrylamide–acrylic acid copolymer and GOx or horseradish peroxidase [27]. The carboxylic group of the copolymer was covalently attached to aminoethylaminomethylferrocene, and GOx was immobilized with an amide group via glutaraldehyde. They have also reported a redox hydrogel obtained by derivatization of polyallylamine with ferrocene and pyridine groups that coordinate iron and ruthenium complexes, and they have prepared the cross-linked redox polymer with GOx via epichlorohydrin [28,29]. However, these redox polymers and GOx may swell because of high hydrophilicity.

Watanabe and co-workers have reported a novel redox copolymer consisting of ferrocenylmethyl methacrylate and methoxy-oligo(ethylene oxide) methacrylate [30]. This redox copolymer and GOx were mixed with carbon paste, and a GOx-immobilized carbon paste electrode was prepared to detect glucose concentration amperometrically. When using this carbon paste GOx electrode, the catalytic current response was highly sensitive and the glucose concentration was determined up to 50 mM. However, the use of carbon paste causes difficulty in evaluating the electrochemical properties of the redox polymer.

Bu and co-workers have reported a redox copolymer of acrylamide and vinylferrocene cross-linked with *N*,*N*'-methylenebisacrylamide, and GOx was entrapped simultaneously in the hydrogel during the polymerization process

[31]. However, this redox hydrogel required a dialysis membrane to be fixed on the electrode.

To improve the above problems, our group has previously proposed a BSA-based immobilization that co-reticulates GOx and a redox polymer with glutaraldehyde on a glassy carbon electrode [32]. A cross-linked copolymer of acrylamide and acrylic acid with 11 bonded side chains containing ferrocene at the ends was used. This immobilization method was highly effective for stabilizing the electrochemical response of the redox polymer. However, synthesis of this redox copolymer was complicated and the yield was low.

In our previous work, we investigated a cross-linked redox polymer that can readily be prepared and we characterized a mediated enzyme electrode using this redox polymer [33]. Polyallylamine was cross-linked with glutaraldehyde and modified successively with ferrocenecarboxylic acid. GOx and the redox hydrogel were immobilized on the electrode surface using glutaraldehyde. To prevent the enzyme-containing redox polymer gel from over-swelling, BSA was also employed for immobilization of the redox polymer and enzyme. The effect of BSA for immobilizing both the redox polymer and GOx was examined by cyclic voltammetry. A time course of the relative peak current of the redox polymer-modified electrodes was investigated. A typical electrochemical response of the ferrocene-containing redox polymer was obtained initially for each electrode. However, continuous soaking in an aqueous solution reduced the redox current of the electrode without BSA. These results indicate that such a decrease in the redox response resulted from swelling of the polymer–protein hydrogel. Therefore, the distance between the redox sites of the polymer was extended and the electron transfer rate among neighboring redox sites decreased. Stable cyclic voltammograms were obtained for the electrode with a greater BSA content than the ratio of redox polymer: GOx : BSA = 150 : 50 : 150 (mg/ml). This result suggests that the addition of BSA prevented the polymer matrix from over-swelling. This paper also showed that the redox response, which was reduced by soaking in a buffer solution, was recovered after drying the electrode.

We also investigated the effect of the redox polymer and enzyme content. Although several studies on enzyme-catalyzed electrochemistry using redox polymers have been reported, an enzyme electrode using a redox polymer at high redox polymer content and/or at high enzyme content has never been examined. Consequently, interesting electrochemical behavior was observed.

The catalytic electrochemistry of the enzyme electrode with the redox polymer was investigated. Figure 4.19 shows cyclic voltammograms of the enzyme electrodes at various redox polymer contents at various glucose concentrations. These figures demonstrate that a catalytic current was observed by the addition of glucose. In Fig. 4.19a, the enzyme electrode at high redox polymer content showed prepeaks at around 260 mV at low glucose concentration. On the contrary, no clear prepeaks were observed at low redox polymer content, as shown in Fig. 4.19b. At high glucose concentration, a

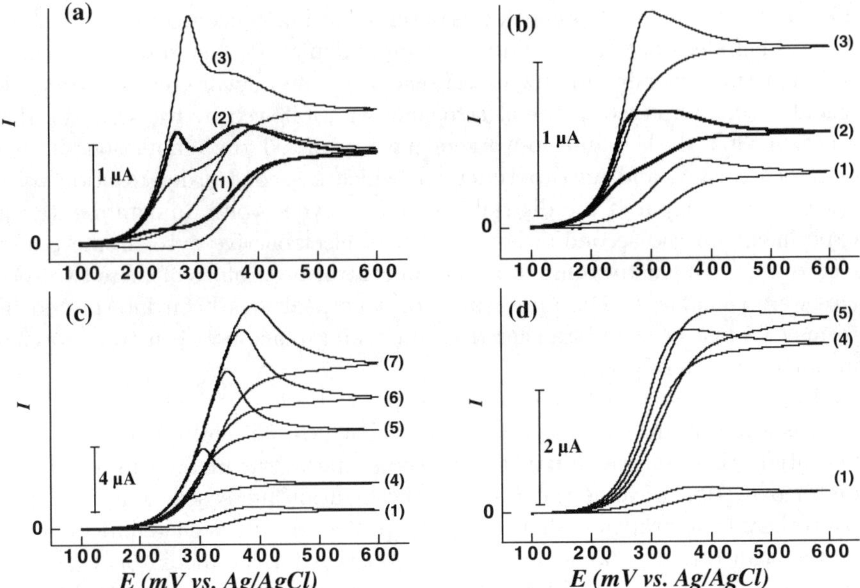

Fig. 4.19. the catalytic electrochemistry of the enzyme electrode with redox polymer: the effect of redox polymer content. Cyclic voltammetry was carried out in a deoxygenated phosphate buffer containing (1) 0, (2) 1, (3) 2, (4) 6, (5) 10, (6) 15, and (7) 20 mM glucose. The enzyme electrode was prepared by loading with a 2.0 µl (BL suspension of a polymer–protein mixture. The ratios of the mixture for immobilization were redox polymer : GOx : BSA = (**a**) and (**c**) 200 : 50 : 150, (**b**) and (**d**) 50 : 50 : 150 (mg/ml). The scan rate was 1 mV/s

typical catalytic current with a sigmoidal shape was observed at low redox polymer content, as shown in Fig. 4.19d. However, peaks were still observed at high redox polymer content in Fig. 4.19c.

The appearance of these prepeaks can be explained by cyclic voltammetric simulation. A cyclic voltammetric simulation for the electrochemically mediated enzyme reaction has been reported by our group [34]. In this chapter, the digitally simulated cyclic voltammogram demonstrates that a prepeak is observed at low substrate concentration, high mediator concentration, and high enzyme activity. This electrochemical behavior can be elucidated from the calculated concentration profiles of the mediator and the substrate. The appearance of the prepeak is due to the depletion of the substrate concentration in the vicinity of the electrode surface. Hence, the recycling rate of the redox site is reduced, and the catalytic current increase is suppressed. The concentration profiles can also elucidate whether the catalytic currents show a sigmoidal shape or a peak at high glucose concentration.

The observation of such prepeaks has also been reported by Coury and co-workers in the enzyme electrochemistry of sulfite oxidase mediated by

$[Co(bpy)_3]^{2+/3+}$ [35]. A prepeak was observed at low concentrations of sulfite. They concluded that this was due to a rapid depletion of sulfate in the catalytic reaction layer by an enzymatic reaction. The appearance of a prepeak has also been observed in the electrochemistry, followed by the second-order reaction [36]. Barker and co-workers have reported the simulation of this electrochemistry, and an experiment in which a second-order reaction took place. They compared the digitally simulated cyclic voltammogram with an experiment on the second-order reaction of electroactive cytochrome *c* and electro-inactive plastocyanin, and the concentration profiles of these two species were calculated. The appearance of a prepeak has been interpreted in terms of a fast electron transfer reaction taking place between two proteins in homogeneous solution.

Figures 4.19a,c also show that the prepeak current increased with an increase in the glucose concentration, and a positive shift of the prepeak potential was observed. Furthermore, the prepeak was merged into the mediator oxidation wave as the glucose concentration increased. Andrieux and co-workers have proposed that a prepeak potential shift occurred, depending on the scan rate, the homogeneous rate constant for the mediator/substrate reaction, and the ratio of mediator to substrate concentration (the excess factor) [37]. By using this proposed theory, Coury and co-workers have deduced that a shift of the prepeak potential is observed at very low substrate concentration at comparatively high enzyme and mediator concentrations, where the rate control should be shifted to the substrate/enzyme reaction [35]. The positive shift of the prepeak potential occurred due to the increasing concentration of the reduced mediator at the electrode as the substrate concentration increased.

The effect of enzyme content was investigated. Figure 4.20 shows cyclic voltammograms of the enzyme electrodes with the redox polymer at various enzyme contents at low glucose concentration. No prepeak was detected at low enzyme content in Fig. 4.20a, whereas a prepeak was observed at high enzyme content in Fig. 4.20b. These results also led us to conclude that these prepeak phenomena occur at high enzyme activity.

The steady state response at fixed potential was investigated using a rotating disk electrode. Figure 4.21 shows the current responses to glucose at three different redox polymer contents. These figures demonstrate that the steady state current was increased by the addition of glucose. At low redox polymer content, the steady state current was saturated at low glucose concentration, as shown in Fig. 4.21a. This is ascribed to the redox polymer content being too low to oxidize the highly concentrated glucose. At high redox polymer content, the response to the first glucose injection was less than that to the second injection, as shown in Fig. 4.21c. This seems strange if only the enzyme kinetics are considered. However, this result can also be elucidated in terms of the cyclic voltammetric simulation. In our previous work [34], at low mediator concentration, a plateau on the cyclic voltammogram

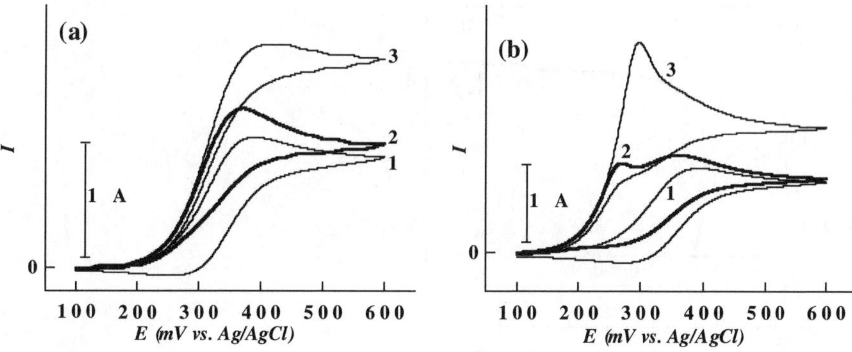

Fig. 4.20. The catalytic electrochemistry of the enzyme electrode with redox polymer: the effect of enzyme content. Cyclic voltammetry was carried out in a deoxygenated phosphate buffer containing (1) 0, (2) 1, and (3) 2 mM glucose. The enzyme electrode was prepared by loading with a 2.0 µl (BL suspension of polymer–protein mixture. The ratios of the mixture were redox polymer : GOx : BSA = (**a**) 200 : 5 : 150 and (**b**) 200 : 50 : 150 (mg/ml). The scan rate was 1 mV/s

was observed at low substrate concentration. On the contrary, at high mediator concentration, a steady state current in the cyclic voltammogram was not obtained at low substrate concentration, and a time-dependent current decrease was observed. Therefore, if the steady state current at high redox polymer content is discussed, there is some possibility that the response to the first glucose injection is less than that for the further injection. Figure 4.22 shows the correlation between the steady state response and the glucose concentration. The response current depended on the redox polymer–enzyme content and increased with an increase in the redox polymer content. This figure demonstrates that it is possible to determine glucose concentration up to 7 mM.

4.3.3 Glucose Sensors on the Market

There are many different blood glucose sensors on the market that are faster and less complicated to use. Manufacturers have developed sensors and test strips that require less blood, conduct the test faster, and offer a variety of functions and data management capabilities to match individual testing needs. This section covers blood glucose sensors for self-testing on the market.

Commercial glucose sensors are based on enzyme-photometric or amperometric biosensor technology. Bayer's Glucometer Encore, LifeScan's One Touch Basic, Profile, and SureStep, and Roche's Accu-Chek Easy use the former technology. Most of the other sensors are based on amperometric detection. The characteristics of the blood glucose sensors on the market are reviewed below.

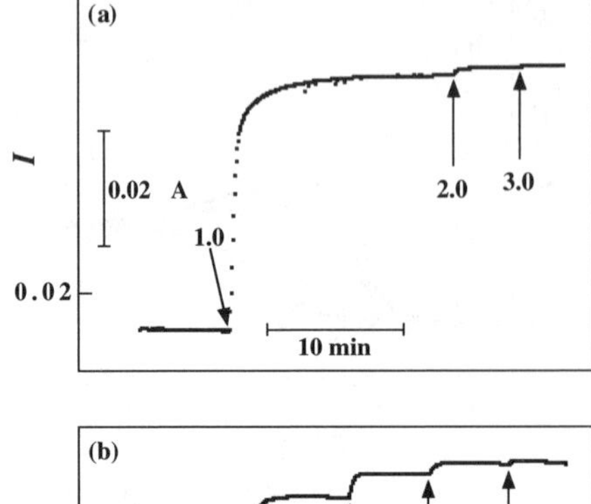

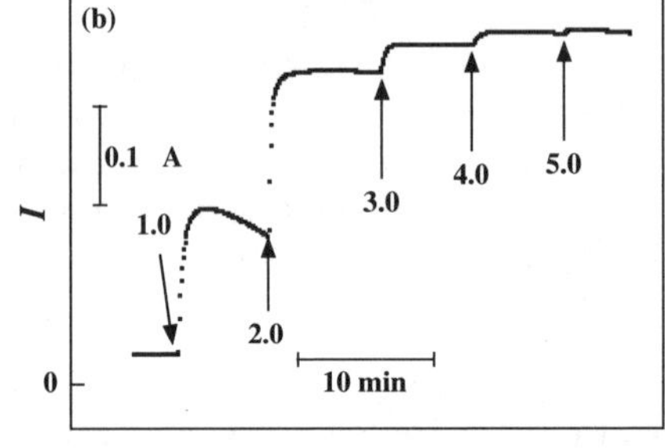

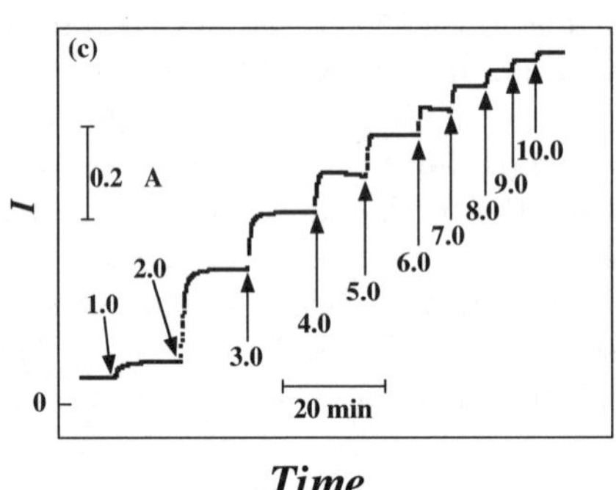

Fig. 4.21. The steady state response of the enzyme electrode with redox polymer to successive additions of glucose. The electrodes were prepared by loading with a 0.5 µL suspension of polymer–protein mixture. The ratios of the mixture were redox polymer : GOx : BSA = (**a**) 50 : 50 : 150, (**b**) 200 : 50 : 150 and (**c**) 300 : 50 : 150 (mg/ml). Numbers on the figures show the glucose concentration (mM) after injection. The enzyme electrodes with the redox polymer were operated at 600 mV and rotated at 1000 rpm

Bayer [38]
Glucometer Elite. Bayer's easiest self-testing system to learn and use, the Glucometer Elite is a small, rugged meter that provides state-of-the-art blood glucose monitoring with a 20-test memory. There are no buttons to push and the automatic sampling provides fast accurate results in just 30 s. User benefits are: MicroEase Sampling action helps strip draw blood from any angle for an accurate result; foil-wrapped strips assure a fresh test result every time; especially beneficial for frequent testers, new patients, those who do not bleed completely automatic – the simplest meter, the simplest way to get accurate results; simple no-button operation minimizes user error; no cleaning required – optimal for hospital blood glucose testing at the bedside; referenced to laboratory methods for virtually lab-accurate results;

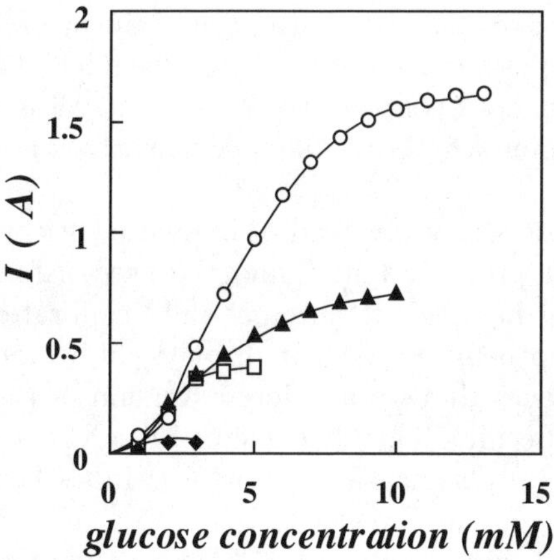

Fig. 4.22. Glucose calibration plots of enzyme electrodes with redox polymer. The ratios of the mixture were redox polymer : GOx : BSA = (○,▲) 300 : 50 : 150, (□) 200 : 50 : 150 and (♦) 50 : 50 : 150 (mg/ml). The electrodes were prepared by loading with (○) 1.0 µl and (▲,□,♦) 0.5 µl suspensions of polymer–protein mixture

new shape, new features – including 20-test memory; fast – results in 30 s; expanded measurement range 20–600 mg/dl (1.2 to 33.3 mmol/l); optimal hematocrit range – 20–60%.

Glucometer Elite XL. The Glucometer Elite XL Diabetes Care System makes blood glucose monitoring for effective diabetes control easier than ever. The Elite XL is simple and convenient to use – needing only the smallest amount of blood. Patients can transfer data from the Elite XL memory to the computer for smart data management. The special WinGlucofacts XL Diabetes Management Software can be downloaded through the web site or patients can order the software and cable from Bayer Customer Service. It's an excellent tool for effective diabetes data organization, analysis, and management. The Elite XL Diabetes Care System is available at most pharmacies and some accessories can be ordered directly from Bayer.

The Elite XL model has a button for manual on/off operation and to go back through stored readings. It stores 120 readings in memory, with date and time taken. It has a data port to download stored readings to a computer via an optional cable. It can delete a reading if a strip has not yet been removed. It can designate a reading as a control before or after obtaining it.

Glucometer DEX. The Dex is somewhat harder to use than the Elites in that the angle at which the strip is held to the drop is less flexible. For accurate results, the meter should be held horizontally and the blood sample should be introduced for 1–2 s after the beep and count-down start, no longer. An insufficient fill will give a falsely lower reading, while an overfill can give a higher than actual reading. The sensor has an open end and an overfill can draw blood into the meter electronics and short them out. Once dry, the meter should work properly again, but do a control test to verify it. Visual inspection of the Dex sensor to determine whether it filled sufficiently is not as reliable as with the Elite strips.

The Dex uses a cartridge of ten test sensors instead of individual strips. Sometimes a sensor does not pop out properly and cannot be used. After use, sensors are ejected without being handled. It automatically calibrates the test sensor cartridge, no code strip being needed. It displays an overall average and four 2 h time-range averages. It uses a different version of the Win Glucofacts software (it is meter-specific). The Glucometer Elite XL, the Elite, and the DEX are referenced to give plasma/serum glucose values when testing capillary whole blood and will compare directly to laboratory results. A meter referenced to whole blood will give results 10–15% lower than the Glucometer DEX and laboratory results.

Glucometer Encore. This competitively priced blood glucose monitor is a great way to obtain accurate readings with an especially easy to use meter. Its fully automatic operation means no hassles, and no complicated procedures to learn. It is the right choice for the infrequent tester who wants to keep things simple.

A wide test range can be achieved, from 10 to 600 mg/dl, for added confidence. The test results are not significantly affected by hematocrit in the range of 20–60%. No battery replacement is needed, with up to a 15 000 test capacity. Visual backup – color chart comparison provides backup security and assurance. There is a ten-test memory with one-button recall. Individual foil-wrapped, easy-grip reagent strips with hexonkinase chemistry are used; there is no oxygen dependence and fewer interfering substances are observed.

MediSense [39,40]. In the UK, MediSense manufactures a range of disposable test strips to support users of consumer and hospital products, including their Precision Plus strip design. This strip delivers accurate, small blood sample, patient friendly monitoring for the Precision family of meter products.

MediSense's family of glucose strip products exploits their unique biosensor technology. The electron flow generated by the specific reaction in the enzyme glucose oxidase with the glucose in the blood sample is coupled by a ferrocene mediator to an electrochemical cell. The current flows along the strip to be measured by the advanced solid state electronics of the sensor, to read out the amount of glucose in the blood. This provides a durable and easy to use system that removes the risk of cross-infection and the need for cleaning.

Precision Q-I-D. The Precision Q-I-D Blood Glucose Test System provides accuracy and convenience for home use testing, to ensure better patient compliance. The precision Q-I-D incorporates the natural testing process of the sensor electrode. The key to the performance of the precision Q-I-D is the sensor electrode test strip design. This advanced biosensor technology limits interference from 50 substances that are commonly found in people with diabetes. It also delivers accurate results in a wide range of temperature, humidity, and altitude conditions.

Precision PCx Point-of-Care System. This fully portable handheld device allows nursing professionals to quickly and conveniently test a patient's blood glucose at the bedside. It combines powerful data management and networking capabilities with the unique MediSense biosensor testing strip technology. This system offers hospitals a complete solution for managing their point-of-care testing needs.

ExacTech. A blood glucose monitoring system that gives accurate results in only 30 s – for easy testing, with no timing, wiping, cleaning, or blotting. No battery replacement is required.

Arkray [41]. Glucocard II Data answers the patient's demand "I want to have full and satisfying data management of my own blood glucose levels." Glucocard II Data is standard equipped with an external output function that stores up to 120 test results. The exclusive cable and data management software (optional) enable the user to transmit the data to a PC for further compilation and processing. For a more scientific and personalized blood glucose management, these are the special features from Arkray, Japan.

Roche [42].

Accu-Chek Advantage. With just two simple steps, the patients get an accurate result in only 40 s. The Accu-Chek Advantage meter has a snap-in code key, making calibration extremely simple. The Accu-Chek Advantage meter has a 100-value memory that lets you automatically store results, time, and date. With the Accu-Chek Advantage meter, the patients can download the test results onto a computer.

Accu-Chek Complete. The Accu-Chek Complete system provides more information, giving users more control over diabetes, and more power to live exactly as they choose.

With just two simple steps, patients get an accurate result in about 40 s. The Accu-Chek Complete meter has a 1000-value memory that lets patients automatically store results, time, and date. The Accu-Chek Complete system records blood sugar, insulin, carbohydrates, exercise, stress, ketones, HbA1c, and much more. The Accu-Chek Complete system features "ATM-like" push-button selection and displays reports in chart or graph form without the need of a computer.

LifeScan (Johnson & Johnson) [43]. The One Touch FastTake Blood Glucose Monitoring System consists of three main products: the FastTake Blood Glucose Meter, FastTake Test Strips, and FastTake Control Solution. These products have been designed, tested, and proven to work together as a system to produce accurate blood glucose test results. The FastTake Meter displays results between 20 and 600 mg/dl (1.1–33.3 mmol/l). If the test result is lower than 20 mg/dL (1.1 mmol/l), "lo" will appear on the meter display. This indicates severe hypoglycemia (low blood glucose). Users should immediately treat hypoglycemia as recommended by their healthcare professional. If your blood glucose test result is above 600 mg/dl (33.3 mmol/l), "hi" will appear on the meter display. This indicates severe hyperglycemia (high blood glucose). Users should seek immediate medical assistance. When the blood glucose test result is between 240 and 600 mg/dl (13.3–33.3 mmol/l), "ketones" will appear on the meter display. This message does not mean that the system has detected ketones, but that testing with a ketone test strip may be advisable. The characteristics are described below: very small blood sample size, 1.5 µl; patients can use a finer-tip lancet, which means less pain; accurate results in just 15 s; plasma calibrated for easier lab comparison; 150-test memory with date, time, and automatic 14-day averaging; small, portable, and easy to use. With One Touch FastTake Test Strips, patients can test on their arm or fingertips. It is entirely up to them. The arm has fewer nerve endings, so testing there feels much better; the FastDraw design automatically pulls blood into the test strip, so patients no longer have to target blood onto the top of the test strip.

4.3.4 Conclusion

Blood glucose sensors are now commercially available. However, there are still some problems, namely, the commercially available glucose sensors are limited to disposable use. Reducing compounds affect the sensor response, and an accurate blood glucose level cannot be determined. Therefore, further improvements of the blood glucose sensors should be necessary.

Except for the glucose sensor, most biosensors have not been used. This is due to the low stability of enzymes except for glucose oxidase. Novel molecular recognition materials are now being developed. Biosensor development in the 21st century will be drastically different.

4.4 Cell and Tissue Monitoring and Their Applications: The Whole Cell Sensor

4.4.1 Overview of Cell-Based Sensor

As also discussed in other sections of this chapter, the biosensor consists of biomaterials for selective sensing and transducers for sensitive signal transduction. Immobilized biomaterials on a transducer recognize specific compounds in the sample solution. The first biosensor started with enzyme, but later studies have extended the immobilized layer to various biomaterials, such as nucleic acids, antibodies, receptors, and cells. Cell-based biosensors employ living cells as immobilized biomaterial.

The most typical cell-based biosensor is a BOD sensor [58]. BOD stands for biochemical oxygen demand, and is an indicator of water pollution. Although the five-day BOD test has remained a standard pollution-monitoring tool since 1936 [44], the test requires a five-day incubation period at 20°C and skill in determination. More rapid and simple BOD measurement methods for estimation of five-day BOD are required for pollution control. A BOD sensor allows rapid and easy sensing. Microorganisms immobilized on an oxygen electrode metabolize organic compounds in sample water that consume oxygen. A decrease in the dissolved oxygen detected by the oxygen electrode indicates the amount of degradable organic compounds. The BOD sensor is already commercially available and the secondary well-produced biosensor has created a market of 400 million yen per year in Japan.

A typical characteristic of cell biosensors is a wide specificity compared to enzyme electrodes such as the blood sugar sensor. The glucose oxidase used in a blood sugar sensor can catalyze only the reaction of glucose. However, the microorganisms used in a BOD sensor metabolize various kinds of organic compounds. This is because many kinds of enzymes that degrade organic compounds are found in the cells. It is difficult to immobilize such enzymes without cells on a transducer. Cells can provide a better microenvironment for proteins as sensing materials than artificial immobilization. Furthermore, cells on a transducer can continuously produce weak proteins as sensing materials, providing characteristics of self-reproducibility.

A signal cascade realized on a transducer by cell immobilization is useful to detect unknown species that show certain effects on them. A cell-based toxicity sensor measures cell activity (usually respiration) as a result of inhibition by toxic compounds in the sample. Not only the normal cell but also the genetically modified cell is designed to produce reporter marker protein in a response to the detection of toxic compounds.

The oxygen electrode is the most common transducer for oxygen measurement. Its large size and complicated structure limit the application of such a conventional electrode to a portable device. The cells' viability or other activity affects the production and consumption of other compounds. Several electrochemical devices as alternatives to the oxygen electrode are now available, such as the microfabricated oxygen electrode, the thick-film electrode, the ion-sensitive field effect transistor, and the surface photovoltage device. In following section, we will describe the recent development and application of cell-based biosensors for BOD measurement.

4.4.2 The Surface PhotoVoltage (SPV) and Its Application

The surface photovoltage (SPV) device, or light-addressable potentiometric sensor (LAPS), is a silicon-based transducer to measure the surface potential of the device, especially the pH of the solution near the surface [45–48]. One side of the device is covered with silicon dioxide and/or silicon nitride layers as the sensor side. The other side is mainly bare silicon and a partially deposited metal layer for ohmic contact. Illumination from the rear of the SPV device induces a photovoltage. When only the sensor side of the SPV device is immersed in an analyte solution, and the potential of the solution is biased against the bulk silicon of the device using ohmic contact, the silicon device acts as a MIS (metal insulator semiconductor) structure. In this case, the solution plays the role of the "M" of MIS due to its conductive nature. Because the depletion layer at the surface of the device has a certain capacitance depending on the surface potential, the modulated photovoltage induces a photocurrent that depends on the sum of the bias potential and the surface potential of the device. Thus, frequently modulated illumination induces a modulated photovoltage and ac photocurrent. The surface potential can be estimated from the typical *CV* characteristics of the MIS structure (Fig. 4.23). Finally, the SPV device gives the same response as the Ion-Sensitive Field Effect Transistor (ISFET).

The SPV device has advantages compared with other pH sensors. The fabrication process is simpler than for an ISFET. The multiplexing of different light sources in different locations allows multisensing (called LAPS in a strict sense) without additional process complexity. SPV devices have been used in some research, such as enzyme-linked immunoassays [49,50], and measurement of taste compounds [51], hydrogen [52], and the metabolism of animal cells [53–57].

The metabolism of cells produces acidic substances such as carbonate ion and organic acid. The microbial metabolism is affected by many factors in a cell's environment. Changes in the biological, chemical, and physical environment of a cell are reflected in the production of acidic compounds. Thus, when microorganisms or cells are immobilized on the sensor side of the SPV device, the system can sense the biological information in the analyte solution.

Recently, many methods have been developed for the measurement of BOD. The BOD sensor consists of microbial cells immobilized on an oxygen electrode and measures the current decrease resulting from a decrease in dissolved oxygen [59–67]. Some other methods have been also reported

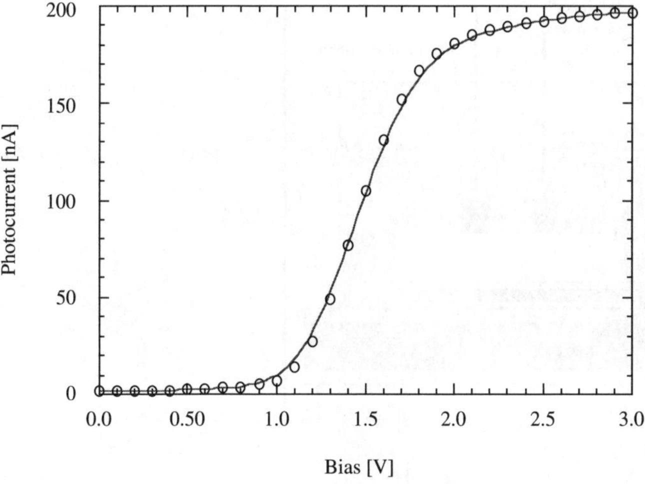

Fig. 4.23. The *CV* characteristics of the SPV device. A phosphate buffer solution of 10 mM (pH 7) without sodium chloride was flowing at 250 µl/min

[68–70]. Both the five-day method and the oxygen electrode method use the oxygen consumption as a parameter of the activity of the immobilized microorganism. The other parameters of the metabolism are potentially useful in constructing a microbial sensor, combined with an appropriate transducer. Therefore, we applied the SPV device to the the fabrication of a novel BOD sensor [71]. In this section, we will discuss fabrication and characterization of a novel BOD sensor combining microorganisms with the SPV device.

The Dependence of the *C–V* Characteristics of the System on pH. Figure 4.24 shows the configuration of a flow cell of the system. The flow cell consists of the SPV device (Shindengen, Japan) and silicone sheets, and was connected to an SPV controller (SE1030, Technologue, Japan) and a computer. Figure 4.25 shows the sigmoidal curves at various pH and calibration curves at various bias potentials. According to the Nernst equation, the rate of potential change against pH is about 58 mV/pH at room temperature, although the experimental value was 46 mV/pH in Fig. 4.25. The difference may due to the existence of other ions in the buffer. The same tendency was reported in the research on ISFETs that had same MIS structure as the SPV. When the sample reservoirs of the flow system were replaced, and changed from pH 7 to pH 4, the photocurrent began to decrease, and then showed a steady response after 6 min. It took about 2 min to replace the buffer in the flow cell after the replacement of the reservoirs. Generally speaking, a slower flow makes the response larger, but it takes a longer time to become stable. The response time for the flow rate of 50 µl/min, about 70 min, was too long, compared to the response time of typical BOD sensor, which is

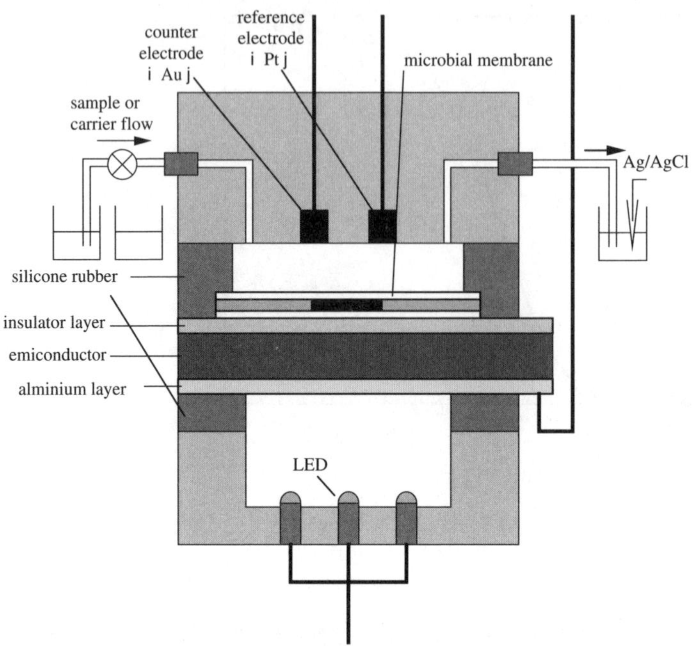

Fig. 4.24. A schematic diagram of a flow cell. LEDs were located at the rear against sample flow side. A platinum electrode and a gold electrode were located inside the wall of the flow cell, and used as reference and counter electrodes, respectively. Two jigs bound a silicon chip and a microbial membrane with silicone rubber sheets as a spacer for the flow cell

about 10–20 min. A flow rate of 250 µl/min was chosen to make the response time about 25 min. The average values of the response from 5 min before replacement to the time of replacement, and from 25 min to 30 min after replacement, was calculated to reduce error in the estimation of the difference as a response value.

Immobilization Method for *T. cutaneum* on the Device. *Trichosporon cutaneum* IFO10466 (AJ4816) was cultured in GP medium at 37°C for 36 h. The cells were collected by centrifugation. Phosphate buffer and glycerol as cryoprotectant were added to the cells. A 2 ml aliquot of the suspension was gradually frozen at −25°C, and stored at −80°C. The cell stock was melted in a water bath at 37°C prior to use. The cell was washed three times with phosphate buffer using centrifugation. A double-sided adhesive tape having a 4 mm hole was put on an acetylcellulose membrane filter (DISMIC-25cs, pore size 0.8 µm, Advantec). Then 1 mg of cells was filtered through the membrane. Another acetylcellulose membrane was put on the membrane. The size of the membrane was adjusted so that it located on the silicon device. Because both sides of the microbial membrane were

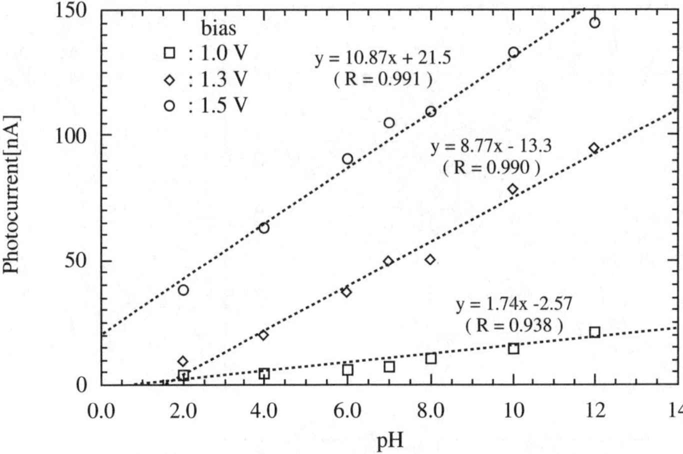

Fig. 4.25. Calibration curves at various fixed bias against various pH levels. The calibration curves were calculated from *CV* characteristics to varied pH solutions at a fixed bias, as described in the figure

made of acetylcellulose, the membrane was easy to handle and store. Storage of a microbial membrane in the 1 mM phosphate buffer (pH 7) containing 0.15 M of NaCl at 4°C kept its activity for more than 14 weeks.

Optimization of the Conditions of the System. Biosensors that measure pH shift have a large dependence on the buffer ability. A high buffer ability inhibits the pH shift at the sensor surface and reduces the sensitivity, but a low buffer ability makes the response unstable. Figure 4.26 shows the dependence of the response on the concentration of the buffer. A lower concentration of the buffer gave a larger response, and even the lowest concentration, 1.0 mM, showed enough stability.

Various concentrations of BOD solution (0–500 ppm) were applied to the system to obtain a calibration curve as shown in Fig. 4.27. A BOD standard solution containing glucose (150 mg/l) and glutamic acid (150 mg/l) was employed as a model wastewater, according to the Japanese Industrial Standard (JIS). The BOD response increased with increasing BOD concentration of the standard solution in a linear relationship between the decrease of the photocurrent (BOD response) and the BOD concentration from 0 to 100 ppm ($r = 0.988$).

Comparison with BOD_S and BOD_5. The SPV-based sensor also measured the BOD value for various kinds of organic substances. Table 4.2 summarizes the substrate specificity of the SPV method, compared with reported data of BOD_5 (five-day method) and BOD_S (sensor method) for the various compounds [61]. For the organic substances, including lactose, soluble starch,

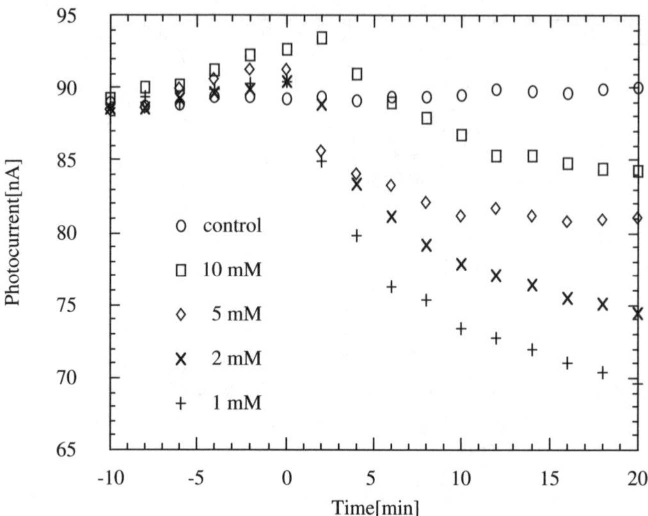

Fig. 4.26. The dependence of the response on the concentration of the buffer. After stabilization of the response with the flow of 1mM phosphate buffer (pH 7) containing 0.15 M NaCl at 250 µl/min, the carrier solution was changed to the sample solution containing 1 g/l glucose in various concentrations of phosphate buffer. The amount of *T. cutaneum* was 1.0 mg. The time when the solution changed is zero on the horizontal axis. The concentrations of phosphate in buffer were 1.0, 2.0, 5.0, and 10 mM, as described in the figure. The control sample was 1.0 mM of phosphate buffer without glucose in the sample

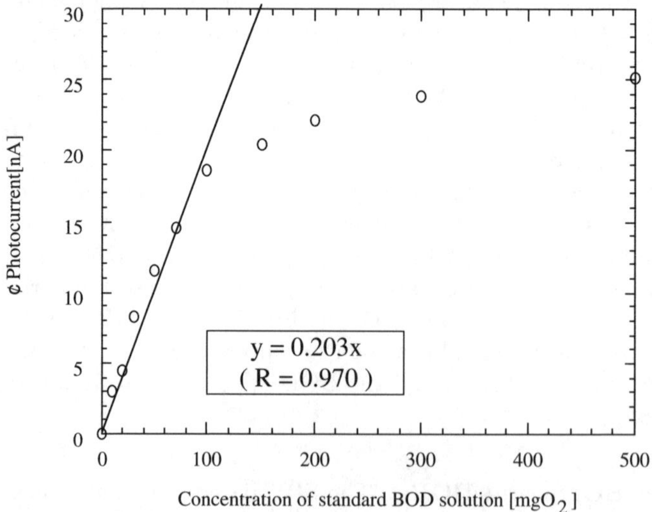

Fig. 4.27. The calibration curve of the SPV-based BOD sensor for the standard BOD solution. The phosphate buffer solution of 1 mM, pH 7, was pumped at 250 µl/min. The amount of immobilized cells was 1.0 mg

Table 4.2. A comparison of BOD values of various organic samples (mg O_2/mg substrate)

Substrate	SPV	Microbial electrode	Five-day method[a]
Glucose	0.66	0.72	0.50–0.78
Fructose	0.73	0.54	0.71
Sucrose	0.45	0.36	0.49–0.76
Lactose	0.04	0.06	0.45–0.72
Soluble starch	0.07	0.07	0.22–0.71
Glycine	0.36	0.45	0.52–0.55
Glutamic acid	0.40	0.70	0.64
Histidine	0.34	0.35	0.55
Acetic acid	0.39	1.77	0.34–0.88
Citric acid	0.18	0.72	0.63–0.88
Lactic acid	0.14	0.17	0.40
Ethanol	0.49	2.90	0.93–1.67
Glycerol	0.44	0.51	70.62–0.83

The SPV method was performed under the conditions of 250 µl/min, 1 mM phosphate buffer, pH 7, containing 0.15 M of NaCl, and 1.0 mg cell weight when wet. The substrate was 100 mg/l, pH 7, in the buffer.
[a] Calculated data from [61].

and lactic acid, the SPV-based sensor and the sensor based on the oxygen electrode show lower BOD values compared with the BOD_5. Because these two methods employ short-time metabolism of the same kind of microorganism, the compounds that are difficult to degrade would have smaller BOD values than BOD_5.

The SPV-based sensor, however, gives lower values for the same compounds, such as acetic acid, citric acid, and ethanol. The most likely explanation is that the system mainly responds to the acidic compounds produced by the microbe in its early metabolism stage. For example, the microbe metabolizes glucose to produce acidic compounds such as pyruvic acid, and lactic acid through glycolysis. Glycolysis is generally faster than the TCA cycle, and the major factor in the SPV response might be glycolysis in the short response time. There is no consumption of oxygen in glycolysis, and the production of only two molecules of NADH requires a small amount of oxygen in the following ATP production reaction. Compared to the TCA cycle, glycolysis makes only a small contribution to oxygen consumption in the oxygen electrode method, although it produces a certain amount of acidic compounds. Thus, a relatively small amount of response might be obtained for the compounds that are metabolized only in the TCA cycle. The mechanism suggests that the method can be used in anaerobic conditions. Microorganisms that metabolize certain materials specifically in anaerobic conditions would be applicable to the microbial sensor when combining it with SPV-based sensor.

Fundamentally, the SPV method measures not the "oxygen demand", or the consumption of oxygen, but the production of acidic compounds, while the other method measures the consumption of oxygen. The acidic compounds produced by the microbe are mainly carbon dioxide and carbonic acid. Oxidation of the substrate produces acidicity. Thus, the production of acidic compounds is closely related to the oxygen consumption. The differing principles can induce different substrate specificities, although they utilize the same microbe. Strictly speaking, BOD_{SPV} has different characteristics from BOD_S and BOD_5. There is also the difference between BOD_S and BOD_5. The five-day method utilizes any kind of microorganism, allowing metabolization for a long time, while the sensor method utilizes the designated microorganism, allowing metabolization for a short time. We have to pay close attention to compare the BOD values to each other when the sample contains compounds of differing specificities. Fortunately, environmental monitoring often treats samples from fixed sites, or the same kind of sample, and one can consider that difference prior to analysis.

Subsequently, the SPV-based BOD values of some wastewater samples were determined. The five-day BOD value of the BOD solution and various wastewater samples were determined by the Japanese Industrial Standard (JIS) method [44] as follows. An incubator flask containing a BOD reagent, the sample, and the microbe was cultured at 20°C for 5 days. After cultivation, a dissolved oxygen electrode measured the dissolved oxygen concentration.

The wastewater samples tested were treated and untreated wastewater collected from a septic tank at our institute and at a brewery. Each of the wastewaters was diluted appropriately with the same solution as the carrier solution prior to use. A good agreement between the SPV-based sensor and BOD_5 methods was obtained for the test sample. The correlation coefficients between BOD_{SPV} and BOD_5 were 0.979 and 0.976 for the sample from the institute and from the brewery, respectively. Although the sample from the brewery may contain redox compounds such as ascorbic acid, the sensitivity without an extra reference electrode is almost the same as that of the sample from the institute.

The results expand the applicability of the SPV method to the field of microbial sensors, although this research only treats BOD measurement. In principle, many microbial sensors measure the activity of immobilized microorganisms, and the method used in this research was less specific to BOD measurement. Thus, the method should be applicable to other microbial sensors simply through the replacement of the immobilized microorganism from *T. cutaneum* with another that is specific to a certain analyte and some optimization. In our recent research, the method has already been applied to process monitoring in a brewery [72].

4.5 Neural Network, Neural, or Brain Analyses: Measurements of Neural Activity and Their Application for Analyses of the Neural Network System in the Living Body

4.5.1 Ultra-Microglutamate Sensors for Brain Analyses

Chemical neurotransmitters play an important role in the brain, since they are the key link in communication between neurons. Glutamate is one of the neurotransmitters that carry nerve signals across the synapse from one neuron to another. Released by the presynaptic neuron, glutamate binds to specific receptors on the postsynaptic neuron. Interactions between glutamate and the glutamate receptor are involved in memory-storage phenomena in the hippocampus, known as long-term potentiation (LTP), and long-term depression (LTD) in the cerebellum [73]. Quantitative analyses of released glutamate are required to further investigate the biomolecular mechanisms of neuron function and the etiology of neural diseases.

Possible approaches to monitoring the glutamate concentration in such small environments are to use the microdialysis (MD) probe sampling system [74] and to develop an "ultra-"microglutamate sensor.

Monitoring of the glutamate concentration released from cultured nerve cells by using an MD probe and an enzyme sensor system was reported by Niwa et al. [75]. The sensor consisted of an MD probe fixed at the manipulator, a small-volume glutamate oxidase enzymatic reactor (0.75 mm i.d. and 2.5 cm long), and an electrochemical reactor in a thin-layer radial flow cell with an active volume of 70–340 nl. The overall efficiency of the glutamate detector with the sensor is 94%. They achieved a sensitivity of 24.3 nA/mM and a detection limit of 7.2 nM. By using this sensor system, they could monitor glutamate concentration changes at the submicromolar level caused by KCl stimulation of a single nerve cell and micromolar glutamate concentration increases caused by electrical stimulation of a brain slice. The combination of the MD probe and the enzyme sensors is useful for not only neural and brain analyses, but also for small-scale bioprocess monitoring. We have reported simultaneous monitoring of glucose and lactate in lactic acid fermentation by using integrated miniature enzyme electrodes and an MD probe sampling system [76].

Recent studies have demonstrated that the in vivo electrochemical sensor is a powerful tool for clinical and neurochemical monitoring [77]. A smaller electrode causes less damage to tissue during its insertion into the brain or nerve tissue. To examine such small environments, a very small electrode with a diameter of the order of micrometers is required. Since microbiosensors based on semiconductor fabrication technology are of the order of millimeters in size [78], novel ultra-microbiosensors should be developed for this purpose.

The carbon-fiber electrode is considered to be one of the most useful transducers for in vivo biosensors, because it can be designed for use as

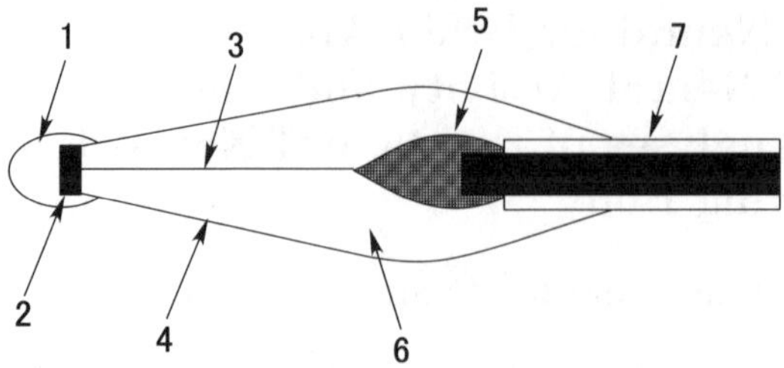

Fig. 4.28. A diagram of the platinized carbon-fiber disk electrode immobilized glutamate oxidase. *1:* Glutamate oxidase enzyme membrane, *2:* electroplated platinum, *3:* carbon fiber, *4:* glass capillary, *5:* silver paste, *6:* silicone resin, *7:* lead wire

an ultra-microelectrode with high strength and low cost. We have used a carbon-fiber electrode as a detector for glucose [79] and acetylcholine [80] determination, but the sensitivity of the carbon-fiber electrode to hydrogen peroxide is not sufficient for the construction of microbiosensors.

In the case of microbiosensors of the order of a micrometer in size, the following problems should be solved. First, since the amount of immobilized enzyme is very small, the electrode should be very active and sensitive. Secondly, special measurement techniques are required, because a small electrode is very easy to inactivate. Thirdly, the reference electrode (counter electrode) should be combined with a working electrode, especially for in vivo applications.

We have developed an ultra-microenzyme sensor by using a platinized carbon-fiber electrode, and it has been applied for the determination of glutamate [81].

The structure of the ultra-microenzyme sensor is shown in Fig. 4.28. A carbon fiber 15 mm long with a diameter of 7 μm was cemented to lead wire with silver paste and then inserted into a prepared glass capillary. Silicon resin was used to fix the electrode inside the glass capillary. After polishing the tip of the electrode, the surface of the carbon-fiber electrode was platinized electrochemically. Glucose oxidase, bovine serum albumin, and photocrosslinking polyvinyl alcohol (PVA-SbQ) were mixed well, the mixture was attached to the tip of the electrode, and it was left for 10 min under a fluorescent lamp. The electrode was dipped into phosphate buffer for 1 h, and then left for 3 min in a glutaraldehyde atmosphere.

Th prepared glutamate microsensor could be characterized as follows. The response was very stable and the response time was within 12 s. The calibration range of the glutamate microsensor was 0.002 to 1.2 mM. The standard deviation at 0.002 mM obtained from 60 measurements was 0.498 pA. Since the actual response current was 8.498 pA, the error for 0.002 mM glutamate measurement was 5.9%.

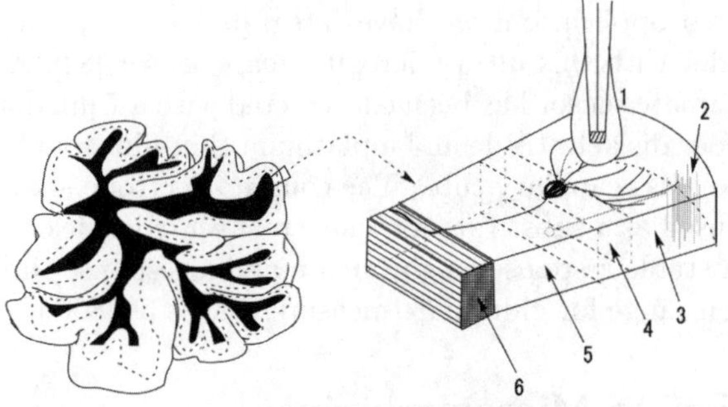

Fig. 4.29. The structure of cerebellar cortex tissues and the measurement point. *1:* Glutamate sensor, *2* parallel fiber, *3:* molecular layer, *4:* Purkinje cell layer, *5:* white matter

Then the glutamate microsensor was applied to in vitro measurement of neural glutamate. Figure 4.29 shows the structure of cerebellar cortex tissue and the measurement point. Cerebellar cortex tissue has four layers: the molecular layer, the Purkinje cell layer, the granule cell layer, and white matter. We inserted the glutamate microsensor onto the molecular layer where synapses occur between parallel fibers, climbing fibers, and Purkinje cells. The presynaptic membrane was depolarized to release neurotransmitters [82]. Then glutamate microsensor responses were recorded before and after potassium stimulation. A rapid increase in glutamate response to a peak concentration value was followed by a gradual decrease to the initial background level, after approximately 90 s. The decay in the appearance of the peak response is attributed to the approximately 20 s required to replace the perfusate solution with high potassium ion solution in the experimental flow chamber. Depolarization-induced release of glutamate, in either the presence or absence of calcium ions, was investigated using our glutamate microsensor. The perfusate solution was changed from normal Ringer solution to a high potassium ion solution containing 5 mM of calcium ions. No glutamate was detected using normal blank Ringer solution, which produced only a slight increase in background current. However, the difference between the depolarization-induced glutamate release under high and low calcium ion concentrations was significant in the cerebellar tissues. These results suggest that voltage-dependent calcium ion channels function prior to the release of glutamate from cerebellar synaptic junctions. To ascertain that the responses obtained by potassium stimulation are due to glutamate, we compared the responses of an enzyme-active sensor with an enzyme-denatured sensor, both without the polyvinyl pyridine membrane. In this experiment, only 5.7% of the sensor response was estimated to be from compounds other than glutamate. By using a calibration curve, the peak current obtained by the potassium stimulation was found to correspond to approximately 400 mM of released glutamate.

Furthermore, for in vivo application, we have integrated the reference electrode (counter electrode) with this ultra-microglutamate sensor [83]. An integrated ultra-microglutamate sensor has been constructed with a 7 µm diameter platinized carbon-fiber disk electrode and a platinum thin-film counter electrode fabricated on the glass capillary tube. The counter electrode shows good stability and can be used as a substitute for a silver–silver chloride electrode. The sensor shows a stable response to glutamate and a response time within 12 s. The calibration range for glutamate measurement is 50–800 µM.

4.6 The Application of Micromachining Techniques to Chemical Sensors, Biosensors, and Microanalysis Systems

4.6.1 Introduction

Considering the application to clinical analyses, the miniaturization of sensors, actuators, and systems is of great importance. In realizing them, semiconductor and micromachining technologies are becoming indispensable. Concurrent advantages accompanying the miniaturization are 1) reduced size, 2) small sample volume, 3) identical, highly uniform, and geometrically well-defined structures, and 4) ease of integration. Among chemical sensors and biosensors based on various detection principles, electrochemical sensors will benefit most from the rapidly advancing technology. In this section, recent advances in the application of microfabrication techniques to electrochemical sensors, biosensors, and micro systems will be discussed.

4.6.2 Basic Technologies

Basically, the technologies used to miniaturize chemical sensors and biosensors originate from semiconductor processes. These include pattern formation by photolithography, thin-film metallization, and chemical etching. In electrochemical sensors and systems, detecting electrodes are formed as patterns by photolithography. As a representative process, photoresist patterns are formed on a metal layer and exposed areas are selectively removed by wet or dry chemical etching (Fig. 4.30a). The metal layers can be deposited by vacuum-evaporation or sputtering. Lift-off is also used in forming patterns of noble metals such as platinum and iridium (Fig. 4.30b). After photoresist patterns are formed on a substrate, a metal layer is deposited. By removing the photoresist layer in an appropriate solvent such as acetone, a faithful reproduction of the patterns on a photomask is obtained. Screen-printing is a technique frequently used to form electrode patterns without expensive facilities. The patterns are literally printed by casting and squeezing an appropriate paste onto a substrate through a mask (Fig. 4.30c). In forming mercury microelectrode arrays or an Ag/AgCl electrode, electrodeposition has also been used supplementally.

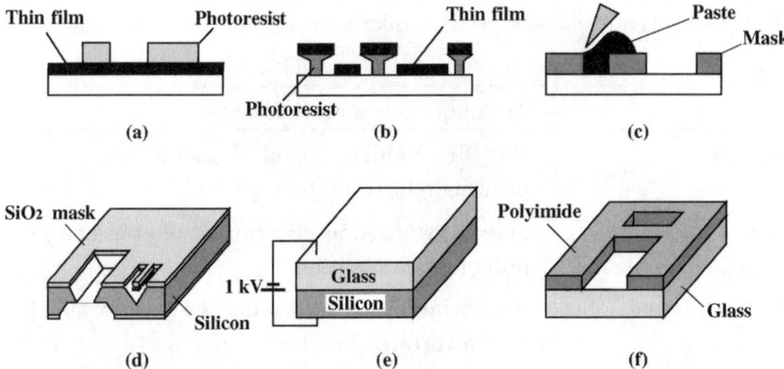

Fig. 4.30. Representative methods for the formation of electrode patterns (**a**–**c**) and microscopic structures (**d**–**f**) used for micro-electrochemical sensors. **a** etching, **b** lift-off, **c** screen-printing, **d** anisotropic etching of silicon, **e** field-assisted bonding, and **f** application of photosensitive polymer

In constructing some kinds of chemical sensors and systems, three-dimensional structures are required. A couple of examples include a container for an electrolyte solution and a flow channel. Anisotropic etching of silicon is promising and is used very often (Fig. 4.30d). The etching utilizes significant differences in etching rates among orientations of crystal lattices. Field-assisted bonding or anodic bonding is also useful in constructing three-dimensional structures (Fig. 4.30e). With a combination of a #7740 glass substrate and a silicon substrate, a negative high voltage is applied to the glass substrate against silicon. Hermetical sealing is instantly achieved even at around 250°C. Photosensitive polymers such as polyHEMA (poly (2-hydroxyethylmethacrylate)), polyimide, and thick-film photoresist such as SU-8^{TM} are useful in forming microscopic three-dimensional structures (Fig. 4.30f). The photosensitive polymers can be patterned following a process similar to that used to produce a photoresist. In addition to these methods, various promising techniques have been used, as summarized in Table 4.3.

4.6.3 Microsensors for Dissolved Gases and Electrolytes

In conducting the measurement of pO_2 and pCO_2 in whole blood, commonly used devises are the Clark-type oxygen electrode and the Severinghaus-type carbon dioxide electrode. The Clark-type oxygen electrode consists of a cathode and anode pair housed in a container. A hydrophobic gas-permeable membrane is placed over the sensitive area. The oxygen electrode measures the current generated by the electrochemical oxygen reduction on the cathode at an overpotential. The structure of the Severinghaus-type electrode is similar to that of the Clark-type oxygen electrode. A pH-glass electrode and a reference electrode are incorporated in a container. A local change in pH caused by the permeation of carbon dioxide through the gas-permeable

Table 4.3. Representative elements used in microsensors and systems, and techniques to form them

Element	Techniques or materials
Electrode patterns	Wet/dry etching, lift-off, masking, screen-printing, electroplating
Electrolyte layer	Screen-printing, application of photosensitive polymers, lamination
Gas-permeable membrane	Sacrificial layer (in forming the membrane over a through-hole)
Flow channel	Anisotropic/isotropic etching, anodic bonding, application of photosensitive polymers, application of a dicing saw
Microcontainer	Anisotropic etching, anodic bonding, application of photosensitive polymers
Liquid junction	Anisotropic etching, porous silicon, sacrificial layer, application of photosensitive polymers
Enzyme-immobilized layer	Lift-off, screen-printing, inkjet printing, electrochemical deposition, application of photosensitive polymers
Diffusion-limiting membrane	Application of photosensitive polymers, dry etching
Through-hole	Anisotropic etching, electrochemical drilling, sandblasting, ultrasonic drilling
Ion-sensitive membrane	Application of photosensitive polymers
Needle-type probe	Anisotropic etching, etch-stop

membrane is measured with the glass electrode. As a result, pCO_2 is related to the potential of the glass electrode against the reference electrode.

Techniques for the miniaturization of the Clark-type oxygen electrode have been established during the past two decades. Structures often seen in oxygen microelectrodes are shown in Fig. 4.31. The structures are very common and can also be seen in other types of electrochemical sensors. At an earlier stage of development, oxygen electrodes were developed whose detecting electrodes are exposed to a test solution (Fig. 4.31a). However, these were soon replaced with the Clark-type cell. In forming the gas-permeable membrane using a precursor solution, an electrolyte solution has been impregnated in a hydrogel such as polyHEMA [84–87]. A pattern of an electrolyte layer can also be obtained using a photocurable polymer. Similarly, the electrolyte layer can selectively be formed by injecting a precursor solution in a micropool formed with polymer such as polyimide [88,89], or in an anisotropi-

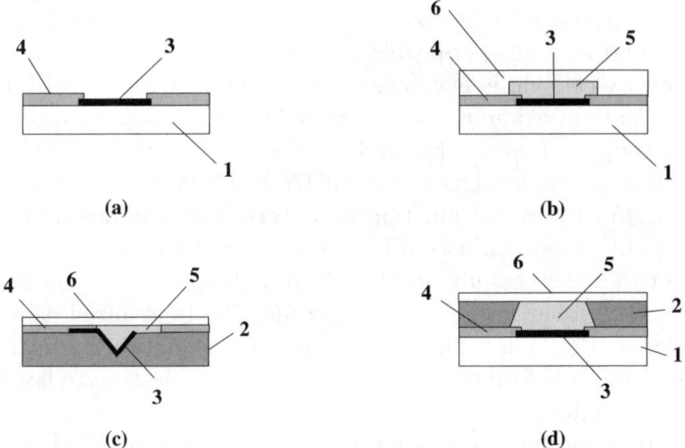

Fig. 4.31. Structures often seen in a micro-electrochemical sensors. **a** A working or indicator electrode exposed to a test solution. **b** A working or indicator electrode covered with a hydrophilic polymer layer containing an electrolyte solution and a hydrophobic membrane. **c** A working or indicator electrode formed in a silicon container covered with a hydrophobic membrane. **d** A working or indicator electrode formed on a planar substrate, to which a container substrate with a hydrophobic membrane is adhered. Liquid electrolyte solution is incorporated just before use. The hydrophobic membrane can be a gas-permeable membrane for the Clark-type oxygen electrode and the Severinghaus-type electrode or an ion-selective membrane for ion-selective electrodes. *1:* substrate 1 (e.g., glass), *2:* substrate 2 (e.g., silicon), *3:* working or indicator electrode, *4:* insulating layer, *5:* electrolyte layer, *6:* hydrophobic layer

cally etched groove on a silicon substrate [90]. Silicone rubber is usually used for the gas-permeable membrane, because it has a good oxygen- or carbon dioxide-permeability, and it shows excellent adhesion to the underlying layer or substrate. The electrolyte layer and the gas-permeable membrane can be formed following a completely dry process. The electrode thus formed can be activated by subjecting it to sterilization in an autoclave [91]. Otherwise, water can be incorporated by osmosis by just immersing the electrode in water at room temperature. Post-incorporation of an electrolyte solution, as is done in conventional macroscopic electrode, is also used. In using the method, an oxygen electrode chip is immersed in an electrolyte solution. The solution is evacuated in an appropriate vacuum chamber.

In carbon dioxide microelectrodes, the pH-glass electrode in its macroscopic counterpart is replaced with a pH-ISFET. Like the oxygen electrode, a commonly used structure is the one shown in Fig. 4.31b. The electrolyte solution in the hydrogel contains KCl and an appropriate concentration of bicarbonate. The gas-permeable membrane can be formed on an electrolyte layer formed as a pattern on the sensitive area [86–89]. In order to avoid the formation of the gas-permeable membrane and to simplify the structure, the internal elements can alternatively be housed in a silicone tube that functions both as a container and as a gas-permeable membrane [92,93]. If the impedance of the indicator electrode is low, signal amplification with an ISFET is not necessary. An iridium oxide indicator electrode can be used instead of the ISFET in the structure shown in Fig. 4.31d [94].

As a representative micro ion sensor, the ISFET has been the subject of hundreds of papers [95]. Its major function is to transduce the membrane impedance to make the sensor unsusceptible to noise. Some groups of ion-selective electrodes can be successfully miniaturized without transducing the signal using an ISFET. Planar ion-selective electrodes can be realized using the structure shown in Fig. 4.31b [96–99]. An internal reference electrode pattern (usually Ag/AgCl) is formed on a substrate and an electrolyte layer and the ion-selective membrane are subsequently formed.

In order to improve mechanical instability, the coated-"film" electrode has been used [100–102]. The coated-film electrode has a membrane–metal interface between which the interfacial potential is not well-defined, and osmotic transport of water to the interface causes noise and potential drift. In practice, if the electrode is calibrated just prior to use, satisfactory accuracy seems to be obtained [100]. In applying the microfabricated ion-selective electrodes to in vivo use, nonspecific protein adsorption and cell adhesion pose a problem. In order to minimize biofouling, physicochemical or biological modification has been tried, using materials such as polyHEMA [103].

Although miniaturization of the indicator electrode including the ISFET has advanced dramatically over the past two decades, the miniaturization and batch-fabrication of a reliable reference electrode has not advanced until very recently. In realizing a miniaturized reference electrode, three lines of approach have been proposed [104]. The first is the so-called pseudo-reference electrode which is usually composed of a metal pattern such as Ag/AgCl that is in direct contact with a sample solution. In potentiometric sensors, it serves as a potential standard in combination with an indicator microelectrode or used as the third electrode in an ISFET-REFET (Reference Field Effect Transistor) configuration [105,106]. Needless to say, the potential of this type of reference electrode varies depending upon the activity of the primary ions determining the electrode potential. Therefore, care must be taken to maintain the activity of the primary ion constant.

The second type is the solid-state reference electrode without a liquid junction. This includes the application of an ion-insensitive membrane often found in differential ISFETs [107]. A perm-selective membrane such as Nafion^{TM} has also been used to suppress the effusion of chloride ions contained in the internal electrolyte solution of a Ag/AgCl reference electrode [108]. However, the interface between the former membrane and the solutions might become polarized by blocking the exchange current. Also, the latter membrane might generate additional potential due to the perm-selectivity itself. In both cases, the situations in which a reliable reference electrode potential can be obtained would be limited.

The third approach is to miniaturize a conventional liquid-junction electrode, which would be most practical considering the accuracy required for potentimetric detection. In some miniature liquid-junction reference electrodes, a silicon microcontainer fabricated by anisotropic etching has been used. A liquid junction was formed on one portion of the container using a restraining material such as porous silicon [109,110], porous glass [111], and polyHEMA [110]. In relation to this, it is necessary to store KCl electrolyte as much as possible to achieve a long lifetime. With a screen-printed electrolyte layer containing KCl in a solid powdered form, a lifetime exceeding 100 h has been achieved [112]. It might be considered that a long lifetime can be achieved by restraining the effusion of internal KCl. However, this is closely related to the liquid-junction potential. If the effusion is restrained too much, it immediately results in a liquid-junction potential that exceeds 10 mV or more. There is a tradeoff between the lifetime and the liquid-junction potential.

4.6.4 Microfabricated Biosensors

Amperometric biosensors have been very actively studied, and blood glucose testing using planar-type glucose sensors has advanced dramatically [113]. Some models have already been commercialized and created a big market. In fabricating amperometric microbiosensors, the working, reference, and auxiliary electrode patterns or cathode and anode pair are integrated. An enzyme is immobilized on the working electrode area. Hydrogen peroxide produced by an enzymatic reaction is often measured with a platinum thin-film or thick-film working electrode.

In immobilizing an enzyme, a method very frequently used for microbiosensors is to immobilize the enzyme in bovine serum albumin with a bifunctional reagent such as glutaraldehyde [114]. In fabricating microbiosensors, the enzyme-immobilized membrane must be formed as a pattern on a necessary portion of the sensitive area of a transducer. Representative techniques for selective pattern formation are summarized in Table 4.4. As seen in the table, a couple of techniques can be combined.

Microbiosensors fabricated by just immobilizing an enzyme are sometimes so sensitive that the linear range of the calibration curves cannot cover the higher range of concentration required in real analyses. The linear range can be expanded using a diffusion-limiting membrane. Representative materials for the membrane include polyurethane, silicone, and polyHEMA.

Electrochemical biosensors whose working electrode is in direct contact with the sample solution are influenced by interfering compounds such as L-ascorbic acid, uric acid, and acetaminophen. In order to minimize the interference, a perm-selective membrane is often used. A couple of examples include a cation exchange polymer such as Nafion^{TM} and cellulose acetate. Another technique often seen is to use a pair of working electrodes with the

Table 4.4. Representative methods used to form patterns of an enzyme-immobilized membrane

Methods
Application of a photosensitive polymer (e.g., PVA-SbQ, polyHEMA) accompanied by spin-coating depending upon the case
Drop-on technique accompanied by spin-coating depending upon the case
Lift-off
Screen-printing accompanied by UV-polymerization depending upon the case
Electrochemical deposition including codeposition during electropolymerization
Physical adsorption on an electropolymerized membrane
Ink-jet printing
Injection of a precursor solution into a micropool
Spraying
Localized immobilization by covalent coupling on a derivatized metal surface

same dimensions and take a differential signal. An enzyme is immobilized on one of the two working electrodes and a deactivated enzyme, or nothing at all, is immobilized on the other. If working electrodes with a very close performance can be obtained, this method will be effective.

Miniature Clark-type oxygen electrodes have also been used as a transducer for microbiosensors. Although this type of sensor is affected by the variation of the dissolved oxygen concentration, it has an excellent selectivity owing to the hydrophobic gas-permeable membrane. As a group of representative potentiometric microsensors, numerous microbiosensors have been fabricated using the ISFETs [115,116]. Because the integration of ISFETs is easy, many integrated biosensors have been fabricated using the method of immobilization listed in Table 4.4. There are few conductometric microbiosensors, but they have been fabricated using interdigitated array microelectrodes [117–119]. Although this mode of detection lacks specificity, the principle is so general that it can be applied to any enzymatic reactions that produce charged species.

4.6.5 The Integration of Sensors and Micro-Electrochemical Systems

A miniaturized system that can perform analyses at or near the site of use would be of great assistance in therapeutic decision-making. In the 1990s, an innovating technology, the μTAS (Micro Total Analysis System), was actively studied. In a completely realized form, the systems will have functions such as sampling, sample pretreatment, separation, and detection along with a high degree of automation and a high level of methodology of analytical chemistry [120,121]. The advantages in using such a miniaturized system are fast response, small sample volume, high throughput analysis, reduction in reagent consumption, reduction in waste disposal, and flexibility in choosing the location of the analysis. Basically, the micro-flow cells can be realized by bonding two substrates, such as glass and silicon, onto a printed circuit board, on which a flow channel along with detecting electrode patterns has been formed [122]. Flow cells can easily be formed with various photosensitive polymers such as a dry film photoresist [123]. A flow cell made from polysiloxane can realize a leakproof cell by placing another substrate on the flow cell and pushing it [87].

The measurement of blood gases, electrolytes, and various metabolites is clinically very important. Several sensor chips have been developed to measure pO_2, pCO_2, and pH [86,88] and six analytes including gases and electrolytes [89]. Some integrated sensor chips or separate ISFETs have been incorporated in a flow cell [87,123–126] or a system [127]. Microbiosensors have also been incorporated in miniaturized systems to measure glucose [128] or glucose and lactate [122]. On-line microsystems have been developed to monitor the concentration of neurotransmitters [129]. With this mode of detection, a lower detection limit can be achieved compared with microelectrode-based sensors.

Although all necessary functions in an analysis system should ideally be integrated on a single chip, this seems to be very difficult considering the present level of technology. Furthermore, it will be time-consuming to integrate different types of components fabricated by different processes. Also, it will not be cost-effective, because the complicated fabrication process decreases the production yield. A good alternative is to decompose the entire system into several modules with different functions [121,130–132].

In realizing a μTAS, an important but yet advanced technology is that of sampling and pumping mechanisms. With the present level of technology, an advanced microsystem uses a simple micromachined pump connected to a flow cell comprised of ISFETs [126,130,132]. Otherwise, macroscopic pumps, valves, and external readout electronics are used to complete the whole system [133]. At present, a few microsystems use a microdialysis tube connected to a syringe pump [134–136]. An advantage of using this technique is that contaminants that may foul the sensors are filtered, which helps to elongate the lifetime of the enzyme electrodes. In these cases, sampling was performed with the help of a syringe pump, and integration of the pump will become a critical issue in the next decade. A couple of trials are beginning to appear to realize a microsampling mechanism. A volume phase transition of poly(N-isopropylacrylamide) gel has been used to fabricate a microsampling mechanism. The mechanism was used in a microsystem for glucose, equipped with a microneedle of 50 μm in inner diameter [137]. The so-called "intelligent mosquito" is promising in realizing a system of analysis that is not highly invasive.

4.7 Microneurography: Measurements and Stimulation of a Single Peripheral Nerve Fiber

4.7.1 Introduction

One of the most important key techniques used in our research (the generation of artificial sensations by microstimulation [138] and the control of an artificial heart system using sympathetic signals [139]) is microneurography or microstimulation. This is probably the only technique that allows recording of the activities of a single sensory nerve fiber or stimulation of a single sensory nerve fiber of a human subject who is awake.

In this chapter, we will briefly explain the techniques for microneurography and microstimulation, which basically use the same procedure.

4.7.2 The History of Microneurography

Microneurography was developed by Hagbarth and Vallbo [140] in Uppsala, Sweden, in the late 1960s. The technique makes use of a very thin tungsten microelectrode that is inserted into a peripheral nerve percutaneously to measure the signals of the nerve fibers attached to the tip of the electrode. Microstimulation of the nerve fiber, but not recording of the signal of the nerve fiber, can be performed with the same microelectrode.

Initially, this technique was used to record the activities of Ia afferent nerve fibers from a muscle spindle, which helped demonstrate the existence of α–γ linkage in voluntary muscle contraction by showing that the activities of the Ia fibers were altered in accordance with the degree of extension of the involved muscle [141]. Next, the technique was applied to record the activities of sensory nerve fibers and those of sympathetic nerve fibers (e.g., Torebjörk and Wallin [142]). This allowed analyses of the response of a single sensory nerve fiber to stimulation of the involved mechanoreceptor unit and contributed to classify the kinds of mechanoreceptors according to the speed of adaptation or the pattern of the response to the given stimulation [143].

With respect to sympathetic nerve activities, the technique allowed for the classification of skin (activities of the vasomotor and sudomotor functions) and muscle sympathetic nerve activity (activities that control the tonus of the vessels in the involved muscle), and a vast amount of research has focused on these two sympathetic activities [142,144].

There have been far fewer studies on "microstimulation," which uses inserted microelectrodes for stimulating nerve fibers [145,146], than on recording nerve activities.

Thus, microneurography has achieved excellent results in the field of basic medical science. In the clinical field as well, many attempts have been made to elucidate the pathophysiology of neurological diseases by recording the activities of the involved nerves, although the technique has not yet been used as a method of medical treatment.

4.7.3 The Technique of Microneurography

As mentioned above, microneurography involves percutaneous and direct insertion of a very thin tungsten microelectrode (needle type) into a peripheral nerve to measure the signal of the nerve fiber attached to the tip of the electrode and to make it possible to stimulate nerve fibers that are attached to the tip of the electrode (microstimulation).

With respect to the tungsten microelectrode, the diameters of the shaft and the tip are around 120 µm and 10 µm, respectively, and when the tip of the electrode (where the electrical insulation by epoxy resin is removed) is properly attached to only one nerve fiber, it is possible to record signals from a single nerve fiber (Fig. 4.32) and to stimulate a single nerve fiber electrically.

The procedure of microneurography that we are currently adopting is based on that of the Nagoya University group under the direction of Professor T. Mano. This research group has played a leading part in the introduction and establishment of this technique in Japan.

The procedure is described below.

Extremely fine tungsten microelectrodes (needle electrodes) are used as working electrodes. One is inserted percutaneously into a peripheral nerve. A reference electrode is attached to the surface of the skin with paste, a few centimeters away from the point of insertion of the working microelectrode.

Although percutaneous electrical stimulation is usually used to search for the direction of the objective nerve, we use ultrasonography to confirm the position of the tip of the needle electrode and that of the target nerve trunk. Using an ultrasonic probe with a comparatively high frequency (we are currently using an annular alley probe with a frequency of 7.5 MHz), normal peripheral nerves at the extremities (in the transverse section) are displayed in a circular area circumscribed by a high-intensity linear structure, and this circular area contains high-intensity microtubular structures (Fig. 4.33). By watching the spatial relationship between the position of the tip of the needle electrode and the objective peripheral nerve, it is easy to make the top of the microelectrode reach the surface of the target nerve. Furthermore, it is possible to set the tip of the needle electrode at any desired position in the nerve trunk (Fig. 4.34). It is easy to judge when the needle electrode is actually inserted into the peripheral nerve, because a subject usually reports a particular sensation when the electrode is inserted.

The measured nerve activities are amplified through a preamplifier, after which they are passed through a band-pass filter (300–5000 Hz), displayed on an oscilloscope, and finally amplified around 50 000 times. At the same time, the nerve signals are connected to a loudspeaker so that they can be audibly monitored (Fig. 4.35).

Electrical stimulation of the nerve fiber can also be performed through the same microelectrodes by changing the circuit from a measurement mode to a stimulation mode using a switch box.

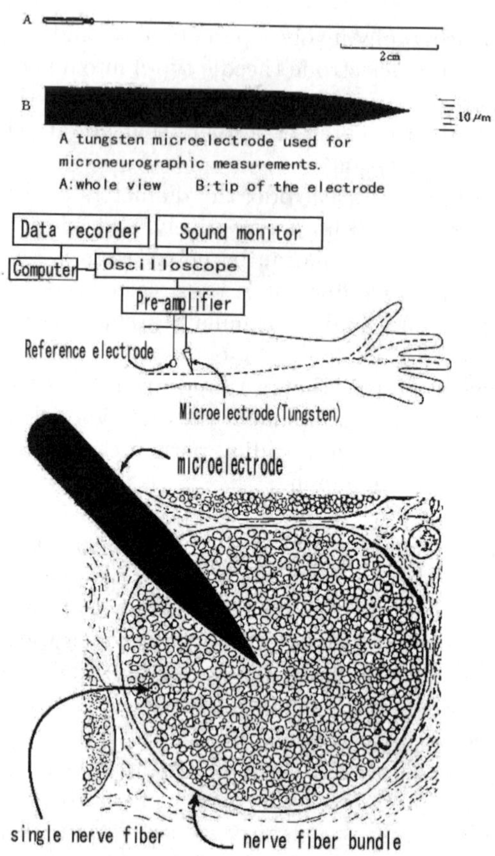

Fig. 4.32. A schematic figure showing the situation when the microelectrode is inserted into a nerve; when the tip of the electrode is attached properly to only one nerve fiber, it is possible to record nerve signals from the single nerve fiber (modified figure of T. Mano)

Although institutions initially manufactured their own tungsten microelectrodes, various types are now commercially available. For example, the tungsten microelectrodes that we are currently using in our experiment have the following properties: 1) the diameters of the shaft, tip, and the top of the tip are approximately 125 μm, 10 μm, and 1 μm, respectively; 2) they are (electrically) insulated with epoxy resin, except for the top of the tip; and 3) the impedance is adjusted between 2 and 12 MΩ.

Although the non-insulated part at the tip of the electrode is larger than the diameter of the sensory nerve fibers (large myelinated nerve fibers), it is

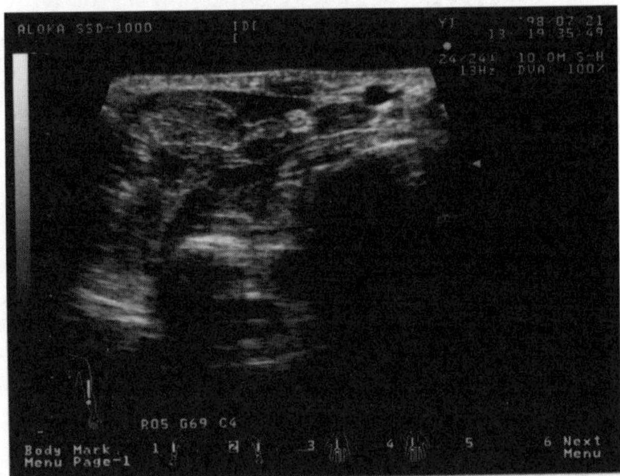

Fig. 4.33. Ultrasonic echography is used to confirm the position of the peripheral nerves (echo-guided microneurography); the photograph shows a median nerve in a wrist observed by ultrasonic echography

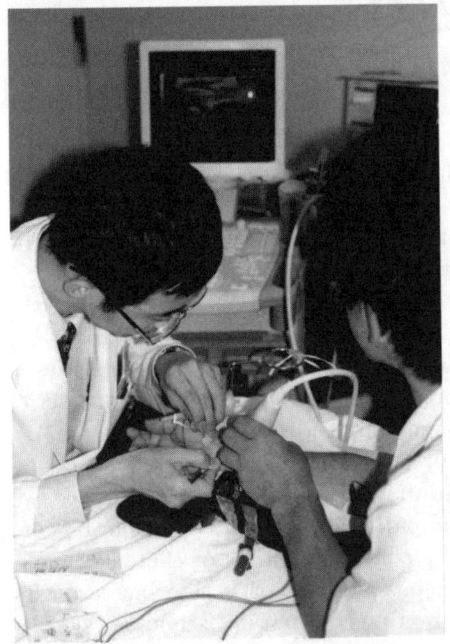

Fig. 4.34. The manner in which microneurography is performed. The microelectrode that is inserted percutaneously into the median nerve of the subject can be observed in an ultrasonic echography probe held by the subject himself in this case

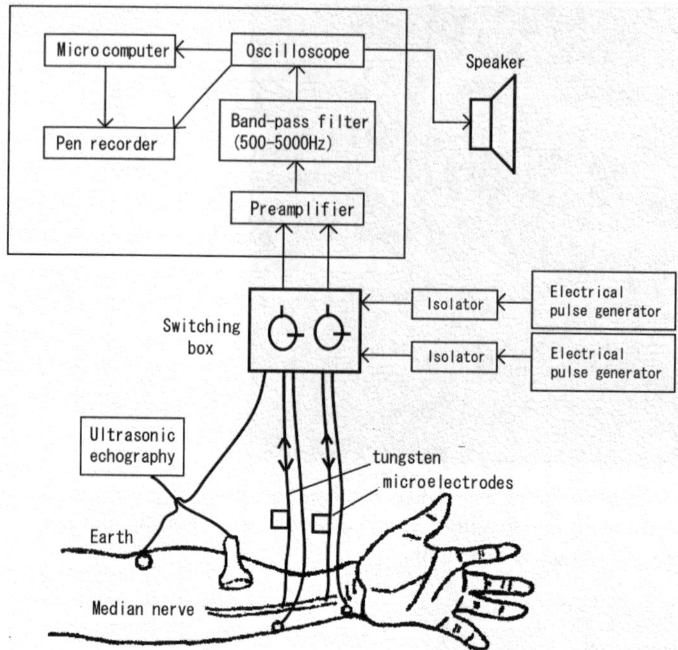

Fig. 4.35. The experimental arrangement and the procedure of microneurography

possible to record the activities of the single sensory nerve fiber using this microelectrode.

Whether the acquired activities originate from sensory nerve fibers can be easily confirmed by checking whether the signal increases when mechanical stimuli (rubbing, touching, or pressing) are applied to a specific area of the skin, which is called a "receptive field" (the area where the sensory receptors are innervated by the involved nerve fibers) (Fig. 4.36). When the activities of a single sensory nerve fiber are recorded, the nerve signal consists of a series of pulses, each having basically the same amplitude and shape (Fig. 4.37).

Sympathetic nerve activity can be recorded as a series of bursts (a combined record of the activities of several unmyelinated fibers). Whether or not the recorded signals originate from sympathetic nerve activities can be confirmed by checking to see: 1) that they consist of bursts of efferent signals, generated simultaneously with spontaneous pulses; 2) that they increase with almost the same latency as mental stress and/or sensory stimulus (such as sound and pain); and 3) that they increase as a result of deep breathing [144]. This sympathetic nerve activity is usually displayed after rectifying and integrating the full wave (Fig. 4.38).

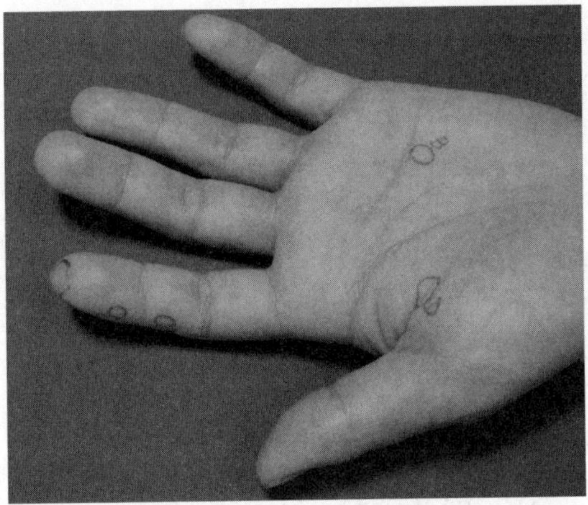

Fig. 4.36. Each area which is encircled by a black line shows a receptive field (the area where the sensory receptors are innervated by the involved nerve fibers)

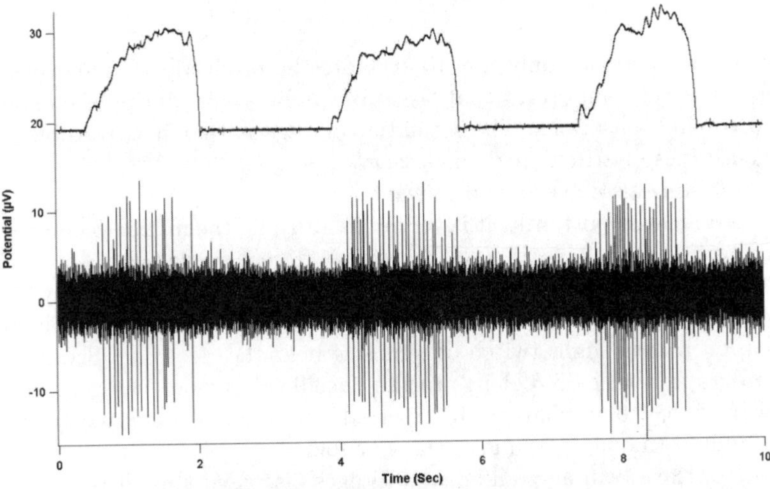

Fig. 4.37. The pressure that is applied to the skin (upper wave) and the response (signal) of the nerve fiber from a single mechanoreceptor unit (lower wave)

4.7.4 The Advantages and Disadvantages of Microneurography

The advantages and disadvantages of microneurography are listed below. Microneurography has two major advantages that cannot be achieved by other methods.

The first advantage is that the technique is not very invasive, and experiments can be performed on subjects who are awake. Sensations are very subjective and, at the moment, we have no good way to evaluate the evoked somatic sensations objectively and quantitatively. Currently, since the sub-

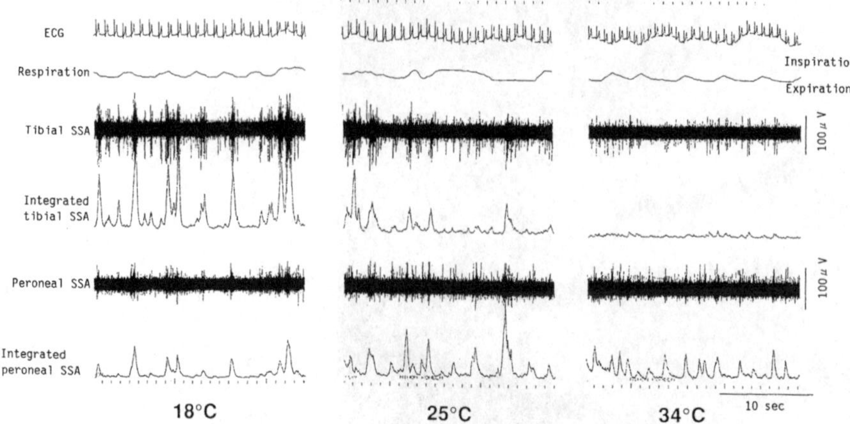

Fig. 4.38. Skin sympathetic nerve activity (SSNA). Sympathetic nerve activities are usually displayed after rectifying and integrating the full wave (quoted from [144]

jects' descriptions are the only way to evaluate the properties of the evoked sensations, it is very important that the subjects be awake during the experiments involving sensations. The second advantage is that it is possible to measure activities or stimulate a single nerve fiber.

This technique also has some disadvantages.

First, and most importantly, it is very difficult to fix the microelectrode at a certain position for a long time so that signals from a single nerve fiber can be recorded under identical conditions. Because the electrode is fixed by the resistance between it and the surrounding tissues, the electrode can easily be dislocated, even by a slight twitch of a muscle near the electrode. Secondly, microneurography basically belongs to the "needle-electrode" category, and the number of electrodes that can be inserted into a nerve simultaneously is basically limited to one or, at most, three or four.

Because of the advantages and disadvantages discussed here, it is obvious that it is difficult to use microneurography as an electrode system for clinical purposes. However, the method is very useful and nearly indispensable for analyzing the relationship between mechanical stimuli, which are given to the mechanoreceptors, and the neural signals evoked by the stimuli, or the relationship between the properties of the electrical stimulation that is given to the nerve fiber and those of the evoked sensations.

4.7.5 Summary

For the development of the next generation of neural interface devices, which will allow bilateral communication between each nerve fiber in the living body and the corresponding electrical (or optical) signal lines from the external devices, two key technical factors will need to be achieved.

First, it will be necessary to develop a multi-channel electrode that can connect each of the sensory nerve fibers contained in the concerned peripheral nerve to an electrical signal line coming from a corresponding sensor located in an artificial hand system. This is explained in a separate section.

Secondly, it will be necessary to develop an algorithm that will permit the modulation of the output signals that are detected by the sensors located on the surface of the robot hand to the signal (sequence of electrical pulses) for electrical stimulation to evoke the desired somatic sensation in the subject.

Microneurography is assumed to be one of the most important and valuable key technologies for dealing with the development of an algorithm.

Microneurography is soon expected to assume a large and indispensable role in "bionic medicine", which involves the connection of external devices to the human body through the nervous system [138,139,147].

4.8 A Microelectrode for the Neural Interface

4.8.1 Introduction

Recently, the direct neural interface between the human nervous system and external equipment has been recognized as a key technology to realize the next generation of prosthetic applications, such as artificial vision and neural control of prosthetic devices [148–150]. This technology is also very important as a tool of basic neuroscience research [151,152].

In order to realize these neural interface applications, two essential problems have to be solved:

1. How to establish intimate contact with the nervous system.
2. How to translate nerve signals.

In other words, the first problem is the development of nerve electrode. Although other techniques, such as magnetic methods, can be used, electrical methods are the most realistic way, at least in the near future. In terms of establishing an intimate and stable interface with the nervous system for the long term, however, the present nerve electrode technology is insufficient in many ways.

The second problem refers to analysis of the nerve signal, or analysis of the coding and decoding rule of nerve signals. For example, if we want to understand the meaning of the command signal given by the circulatory center to the natural heart, we have to know the method of interpreting it. If we want to make patients feel some touch sensation, we have to know what kind of signal to input into their sensory nerve fiber.

As the wide and important application of nerve electrodes has been recognized, and the recent rapid development of silicon process technology has been encouraging the study, a lot of groups around the world are eagerly untertaking research in this area.

In this section, we review recent developments in nerve electrode technology, including the work of our research group.

4.8.2 The Nerve Electrode (Handmade)

An Electrode for a Nerve Bundle. First, nerve electrodes for compound signals are shown:

- the wire electrode [153]
- the cuff electrode [154]
- the collagen electrode [155]

All of these can be used chronically, and Fig. 4.39 shows the schematics of the wire electrode and the cuff electrode.

Wire electrodes made from various metals, such as SS, Pt–Ir, and silver, have been used for long-term measurements. Their impedance is determined by the dimensions and the surface status of the exposed area (for example, whether or not platinum black is deposited). By adapting their impedance, they can be used for measurement of single nerve fibers or a small number of fibers. A wire electrode and nerve bundle are embedded in two-component silicone rubber for fixation [153].

Also, cuff electrodes are widely used for recording from and stimulating peripheral nerves. They fix a nerve bundle by covering it with their silicone tubes. The two wire electrodes arranged inside the tube work as electrodes.

Collagen electrodes were developed by Ninomiya and co-workers, for circulation physiological research. The nerve signal is measured through the skin electrode, which consists of collagen fiber. Stable long-time measurements can be realized, because the nerve bundle and the metal electrode do not contact each other directly [155].

In general, the influence of the connective tissue or the exudate from the surrounding tissue limits the recording time of these kinds of electrodes for nerve bundles.

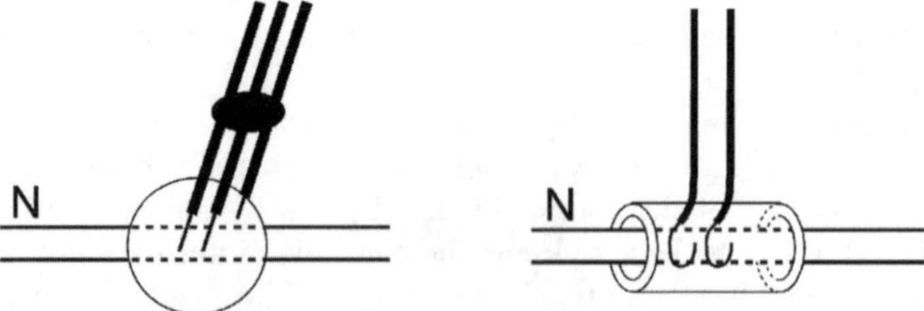

Fig. 4.39. **a** A wire electrode (embedded in two-component silicone rubber for fixation). **b** A cuff electrode (fixed to the nerve bundle by a silicone tube). *N* : Nerve bundle

The Intrafascicular Electrode. Lefurge and co-workers [156] reported their measurement of signals from the radial nerve of cats by intrafascicular electrodes for six months. The intrafascicular electrode is a type of wire electrode. Its recording site is created by removing some insulation, by means of heat, from the wires at some distance from one end, and then platinum black is deposited in the uninsulated region. A tungsten needle is used to insert the electrode into the nerve fascicle (Fig. 4.40).

The Tungsten Microelectrode. In the method of microneurography [157,158], insulated tungsten microelectrodes are used. The most typical type of electrodes is 120–200 micro meters in diameter, and only the tip of the electrode is uninsulated. They can be inserted percutaneously into the median nerve of awake human subjects, because the shaft is sufficiently rigid. Thus they are capable of recording from and stimulating a single nerve fiber. We use this type of electrode to stimulate the sensory nerve fiber that comes from a cutaneous receptor and generate an artificial touch sensation [159]. However, this type of electrode is only good for acute experiments, and is difficult to use in multi-channel interfaces.

A Summary of the Handmade Electrodes. The handmade electrodes mentioned above are still used as fundamental tools in a lot of basic research areas. There are also some applications in which compound signals from the nerve bundle are useful. Moreover, if the number of neurons in the compound signal could be made small enough by selecting the implant position, useful information could be extracted by suitable signal-processing methods. On the other hand, the tungsten electrode has sufficient properties to record single unit activity, although it is limited to acute experiments. However, these handmade electrodes have limitations in the feasibility of multi-channel chronic interfacing, which is necessary to realize the applications mentioned

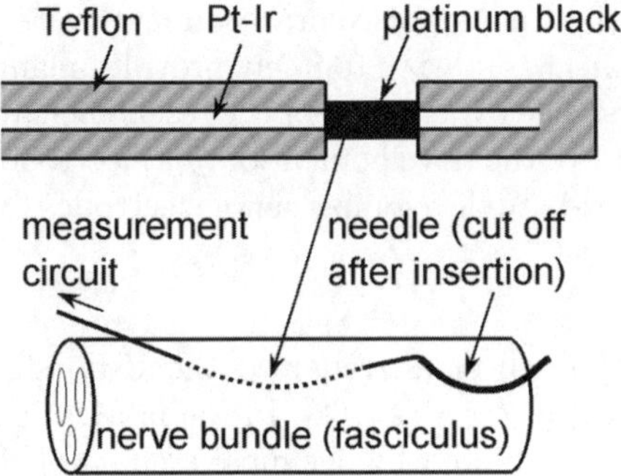

Fig. 4.40. The intrafascicular electrode. A needle is used to insert the electrode into the nerve fascicle

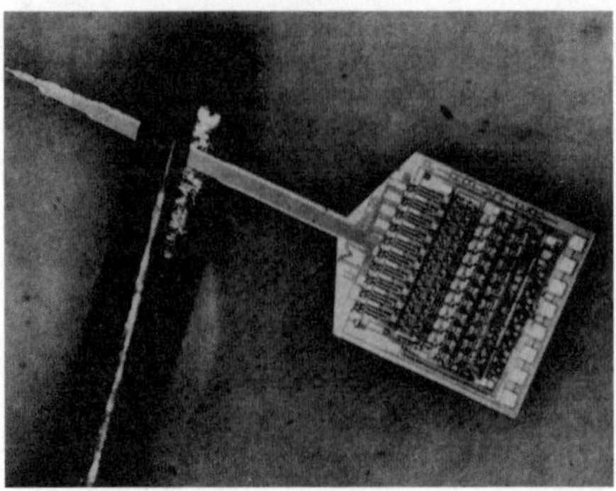

Fig. 4.41. A multi-channel multiplexed intracortical microprobe reported by Wise and Najafi [161] (with permission: © 1991 American Association for the Advancement of Science)

above, such as the control of prosthetic devices and the display of artificial sensations.

In the next section, we review new microelectrode technologies which are encouraged by the development of silicon microtechnology. In these technologies, the recording method is basically extracellular recording, as is the case of the tungsten electrode.

4.8.3 A Microelectrode for the Neural Interface

The Probe Electrode. Wise has reported many types of microelectrodes (probe electrodes) which have multiple recording sites [160]. They have also developed electrodes that have buffer amplifiers on their carrier shank (Fig. 4.41). Their electrode is designed mainly for recording from cortical neurons. But successful measurement is not accomplished easily. One of the reasons may be that this electrode has its recording sites on the shank, so it can record only from the neurons that are damaged by the passage of the electrode [150]. Recently, CNCT (the Center for Neural Communication Technology at the University of Michigan) has started to freely provides many kinds of probe electrodes for researchers. The experiences of the researchers in implantation and recording should improve the development of this electrode. Nowadays, many groups around the world are developing nerve electrodes by using a silicon microprocess.

The Needle Array Electrode. Campbell et al. reported the Utah silicon needle array, which has 100 (10×10) electrodes [162]. In the fabrication process, they used a diamond dicing saw followed by chemical etching. The array was insulated with silicon nitride. Some types are currently commercially available as a recording tool from cortical neurons for neuroscience research. Figure 4.42 shows the typical type of electrode, which has a 100 µm

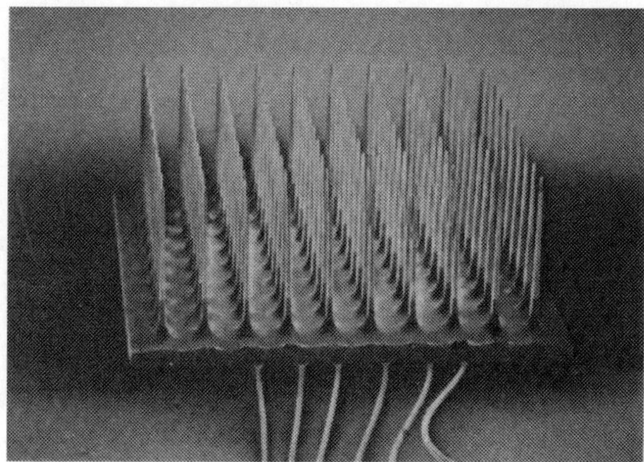

Fig. 4.42. The Utah silicon needle array. The center-to-center spacing between electrodes is 400 μm for scale [162] (with kind permission from IEEE: © 1991 IEEE)

shaft diameter, a 1.0 or 1.5 mm needle length, and the distance between the needles is 400 μm. The size of the whole electrode is 4 mm × 4 mm. Because of the large number of electrode needles, the whole friction is too large, so they have developed a special pneumatic inserter.

Recently, the same group has reported a new shape of needle array in which the length of the needle changes gradually. They call it the Utah Slant Array (U.S.A.) [163]. It was able to interface with many individual sensory and motor nerve fibers in peripheral nerves.

The Regeneration Electrode. The principle of the regeneration microelectrode is as follows. Peripheral nerves of vertebrates will regenerate after they are severed. In this process, the axons in the distal portion will degenerate, but the axons will regenerate again from the proximal portion and will reach the distal end of the severed nerve bundle. Therefore, if we implant a device consisting of many microelectrode holes between the severed stumps, axons can regenerate through the holes, and the action potential of the regenerating axons can be measured by the electrode. This regeneration electrode has many advantages theoretically. It is possible to:

- achieve intimate stable contact due to good electrical and physical compatibility between the electrode and the nerve fibers
- measure signals from only one or a very small number of nerve fibers, with a proper choice of the hole diameter
- achieve a great number of channels

Regeneration electrodes require transection of a nerve and subsequent successful regeneration. These restrictions may limit the range of application

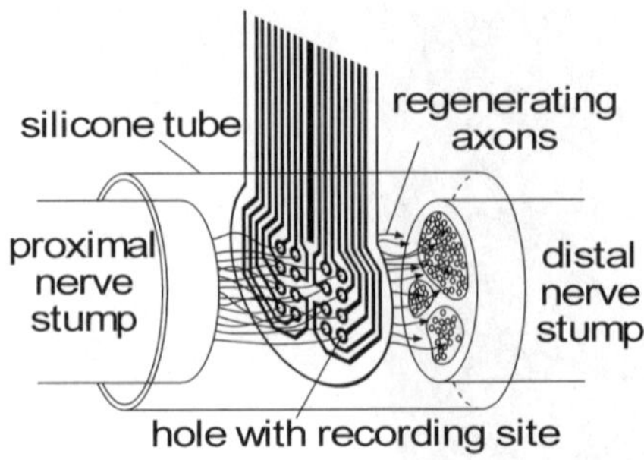

Fig. 4.43. A schematic of the regeneration electrode

to prosthetic devices, such as the control of artificial limbs for amputees or the control of an artificial heart by information from the autonomic nerve.

The development of the regeneration electrode has been undertaken using many kinds of materials since the 1960s. After Matsuo reported their fabrication on a silicon substrate, many kinds of silicon regeneration electrode have been proposed. Kovacs reported a micromachined regeneration electrode [164]. Most of the recent successful reports of measurements of nerve signals by regeneration electrodes have been based on silicon substrates. However, the weakness of the substrate becomes a problem when the implantation procedure is put to real use.

We have designed and fabricated a flexible regeneration microelectrode on a polyimide substrate using ordinary Flexible Printed Circuit (FPC) technology [165]. Polyimide has a good biocompatibility and is easily workable by an excimer laser, but the greatest advantage of polyimide is its flexibility, which aids physical compatibility during long-term implantation.

The schematic of the flexible regeneration microelectrode that we fabricated is shown in Fig. 4.43. The electrode was composed of three layers. A metal electrode pattern layer was sandwiched between two polyimide layers. Silicone tubes fixed the nerve stumps to the electrode. The other terminal of the electrode was connected to the connector during signal recording.

We implanted the electrode between the ends of the surgically severed sciatic nerve of rats under pentobarbital anesthesia. We cut the nerve bundle, inserted the nerve ends into each of the silicon tubes and fixed them with one 10-0 nylon suture at each end. It was possible to record the spontaneous action potential from the regenerating nerves.

At present, we are improving the electrode. We are using Pt as the metal layer and the polyimide layer is thinner (approximately 8 µm).

Summary of Microelectrodes. To summarize the microelectrode technology mentioned above, it is true that the technology has been developed remarkably, especially with regard to the number of the channels (electrodes) and the size of the whole electrodes, as compared with the handmade electrodes. However, in the present technology, the highest number of electrodes is around 100, which is rather insufficient for an application such as artificial sensation. (The future increase of the number of electrodes may cause wiring problems, which should be solved by a built-in signal processing unit such as a multiplexer.) The distance between recording sites is too wide to use as a tool for analyzing the activities of multiple neurons. Furthermore, the solid structure of these electrodes is invasive and causes easy slipping out of an adequate recording position. Thus the development of an ideal electrode needs another breakthrough in the technology, especially with regard to bio-compatibility and flexibility, as well as further micromachining.

4.8.4 Conclusion

In this section, we have reviewed the technology of electrodes for interfacing with the nervous system, along with promising applications in the field of prosthetic devices. It is true that the innovation of micromachining technology has brought about considerable development of the nerve electrode, but another breakthrough is necessary to realize an ideal electrode capable of a one-to-one interface between each electrode and nerve fiber. The key concepts for this breakthrough are bio-compatibility and flexibility, as well as further micromachining. These technologies will be developed in cooperation with other fields of technology.

4.9 Regulation: On-Line Measurements of Humoral and Neural Information from the Living Body and Their Application for the Control of Artificial Organs and Limbs

4.9.1 The Control of Artificial Hearts Using Humoral Information

On-Line Measurement of Humoral Factors by Electrochemical Methods. In addition to the autonomic nervous system, the humoral system also plays an important role in the control of the human circulatory system. Although artificial heart systems have now become widely used, the control algorithm and control systems for artificial hearts have not been sufficiently studied. So far, several studies have been performed to control the driving condition of artificial heart systems on the basis of the physical demands of the subjects or experimental animals [166–169]. However, most studies have attempted to use hemodynamic parameters as feedback parameters, while humoral factors, which play an important role in the control of the actual circulatory system of the living body, have never been used, even though they could provide very useful information for the control of artificial heart systems.

The main reason for not using humoral factors is the difficulty (or impossibility) of performing real-time measurement of their concentrations in the blood or in tissues.

One method that would permit real-time measurement of the humoral factors is measurements based on the electrochemical measurement technique [170,171]; alternatively, a sensor such as the ion-sensitive field effect transistor (ISFET [172]) or the enzyme field effect transistor (EnFET [173]) could be used. Both of these are basically electrochemical measurements.

In this section, we would like to introduce our attempt to develop a system that allows real-time sensing of catecholamine concentration, which is one of the most important humoral factors involved in the control of the circulatory system of the living body, as well as providing feedback on the factors involved in the control of an artificial heart system.

As mentioned above, we adopted an electrochemical technique in this study to measure catecholamine concentration. The main reason is that this technique permits real-time measurement of the electrolyte concentration in solution; in addition, the technique has already been used for the measurement of catecholamines in the brain.

Figure 4.44 shows the principle underlying the method of electrochemical measurement. If only electrode A (or B) is dissolved in the solution, the current increases by H_A (H_B) due to the oxidation of electrolyte A (B) when the applied voltage is greater than ε_A (ε_B) (the oxidation potential for electrolyte A (B)). Similarly, if both A and B are dissolved in the solution, the electric current increases by H_A and ($H_A + H_B$) when the applied voltage is greater than ε_A and ε_B, respectively. Thus, the increase in electric current due to the oxidation of electrolyte B is equal to the difference between the electric currents when the electric potentials are ε_A and ε_B.

Figure 4.45 shows the experimental setup of our experiment for measuring catecholamine concentrations and providing subsequent feedback to the artificial heart driving system. The sensing system is composed of a working electrode consisting of a carbon fiber, an Ag–AgCl reference electrode, and a potentiostat. The electrodes were immersed in phosphate-buffered solution or goat plasma, and the catecholamine concentration of the solution (or plasma) was then varied. Changes in the electric current resulting from the amperometric measurement of the catecholamines were determined by the system, and the information was transferred to a personal computer.

The operating parameters of the pneumatically driven artificial heart system were altered in accordance with the algorithm for changes in the catecholamine concentration (the catecholamine concentration and the alteration of the driving parameters were measured every 2 s).

The results of this study indicated that the system worked well for the measurement of catecholamine concentration in phosphate-buffered solution.

Figure 4.46 shows the correlation between the electric current and the adrenaline concentration in the phosphate-buffered solution. The minimum

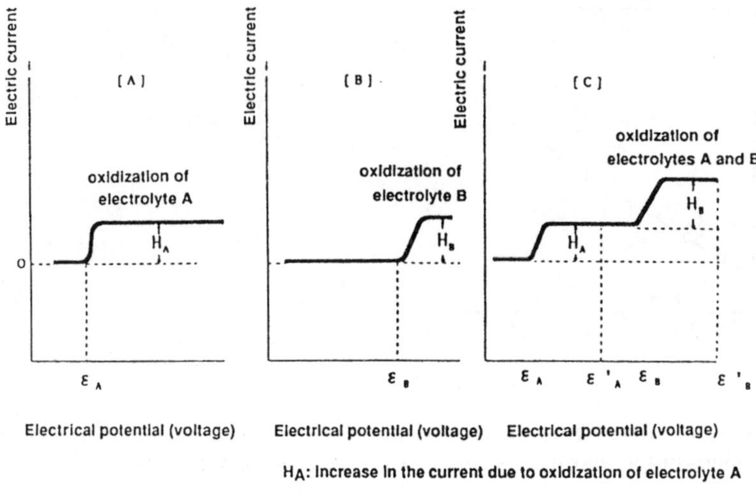

Fig. 4.44. The principle behind the method of electrochemical measurement (quoted from [200])

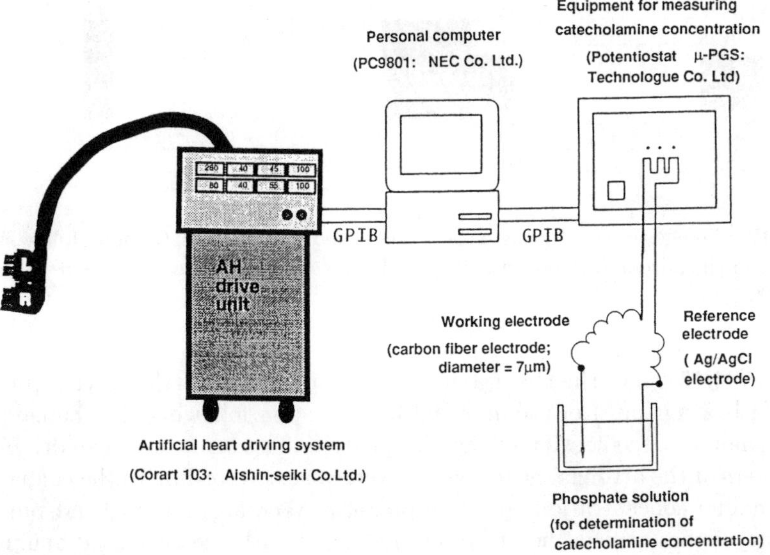

Fig. 4.45. The experimental arrangement for measuring catecholamine concentrations and subsequent feedback to the artificial heart driving system (quoted from [200])

detectable concentrations of both adrenaline and noradrenaline in the phosphate-buffered solution were approximately 1–2 ng/ml (10^{-8} mol/l).

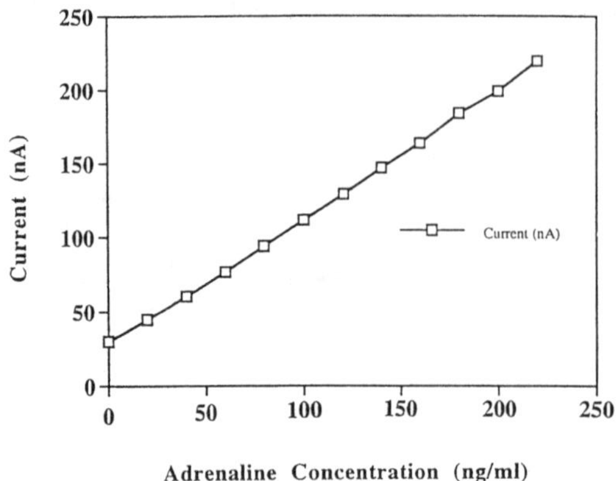

Fig. 4.46. Correlations between the oxidation current and the concentration of catecholamines in the phosphate-buffered solution (quoted from [200])

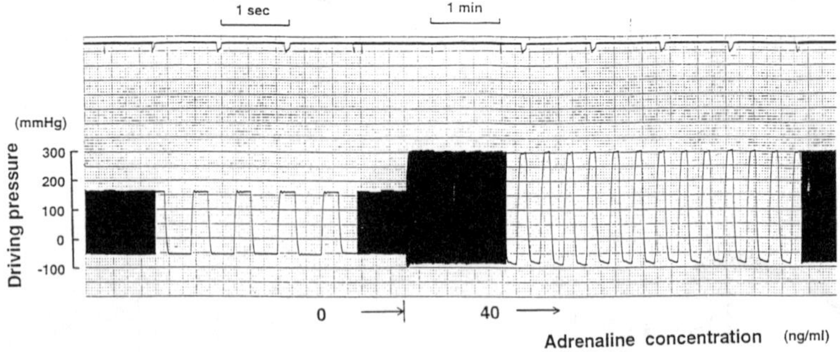

Fig. 4.47. Changes in the driving parameters of the artificial heart system following a change in the adrenaline concentration of the phosphate buffered solution (quoted from [200])

Figure 4.47 shows the changes in the driving condition (the driving pressure and beating rate) of the artificial heart system following the change in the adrenaline concentration of the phosphate-buffered solution. The driving parameters of the artificial heart system were varied according to the changes in adrenaline concentration (positive pressure, vacuum pressure, and pumping rate of the artificial heart system increased with the rise in adrenaline concentration). There was almost no time delay between the addition of the catecholamines and the changes in the operating parameters.

As for the resolution of the catecholamine concentrations in the solution, results of a previous study by the authors [174] on the concentrations of plasma catecholamines before, during, and after treadmill exercise suggested that the minimum resolution of catecholamine concentrations must be improved to at least 10^{-9} mol/l or hopefully 10^{-10} mol/l, so that the system

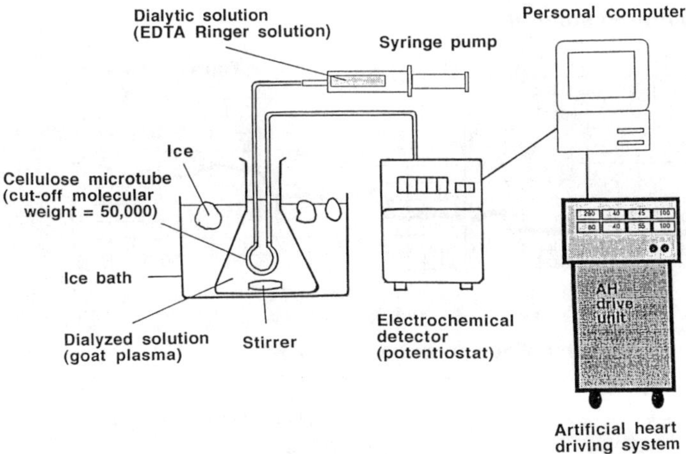

Fig. 4.48. The experimental arrangement for measurement of the catecholamine concentrations using an electrochemical detector in conjunction with the microdialysis technique

can feed back the concentrations of catecholamines and control the driving parameters optimally during the exercise.

Thus, except for the resolution of the catecholamine concentration, the system worked satisfactorily when buffered solution was used instead of actual blood or plasma. However, when the sensing system was used in goat plasma, the sensitivity of the sensor decreased greatly with time after immersion in plasma, and the electrode came not to respond to the catecholamines in it immediately.

It was observed that plasma proteins were adsorbed onto the electrode, which was suspected to be the main cause of the decrease in sensor activity.

One of the possible solutions for preventing blood proteins from being adsorbed onto the electrode is to adopt a microdialysis technique [175,176] in conjunction with the above-mentioned electrochemical method. In our experiments, a cellulose microtube (cutoff molecular weight = 50,000) was used for dialysis; the microtube was immersed in the goat plasma, and a dialytic solution (phosphate-buffered solution) was flushed through the microtube. The catecholamines in the goat plasma were dialyzed into the dialytic solution through the cellulose membrane, and catecholamine concentrations in this dialytic solution were measured using the electrochemical detector described above (Fig. 4.48). With this system, the ratio between catecholamine concentrations in the dialytic solution and that in the plasma was approximately 20% when the flow rate of the dialytic solution was 10 µl/min, with a delay of approximately 90 s between the time when the catecholamine concentration in the goat plasma was altered and that when the detector identified the changes of the catecholamine concentrations as those of the dialytic solution

(Phosphate-buffered solution)

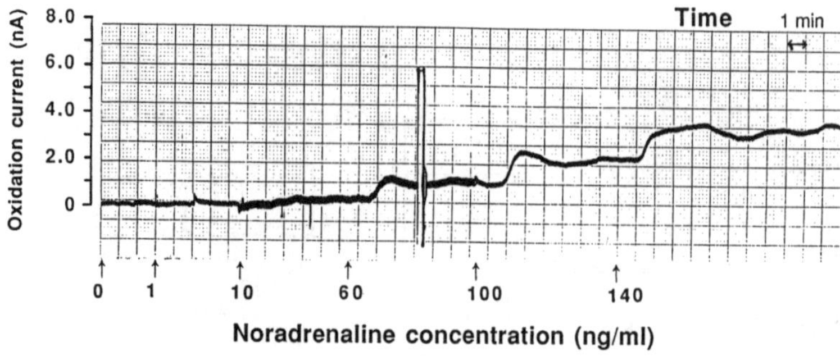

flow rate = 10μl/min,
applied voltage = 400 mV

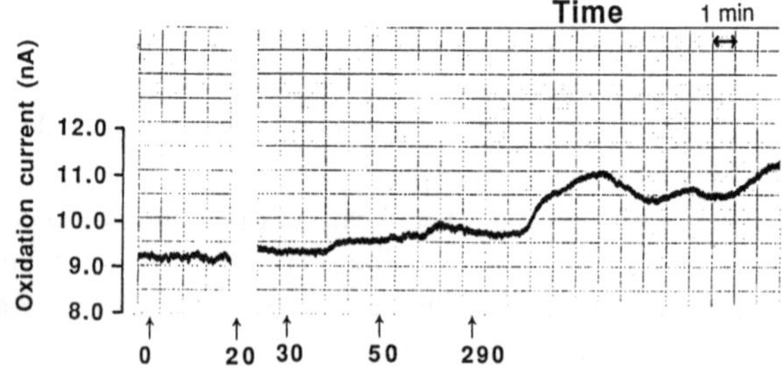

Noradrenaline concentration (ng/ml)

flow rate = 10μl/min,
applied voltage = 400 mV

Fig. 4.49. Changes in the oxidation current that are detected by the sensing system which is shown in Fig. 4.48. The oxidation current of the dialytic solution increased (with fluctuations) with a time delay of approximately 90 s after catecholamines were added to the original (dialyzed) goat plasma or phosphate-buffered solution

(Fig. 4.49). The ratio and the time delay generally decreased in inverse proportion to the increase in the flow rate of the dialytic solution. Using this system, it was possible to alter the driving air pressure and pulse rate in accordance with the algorithm for changes in the catecholamine concentration of the plasma in which the sensor was immersed, although the minimum detectable concentration of both plasma adrenaline and noradrenaline decreased to 10^{-7} mol/l.

Thus, the use of an electrochemical detector in conjunction with the microdialysis method shows the future possibility of developing a feedback

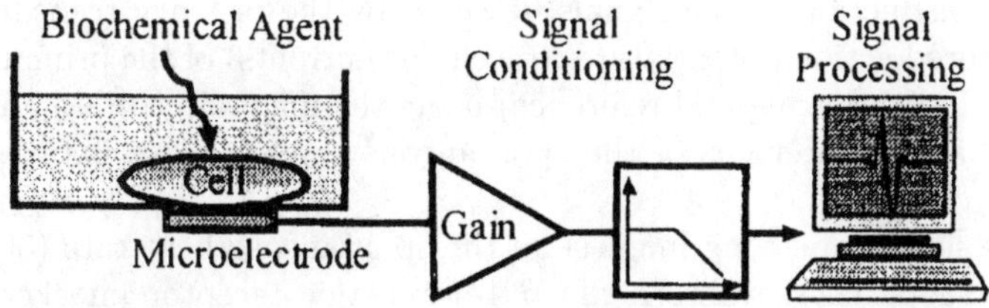

Fig. 4.50. The concept of the "whole cell sensor" [201]

control system for artificial hearts using blood catecholamine concentrations. However, in order to use this system for actual clinical use, it is necessary to both improve the sensitivity of the sensor system (for the detection of catecholamines) to at least 10^{-9} mol/l and decrease the time delay to no more than 10 s or so.

On-Line Measurement of Humoral Factors by Whole-Cell Sensors. The other method that can achieve stability for the measurement of the chemical factors in the blood and high sensitivity for specific biochemically active substances is to use a living cell with high sensitivity for biochemical substances as the detector (cell-based biosensor – whole-cell sensor [177]).

In the case of the measurements of plasma catecholamine concentrations, myocardial cells, which have adrenergic receptors, can be used as a primary transducer.

DeBusschere et al. (Stanford University) have reported that they have succeeded in measuring catecholamine concentrations in a Hepes-buffered culture medium using a cell-based biosensor [178].

They used spontaneously beating mouse myocardial cells as the primary transducers. In order to measure a specific bioactive chemical agent, they used wild-type cells and genetically engineered cells that lack a specific receptor for the chemical agent (by developing a knockout mouse of the receptor) as the primary transducer and made differential measurements between the results of these two cells (Fig. 4.50).

Specifically, they used wild-type myocardial cells as the working primary transducer for measuring β-adrenergic agents, and they developed genetically engineered knockout cell lines (null mutants) of either the β 1-adrenergic receptor or the β 2-adrenergic receptor as the reference primary transducer. The former cells respond to β-adrenergic agents (e.g., isoproterenol) and alter their beating rate; in contrast, the latter are not affected by the β-adrenergic agents. It is therefore possible to evaluate the concentration of the β-adrenergic agents by comparing the difference in the beating rate between these two kinds of cells.

These authors used a 6 × 6 array of circular platinized iridium, gold, or platinum microelectrodes fabricated on either glass or silicon wafers for the secondary transduction of the signals detected by the primary transducers. They monitored action potentials (extracellular currents) of the primary transducer cells (both working and reference) using the microelectrodes, and made a differential measurement of the spontaneous beating rates between the two types of cells.

Their results showed a strong increase in the spontaneous beat rate (50–80%) of the wild-type cells, whereas the β 1-adrenergic receptor knockout cells exhibited only a minor response (1–5%) when a 10 µM concentration of isoproterenol (which is expected to saturate the receptors) was added to the medium.

The main problem for the development of this system is that the knockout of the target receptor would result in a lethal mutation in many cases and, as a result, it would become impossible to develop the target cell lines, as these cells are derived from the primary culture of the tissues of the knockout mouse.

However, this technique is expected to allow on-line measurements of the concentration of specific biochemical agents that are active to the receptors that are made knockout. It is also expected that this measurement can be effectively and reversibly performed in the blood or plasma, with a very high resolution.

The authors are also developing a similar whole cell sensor system using genetically engineered cells derived from embryonic stem cells.

Summary. At the moment, minimal detectable concentrations of both adrenaline and noradrenaline with respect to the sensing system are still inadequate for practical use. Stability of the measurement in blood or in plasma is also necessary for the implementation of this sensing system.

However, novel techniques such as "whole cell sensors" and "microdialysis" will allow the development of a control system for artificial heart systems using feedback of catecholamine concentrations in blood in the near future.

4.9.2 Control of Artificial Hearts Using Autonomic Nervous Signals

Introduction. The nervous system, together with the humoral system, plays a significant role in controlling the human body. Not only does it control our voluntary movements, but it also plays an indispensable role in governing the autonomic nervous control, which regulates the functioning of internal organs to optimum levels in accordance with external conditions. The development of a man–machine interface, which can allow information sent by the human nervous system to control external equipment, is therefore extremely important for the development of the next generation of prosthetic devices, such as artificial organs, hands, and legs. On the other hand, with regard to the control method of artificial heart systems, a definitive and reliable method has not yet been developed, although several methods have so far been proposed and tried [179,180].

In this section, we discuss the possibility of developing an artificial heart system that takes information from the autonomic nervous system of the living body and can control the driving conditions of an artificial heart system using autonomic nervous activity, so that the artificial heart works optimally.

In Vitro Experiment to Control an Artificial Heart System Using Skin Sympathetic Nervous Signals. It has been reported that emotional stress or various kinds of nociceptive stimuli to the sensory receptors at the skin or viscera provoke increases in heart rate and peripheral vascular resistance, and augmentation of the contractility of the heart through changes in the activities of the sympathetic nervous system (the cardiac sympathetic nerve-accelerator) and the parasympathetic nervous system (the cardiac branch of the vagal nerve-suppressor) by reflex.

In this study, we attempted to reproduce the change that takes place in the actual hemodynamics of the living body when nociceptive stimuli are given to a subject. To that end, we used a mock circulatory system connected to an artificial heart system, and tried to control the driving condition of the artificial heart system by using information from the sympathetic nervous activity.

In this experiment, we used skin sympathetic nerve activity (SSNA) as the parameter to determine the condition of the artificial heart system. SSNA includes the activities of the sympathetic nerve fibers that are contained in a peripheral nerve and involve the vasomotor and sudomotor functions at the skin area.

Although this activity is directly concerned with cardiac function, the first and most important reason for selecting SSNA is that, at the moment, it is the only sympathetic nervous activity that can be measured from awake and nonanesthetized human subjects. The second reason is that, as SSNA concerns the activity of a vasoconstrictor, a change in SSNA seems to reflect a change in the peripheral vascular resistance or peripheral blood flow.

It has generally been accepted that the discharge of a vasomotor nerve fiber burst accelerates the constriction of peripheral blood vessels, and that the increase in this activity causes an increase in the peripheral resistance and consequently in the peripheral blood pressure [181,182]. The number of SSNA bursts counted using a pulse counter was therefore used as the parameter to determine the driving condition of the pneumatically driven artificial heart system connected with a mock circulatory system with a peripheral resistance.

The Experimental Arrangement. The experimental arrangement is shown in Fig. 4.51. The subject sat in an armchair in a fairly relaxed posture throughout the experiment. The measurements included blood flow rate in the skin, digital arterial blood pressure, and skin sympathetic nerve activity (SSNA). The measurement of SSNA was carried out from the median nerve by means of microneurography [183]. Details of the technique of microneurography will be presented in another section. Skin blood flow was measured using a laser Doppler blood flowmeter in the regions innervated by the nerve from which SSNA activity was recorded. The blood pressure of a digital artery was measured by a noninvasive continuous monitoring system for arterial blood pressure.

SSNA and other physiological parameters were measured using the above-mentioned system when the subjects were sitting in a fairly relaxed manner and when they were exposed to a series of mental or physical stresses (electrical stimulations, unpleasant white noise, and the sound of a firing revolver) and recorded using a data recorder.

The Development of a Program for Controlling the Driving Parameters of an Artificial Heart System and the Results of the Control. From the changes with respect to time in the above-mentioned measured parameters, the lapses between the time when a stimulus was given to the sub-

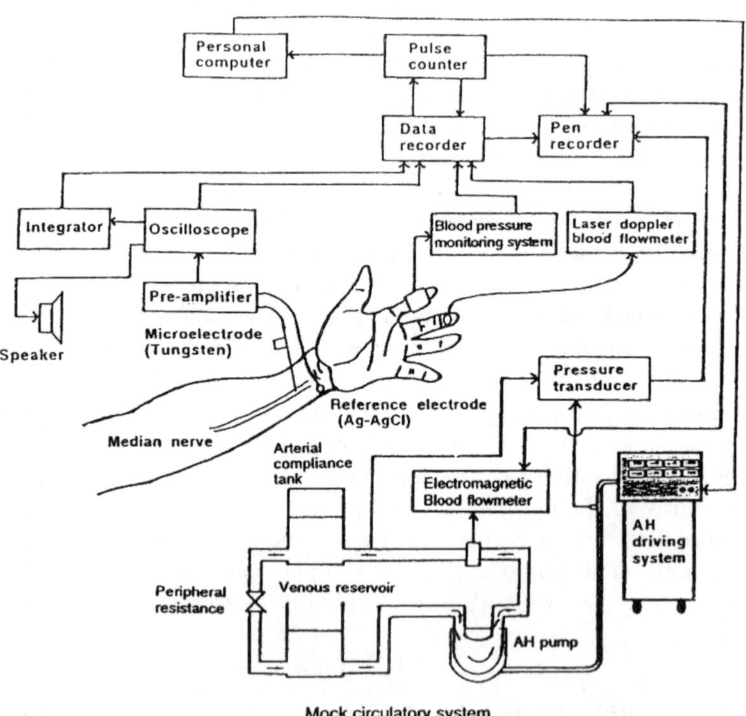

Fig. 4.51. The experimental arrangement

ject and the onset of an SSNA burst and that between the onset of an SSNA burst and the time when the hemodynamic parameters began to change were studied. Based on these results, a program was developed to estimate the vascular response caused by the change in the number of SSNA pulses, to alter the driving condition of an artificial heart system in accordance with the change in the estimated vascular response, and consequently to change the hemodynamic parameters of the mock circulatory system so that it mimics the changes in the subject's actual hemodynamics.

Next, the time series data recorded for the SSNA under various stimuli were inputted into the computer, and the driving parameters of the artificial heart system (the values of beating rate, positive pressure, and vacuum pressure) were determined and altered using this program in accordance with the deviation of the counted SSNA number from the control value, and the hemodynamic parameters of the mock circulatory system were compared with the actual change in the hemodynamic parameters of the subjects when the subjects were exposed to the various stimuli.

Here, the control program is not described in detail for the sake of brevity, and the method of determining the driving parameters of the artificial heart system (the values of pumping rate, positive pressure, and vacuum pressure) is briefly outlined below.

First, SSNA(T), the index of the SSNA level to determine the driving conditions of the artificial heart at time T (s), was determined as the weighed summation of the SSNA number from 2.5 s to 7.5 s before time T, giving consideration to the lapse until the onset of the SSNA burst. Next, the driving parameters of the pneumatically driven artificial heart system (the values of the beating rate, positive pressure, and vacuum pressure) were determined and altered in accordance with the deviation of the SSNA index from the control value (an average value of SSNA(T) when the subject was sitting in an armchair in a fairly relaxed manner), and the driving condition of the artificial heart system was altered every 3 s.

Figure 4.52 shows an example of the results of this control. The upper part of Fig. 4.52 shows the changes with respect to time in SSNA (discriminated wave, integrated wave, and number of pulses), the skin blood flow at the finger, and the digital arterial pressure of a subject when the subject was repeatedly exposed to unpleasant white noise. The time at which each stimulation was given to the subject is indicated at the top.

As can be seen in this chart, an increase in SSNA bursts (activation of the sympathetic nervous system) was observed after the subject received auditory stimulation due to the unpleasant white noise; consequently, the digital arterial pressure increased, and the skin blood flow at the finger decreased. The lower part of Fig. 4.52 shows the driving condition of the artificial heart system and the arterial pressure of the mock circulatory system. The positive pressure, vacuum pressure, and pumping rate of the artificial heart system

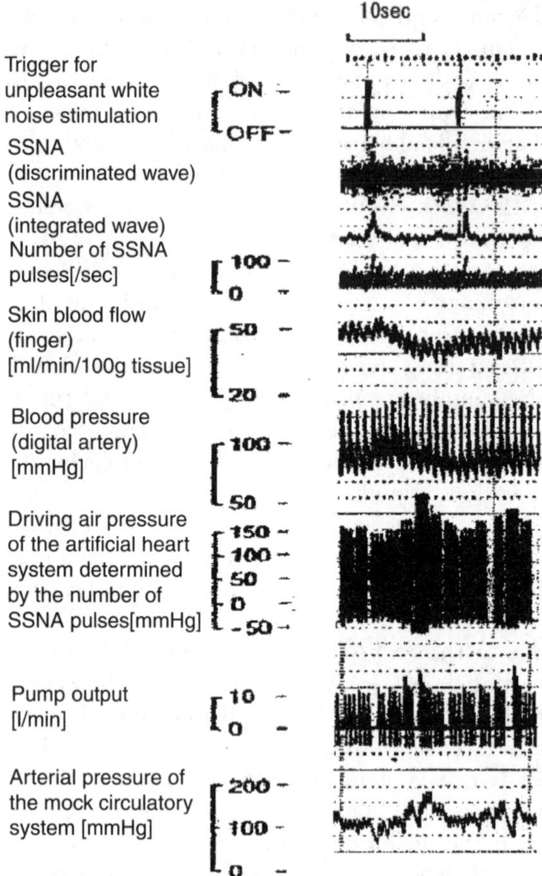

Fig. 4.52. *Upper part:* the changes with respect to time in the SSNA, skin blood flow at the finger, digital arterial pressure of a subject when the subject was repeatedly exposed to unpleasant white noise. *Lower part:* the driving condition of the artificial heart system and the arterial pressure of the mock circulatory system

also increased in response to this change in the number of SSNA pulses in accordance with the algorithm for changes in the driving parameters.

In order to reproduce all the responses of the cardiac and circulatory system in the living body in the mock circulatory system, it is necessary to change not only the driving condition of the artificial heart system but also the passage resistance of the mock circulatory system on-line in accordance with the change in the actual total peripheral vascular resistance of the living body. However, in this experiment, it was impossible to achieve this (the resistance was maintained at a constant value), and we controlled the artificial heart system so that the changes with respect to time in the blood pressure of the mock circulatory system mimicked the changes that actually occurred in the actual living body.

Thus, it seems possible to alter the driving condition of an artificial heart system so that the artificial heart mimics the change in the function of the natural heart through the information of the autonomic nervous system. Our experiment demonstrates the possibility of controlling the driving conditions of an artificial heart system using sympathetic nervous activity. (Further investigation is, of course, needed with regard to whether the activity accurately reflects the systemic hemodynamic conditions.)

An In Vivo Experiment to Control an Artificial Heart System Using Autonomic Nervous Signals. Although the in vitro experiment employed SSNA as the signal for controlling an artificial heart system, it would of course be desirable to employ the activity of cardiac nerves instead of SSNA for the purpose of controlling an artificial heart system. As this is impossible to perform without invasive procedures, we are currently employing this procedure in in vivo experiments on animals (goats) that are equipped with a left ventricular assist device (an artificial heart system). We are also trying to control the driving condition of this artificial heart system by using the signals from the autonomic nervous system, so that it can mimic the actual changes in the pumping conditions of a natural heart.

As indicated above, goats were equipped with a left ventricular assist device. The electrode used in these experiments was a wire electrode made of stainless steel. It would be desirable to set and fix the electrode to the cardiac sympathetic nerve; however, in the experiments currently under way, the electrodes have been fixed to the left vagal and sympathetic nerve trunk at the neck for technical reasons (Fig. 4.53).

The artificial heart system used in these experiments is the same as that used in the in vitro experiments (Fig. 4.54). At present, it is not always easy to measure and record signals of the autonomic nervous system stably from a goat without preventing it from moving freely, and there still remain several problems such as:

1. working out a countermeasure against noise;
2. improving the stability and the sensitivity of the nerve electrodes;
3. separating the efferent sympathetic (or parasympathetic) signals from the afferent signals.

We are currently working on solving these problems.

4.9.3 The Control of Somatic Sensations and the Generation of Artificial Sensations by Direct Stimulation of the Neural System

Introduction. In this section, we discuss a method to evoke artificial somatic sensations by stimulating sensory nerve fibers. The technique is currently

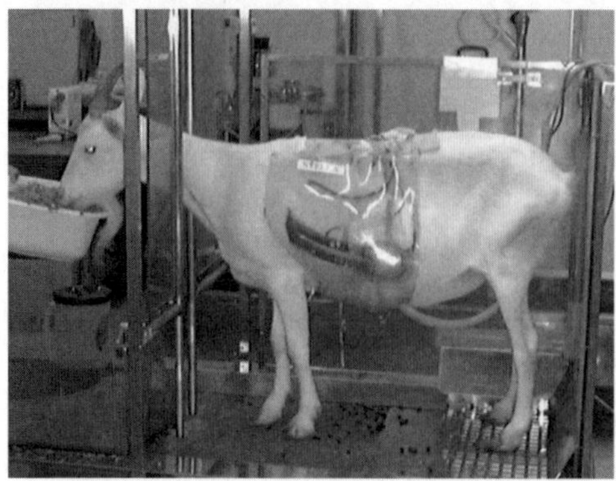

Fig. 4.53. An in vivo experiment to control a left ventricular assist device (an artificial heart system) using information from the sympathetic nervous system. The electrode is expected to be fixed to the cardiac sympathetic nerve. However, at the moment, the electrodes have been attached to the left vagal and sympathetic nerve trunk at the neck for technical reasons

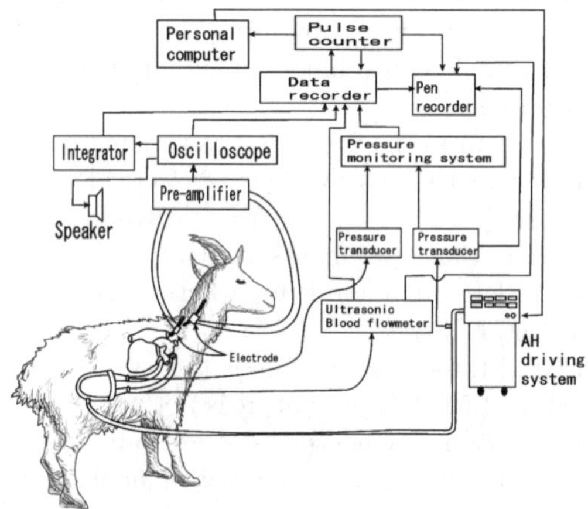

Fig. 4.54. A schematic diagram of the in vivo experiment to control the driving condition of an artificial heart system using information from the autonomic nervous system

being used in our project to develop an artificial arm (or leg) system that is capable of sensing mechanical stimuli and then transferring these stimuli to the subject, so that the subject can experience the stimuli as the corresponding somatic sensations. First, we will explain the concept and technique used to evoke somatic sensations artificially by electrically stimulating a sensory nerve fiber from a single mechanoreceptor unit. Next, we will discuss how the properties of an evoked sensation are affected when electrical stimulations

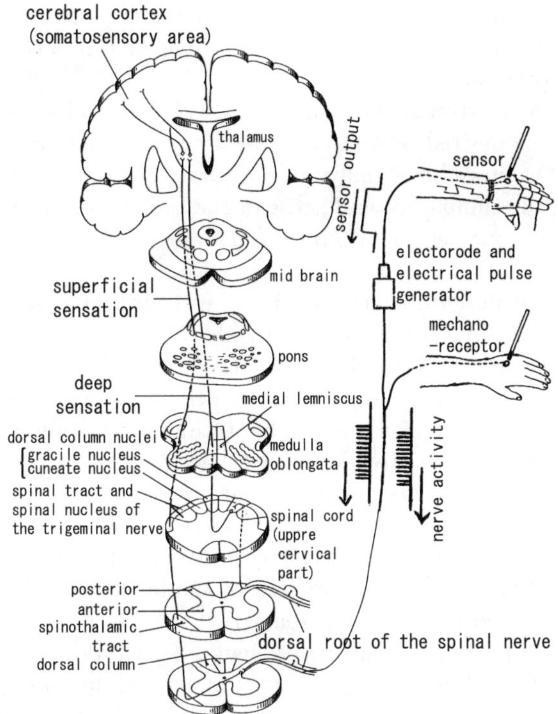

Fig. 4.55. The concept for generating artificial somatic sensations by the electrical stimulation of afferent sensory nerve fibers

are given simultaneously to multiple sensory nerve fibers which separately innervate different single mechanoreceptor units. Finally, we will describe the prototype of the artificial arm system mentioned at the beginning of this section.

The Concept and Method for Generating Somatic Sensations by Intraneural Microelectrical Stimulation. The concept for generating somatic sensations is shown in Fig. 4.55.

When we touch something with our hands and a somato-sensory receptor (mechanoreceptor) is activated by a stimulus such as pressure, heat, or vibration, or by nociceptive stimuli, the nerve signals (which consist of a series of pulses of evoked potential) generated at each mechanoreceptor are transferred to the somato-sensory area of the cerebral cortex through a peripheral nerve and the spinal cord, which evoke the corresponding somatic sensation to the subject.

The underlying idea is that if each single peripheral nerve fiber could be supplied with appropriate electrical stimulation, and if we could thereby generate the same afferent signal (the same series of evoked potential pulses) to the nerve fiber, we would then be able to generate in the subjects the same somatic sensations as would be evoked by stimulation of the actual corresponding mechanoreceptor unit.

We have been using the micro-stimulation (microneurographic) technique [184,185] in order to evoke the desired sensation by electrical stimulation to the sensory nerve fiber, and we have been analyzing the relationship between the properties of the electrical stimulation and those of the evoked sensation generated by stimulating only one sensory nerve fiber coming from a single mechanoreceptor unit.

Microneurography involves percutaneous and direct insertion of a very thin tungsten microelectrode (needle-type) into a peripheral nerve, so that the signal of the nerve fiber attached to the tip of the electrode can be measured: it is also possible to stimulate nerve fibers which are attached to the tip of the electrode (microstimulation) using this technique. Details of the technique of microneurography are provided in Sect. 4.7.3.

In our experiments, we inserted a microneurographic electrode into the median nerve of each subject. The position of the electrode was adjusted and fixed so that the nerve signal from a single mechanoreceptor unit could be recorded (Fig. 4.56), and the type and receptive area of the involved mechanoreceptor unit were determined from the response of the receptor to the mechanical stimuli [184,185]. (In a set of our experiments, SA I type mechanoreceptors are basically selected as the object for electrical stimulation.) Then, we gave electrical stimulation to the nerve fiber in a series through the same microelectrodes by changing the circuit from the measurement mode to the stimulation mode using a switching box. Subsequently, we investigated the properties of the evoked sensation and the relationship between the area and the subjective intensity of the evoked sensation and the frequency, amplitude, and temporal profiles of the electrical stimulation. The relationship between the intensity of the evoked sensation (pressure sensation) and the frequency or amplitude of the electrical stimulation was quantitatively evaluated using the so-called "magnitude estimation method" (we used a positive square-wave pulse for 250 μs; the frequency was fixed at 20 Hz).

Note 1: Six kinds of mechanoreceptors (or endings of the nerve fiber) have been reported to exist on the skin [185–187] (Fig. 4.57): Merkel disks; Pinkus corpuscles; Ruffini endings; Meissner corpuscles; Vater–Pacini corpuscles; and hair follicle receptors. The response of these receptors is classified into four patterns: SA I (slowly adapting type I), SA II (slowly adapting type II), RA (or FA I – rapidly adapting type), and PC (or FA II – Pacinian type) according to their properties with respect to the adaptation of the response and the area of the receptive field (refer to Fig. 4.58).

The slowly adapting types correspond to the signals from Merkel disks or Ruffini endings, which detect deviations of the skin; these two receptors are suitable for detecting continuous pressure or touch.

The rapidly adapting type corresponds to the signals from Meissner corpuscles or hair follicle receptors, which detect the velocity of the skin

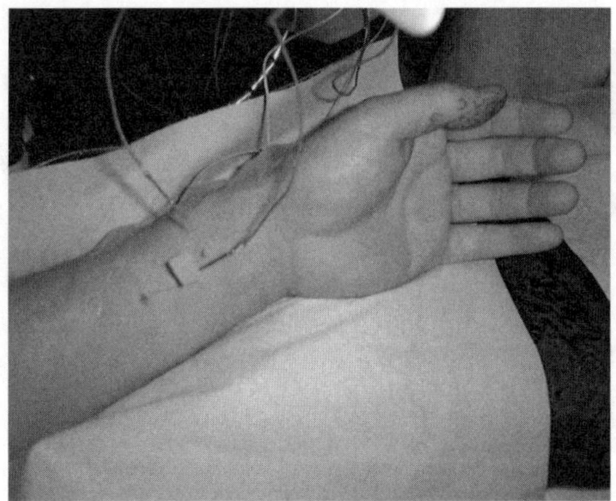

Fig. 4.56. The manner in which the experiment for the measurements and stimulations were performed. The position of the inserted electrode was fixed so that it allowed recording and stimulation of a single nerve fiber from a single mechanoreceptor unit

movement; these receptors are suitable for detecting slow vibrations whose frequency is less than 40 Hz.

The Pacinian type shows very sensitive responses and very fast adaptation to the stimuli and corresponds to the signals from Pacinian corpuscles, which detect the acceleration of the skin movement; this type of receptor is suitable for detecting vibration.

Ochoa et al. have reported that a pressure sensation is evoked by stimulating an SA I mechanoreceptor unit, and that the magnitude of the evoked sensation is influenced by the frequency of the stimulus but not by the amplitude of the stimulus [188].

Our results basically confirm the results reported by Ochoa et al. With respect to sensory nerve fibers from the SA I mechanoreceptor unit, our results showed that the electrical microstimulation evoked pressure sensations, and that the magnitude of the evoked sensation increased with the stimulus frequency and seemed to be a positive power of the frequency (Fig. 4.59). This type of response is often observed in the field of sensation, as in the case of the relationship given by the Weber–Fechner law or Stevens's law.

Combined Evoked Sensation due to the Simultaneous Electrical Stimulation of Multiple Single Mechanoreceptor Units. So far, we have mentioned the methods of evoking artificial sensation by electrically stimulating a single mechanoreceptor unit. However, when we actually touch something, it is impossible for only one mechanoreceptor unit to be stimulated; in fact, both a large number and various kinds of nerve fibers are

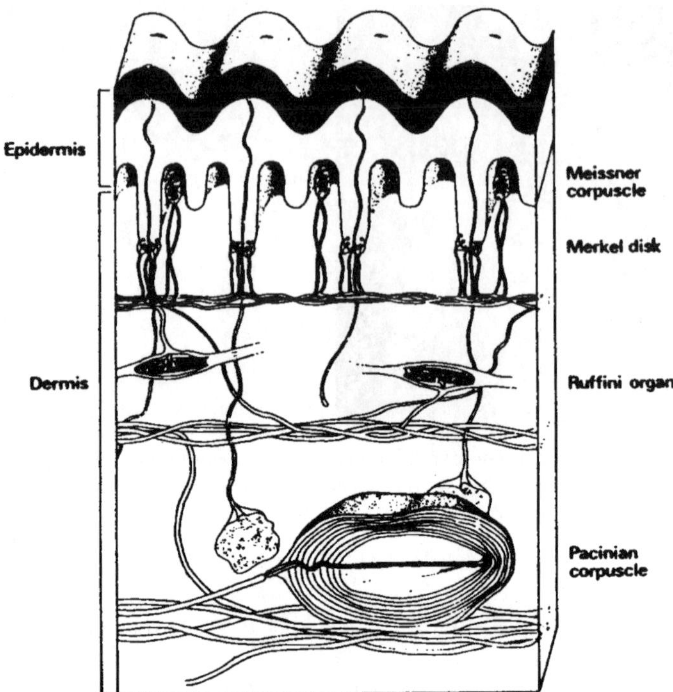

Fig. 4.57. Mechanoreceptors located on the skin (quoted from [186])

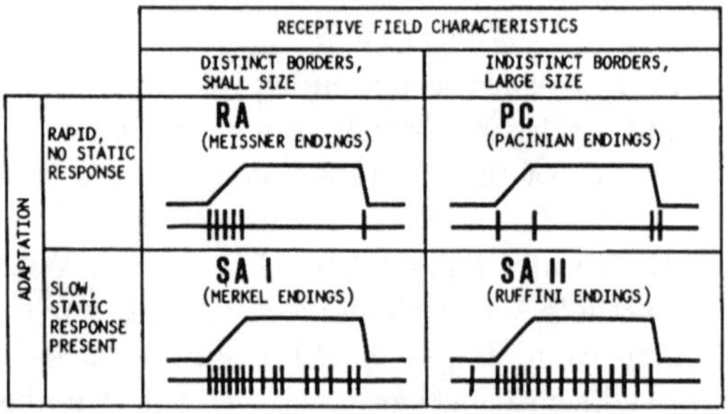

		RECEPTIVE FIELD CHARACTERISTICS	
		DISTINCT BORDERS, SMALL SIZE	INDISTINCT BORDERS, LARGE SIZE
ADAPTATION	RAPID, NO STATIC RESPONSE	RA (MEISSNER ENDINGS)	PC (PACINIAN ENDINGS)
	SLOW, STATIC RESPONSE PRESENT	SA I (MERKEL ENDINGS)	SA II (RUFFINI ENDINGS)

Fig. 4.58. The response of the mechanoreceptors is classified into four groups according to the firing pattern (type of the adaptation of the response) to the pressure stimulation which is given to the receptor, and the area of the receptive field (quoted from [184])

stimulated simultaneously. Therefore, in order for this technique to be acceptable for actual clinical use, it is necessary to clarify how the properties (type, projected area, subjective amplitude, etc.) of the evoked sensation affect each other when electrical stimulations are given simultaneously to multiple sen-

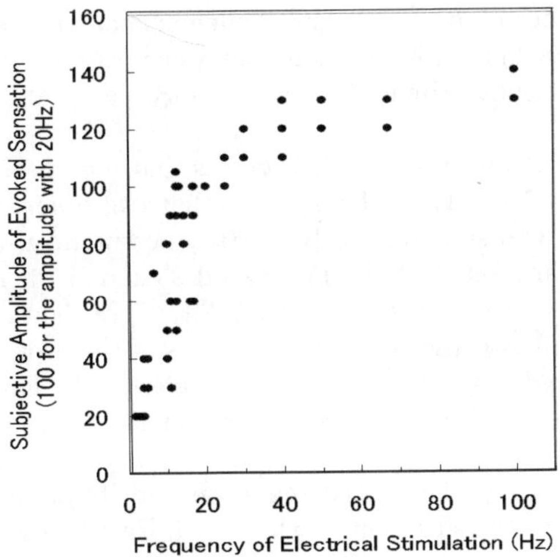

Fig. 4.59. The elationship between the frequency of the electrical stimulation to the nerve fiber from the SA I mechanoreceptor unit and the subjective magnitude of the sensation which was evoked by the electrical stimulation

sory nerve fibers that separately innervate different single mechanoreceptor units.

In the case of direct stimulation to the skin, several kinds of nonlinear phenomena or interactions between two evoked sensations such as "masking effects" and "phantom sensations" have been reported [189–192] to occur when multiple stimulations are given to a subject simultaneously.

Note 2: The "masking effect" represents the phenomenon in which the subjective amplitude of the test stimulus decreases or becomes subliminal when another stimulus is presented to the skin simultaneously, or with a very short time interval between the two stimuli. The "phantom sensation" represents the phenomenon in which two separated transient stimuli to the skin come to be felt as only one single stimulus at a point between the two actual stimuli. This occurs when these two stimuli are presented simultaneously and of almost equal intensity, and the distance between the two stimuli is adjusted within a certain range.

In addition to these two phenomena, a few nonlinear phenomena such as "cutaneous saltation" or "apparent tactile movement" have been also reported [193,194].

Theoretically speaking, it seems that the same phenomena can also occur when the two sensations are evoked by electrical stimulation to the sensory nerve fibers as well as by mechanical stimuli to the mechanoreceptors at the skin, and we investigated whether or not a nonlinear phenomenon such as the "phantom sensation" or some kind of interaction may occur between the two sensations evoked by electrically stimulating two sensory nerve fibers simultaneously [195].

The experimental arrangement was basically the same as that used in the experiments described in Sect. 4.7.3.2. The difference was that two tungsten microelectrodes were separately and simultaneously inserted percutaneously into the median nerve at two points on the forearm (around 2 cm and 10 cm proximal to the wrist joint, respectively) of each subject. After investigating the type of mechanoreceptor unit and the physical properties of the evoked sensation (quality, magnitude, and projected area) by giving electrical stimulation (a series of electrical pulses) to each electrode individually, electrical stimulation was then given to these two electrodes simultaneously, and the resulting changes in the properties of the evoked sensation and its projected area were then compared with the situation in which each electrode was stimulated individually.

In our experiments, combinations of types of mechanoreceptor units stimulated simultaneously were slowly adapting I (SA I) and SA I in most cases. In some cases, however, combinations of types of mechanoreceptor units between Pacinian (PC) and SA I, and between PC and PC were also examined.

From the results obtained to date, it may be concluded that the combined evoked sensation when two nerve fibers are stimulated simultaneously became a simple algebraical summation of two evoked sensations evoked by the electrical stimulation of the individual mechanoreceptor unit. It may therefore not be necessary to pay attention to nonlinear interactions between the properties of the evoked sensation of each sensory fiber when we give electrical stimulation to each sensory nerve fiber in accordance with the output signal from the sensor system of an artificial arm system.

However, further research is needed in order to reach conclusive results, because the number of the experiments is still too small, mainly due to the difficulty in fixing the positions of the microelectrodes for a long time, so that a nerve signal from a single mechanoreceptor unit can be recorded by each electrode.

The Development of a System of Interpreting Somatic Sensations for Use with Artificial Hands and Limbs. Based on the results of the abovementioned studies, we developed a prototype of an artificial arm system capable of sensing mechanical stimuli and then transferring these stimuli to the subject, so that the subject may experience the stimuli as the corresponding somatic sensations [196].

The system consists of: 1) an artificial arm system whose palm and fingers were covered with pressure-conductive rubber [197] in order to detect pressure applied to the artificial hand (Fig. 4.60); 2) an electrical stimulator (pulse generator) which can provide microelectrical stimulation to the nerve fiber; 3) a controller (a personal computer system) which can calculate the frequency of the electrical stimulation (a series of electrical pulse sequences) that should evoke in the subject the same magnitude of pressure sensation as is given to the pressure-conductive rubber of the robot hand, and which can control the electrical stimulator so that it generates an electrical pulse sequence with that frequency; and 4) a tungsten microelectrode which, when inserted into the subject's nerve, can stimulate a single nerve fiber coming from a mechanoreceptor unit. The schematic diagram is shown in Fig. 4.61.

When pressure was applied to the pressure-conductive rubber, the personal computer (controller) modulated the output voltage of the pressure-conductive rubber to the frequency of the electrical pulses for stimulating the sensory nerve fiber; it controlled the electrical stimulator so that it could output the above-mentioned electrical pulses. The sensory nerve fiber was stimulated by the electrical pulses via the tungsten microelectrode, and the same afferent signal (as should be evoked when the pressure was applied to the mechanoreceptor unit itself) was evoked in the nerve fiber, making the subject evoke a pressure sensation at the projected area.

Just before the experiment to convey sensation via the artificial arm system, we evaluated the quantitative relationship between the frequency of the electrical stimulation and the intensity of the evoked pressure sensation using the magnitude estimation method with the same subject; in addition, we determined the equation to calculate frequency of the electrical stimulation from the output voltage of the conductive rubber.

The subjective intensities of the sensations evoked by the electrical stimulation were described by the subjects using a slide volume (Fig. 4.62), and these subjective evaluations were then compared with the strengths of the stimuli that were applied to the pressure-conductive rubber of the robot hand.

The system worked satisfactorily. The subjects were able to feel pressure sensations resulting from the pressure applied to the conductive rubber of the

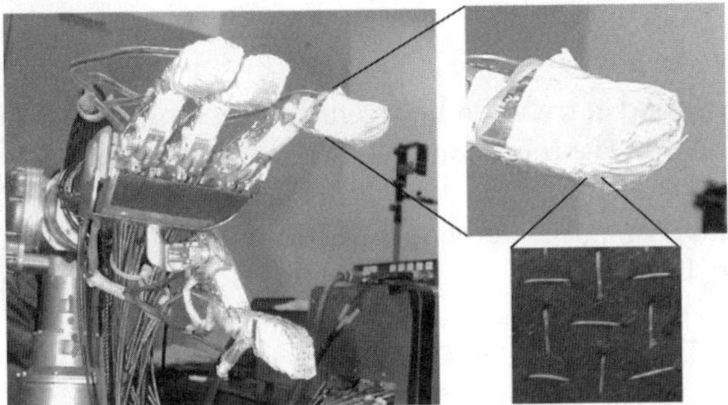

Fig. 4.60. The prototype of the artificial arm system developed in our study; the palm and fingers of the arm system were covered with pressure-conductive rubber in order to detect pressure applied to the arm system

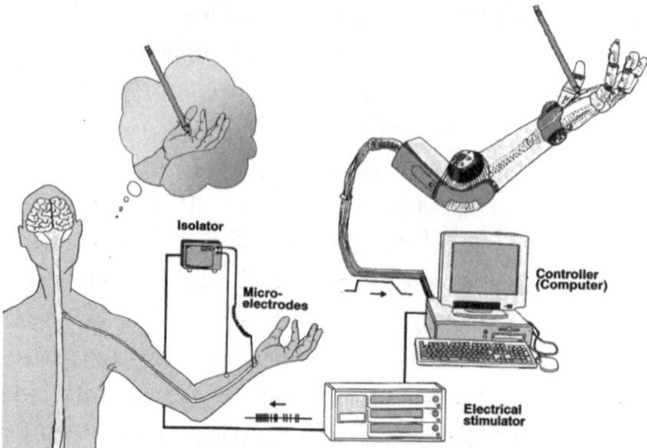

Fig. 4.61. The schematic diagram of the technique with which the artificial arm system senses mechanical stimuli and transfer these stimuli to the subject, so that the subject experiences the stimuli as the corresponding somatic sensations

Fig. 4.62. The slide volume which was used in our study in order to describe the subjective magnitude of the evoked sensation

robot hand. There was also a good correlation between the pressure applied to the pressure-conductive rubber of the robot hand and the subjective intensity of the evoked sensations.

The upper part of Fig. 4.63 compares the changes with time in the pressure applied to the finger (upper line) and the subjective changes in the intensity of pressure sensations felt by the subjects.

On the other hand, the lower part of Fig. 4.63 shows the changes with time in the pressure applied to the conductive rubber (upper line), and the changes with time in the subjective intensity of the evoked sensation when the pressure on the robot hand was transferred to the subject by the system.

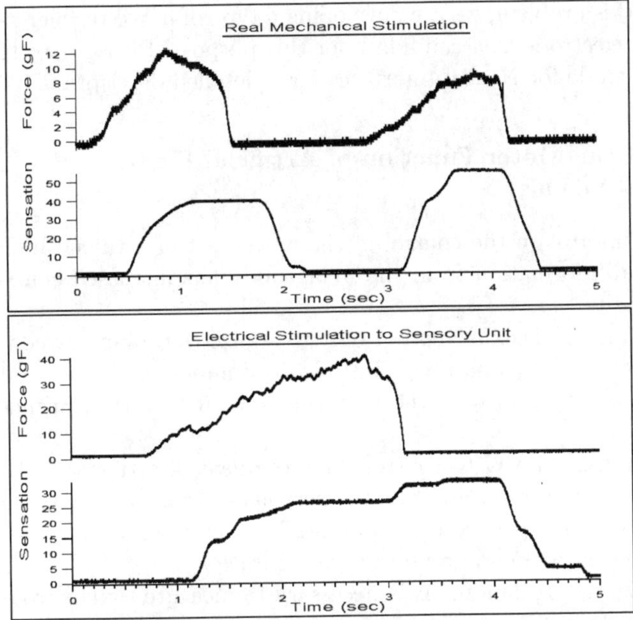

Fig. 4.63. *Upper:* changes with time in the pressure that were applied to the finger (*upper line*), and those in the intensities of the evoked pressure sensations felt by the subjects. *Lower:* changes with time in the pressure applied to the conductive rubber (*upper line*) and those in the subjective intensities of the evoked sensation, as described by the subject (*lower line*) when the pressure was transferred to the subject by the system

In both cases, although some motion delay can be observed in the chart, similar changes in the evoked sensations were observed, corresponding to the changes in the pressure applied to the hand (former case) or to the pressure-conductive rubber (latter case); furthermore, stimuli detected by the sensors of an artificial arm system were thought to be successfully conveyed to the subjects via this artificial arm system, which produced the same somatic sensation as the original stimuli and with corresponding magnitude.

However, several obstacles still remain to be overcome in order to improve this prototype system so that it can be used in actual clinical cases.

The main problem lies in the development of a multi-channel microelectrode that is capable of connecting each nerve fiber to an electrical signal line from external devices, so as to enable actual clinical use. Basically, the number of lines that can be connected using the microneurographic technique is limited to at most two or three. However, in order to connect all fibers contained in a peripheral nerve to each corresponding electrical signal line, it will be necessary for the electrode to have a minimum of more than 10,000 channels.

With regard to this problem, we are developing a type of nerve-regenerating (multi-channel) electrode as a candidate for this purpose. Please refer to Sect. 4.8 (Microelectrode for Neural Interface) for a detailed description.

4.9.4 Control of the Motor Function of Artificial Limbs (by Neural Signals)

In the recent past, interest in the control of the movement of artificial limb systems using neural information from the living body has increased considerably.

This kind of technology is expected not only to realize the voluntary control of artificial limbs used by amputees, but also to improve the control of paralyzed limbs in combination with the FES (Functional Electrical Stimulation) technique.

Although studies using EMG to control the movement of artificial limbs have the same goal, using information from the nervous system for the control of prosthetic devices seems to have more potential to achieve more natural control with more accuracy and a greater degree of freedom.

In order to realize this technique, it is necessary to measure signals from the motor nervous system using nerve electrodes. Concerning the location for implanting the electrode, three candidates are proposed and are being tried at the moment: the cerebral cortex, the spinal cord, and the peripheral nerve trunk, which contains efferent motor fibers.

Very recently, Chapin et al., researchers at the SUNY (State University of New York) Health Science Center in Brooklyn, reported that they had succeeded in experiments to control the movements of a robot arm using information derived from the brain of a rat [200].

The research group reported that they could make the rat control the robot arm voluntarily by decoding nervous activities recorded from neurons in the brain to the control signal of the robot arm. In this experiment, multi-channel electrode arrays were chronically implanted in the MI cortex (16 channels) and VL thalamus (eight channels) of rats. As each channel has two pulse discriminators, a maximum of 48 single neuron activities were recorded to obtain control information.

First, rats were trained to use their forelimbs to push a bar that controlled a robotic arm to bring water to them. Then, the research group derived a neural population function (NPF) capable of predicting the movement of the robot arm from the neural information when the rats pushed the bar. Finally, the control mode was switched in accordance with the NPF so that the position of the robot arm was controlled only by the information extracted from neural activities.

Wessberg (Duke University, a co-worker of I. Chapin) et al. also performed a similar experiment, in which they implanted multiple microelectrode arrays (16–32 channels, made from Teflon-coated stainless steel microwires) at different cortical areas (the premotor, primary motor, and posterior parietal cortical areas) of an owl monkey and reported similar results [199].

With regard to the attempts to utilize the signals of motor nerve fibers measured from the spinal cord or peripheral nerves to control the movements of a robot limb, no fruitful results have been achieved up to the moment, although several studies are currently under way.

For this technique, it is difficult to determine the location at which the electrode is to be implanted. This is because each system (the cerebral cortex, the spinal cord, and the peripheral nervous system) has its own advantages and disadvantages in this respect. When the electrode is implanted in the peripheral nervous system, it may be easier to understand the meaning of the recorded nerve signals; invasiveness and the risks of implantation are also lower than when the electrode is implanted in the motor cortex. On the other hand, the stability of the recording and the signal-to-noise ratio of the signal would worsen because peripheral nerves are usually circumscribed by moving muscles. A system utilizing the signal of the motor cortex has the advantage that the system would work well regardless of the position of the paralysis or amputation, as long as the motor cortex or the upper system is being kept intact, thus guaranteeing a wider area of application.

In conclusion, the development of a multi-channel electrode that allows long-term and stable recording of neural activities of a motor function is thought to be the key technology to realize this application.

References

1. I. Karube, T. Turner, eds.: *Biosensor-Fundamental and Applications* (Oxford University Press, 1987)
2. F. Scheller, R.D. Schmid, eds.: *Biosensors, Fundamentals, Technologies and Applications*, GBF Monographs, vol. 17 (VCH Publishers, 1992)
3. G. Costa, S. Miertus, eds.: *Trends in Electrochemical Biosensors* (World Scientific Publishing Co., 1992)
4. A. Berg, W. Olthuis, P. Bergveld, eds.: *Micro Total Analysis Systems 2000* (Kluwer Academic Publishers, 2000)
5. K. Hashimoto, K. Ito, Y. Ishimori: Anal. Chem. **66**, 3830 (1994)
6. T. Ihara, Y. Maruo, S. Takenaka, M. Takagi: Nucl. Acids Res. **24** (21), 4273 (1996)
7. P. Belgrader, S. Young, B. Yuan, M. Primeau, L.A. Christel, F. Pourahmadi, M.A. Northrup: Anal. Chem. **73**, 286 (2001)
8. L.C. Waters, S.C. Jacobson, N. Kroutchinica, J. Khandurina, R.S. Foote, J.M. Ramsey: Anal. Chem. **70**, 5172 (1998)
9. I.A. Nazarenko, S.K. Bhatnagar, R.J. Hohman: Nucl. Acids Res. **25**(12), 2516 (1997)
10. J.A. Parkinson, J. Barber, K.T. Douglas, J. Rosamond, D. Sharples: Biochem. **29**(44), 10,181 (1990)
11. S.O. Kelley, E.M. Boon, J.K. Barton, N.M. Jackson. M.G. Hill: Nucl. Acids Res. **27**(24), 4830 (1999)

12. M.S. Ibrahim, R.S. Lofts, P.B. Jahrling, E.A. Henchal, V.W. Weedn, M.A. Northrup, P. Belgrader: Anal. Chem. **70**, 2013 (1998)
13. G. Bardeletti, F. Sechaud, P.R. Coulet, in: *Biosensor Principles and Applications*, ed. by L.J. Blum and P.R. Coulet, Chap. 2 (Marcel Dekker, 1991)
14. A.E.G. Cass, G. Davis, G.D. Francis, H.A.O. Hill, W.J. Anson, I.J. Higgins, E.V. Plotkin, L.D.L. Scott, A.P.F. Turner: Anal. Chem. **56**, 667 (1984)
15. T. Ikeda, I. Katasho, M. Kamei, M. Senda: Agric. Biol. Chem. **48**, 1969 (1984)
16. T. Ikeda, H. Hamada, M. Senda: Agric. Biol. Chem. **50**, 883 (1986)
17. K. Yokoyama: Electrochem. **63**, 1193 (1995)
18. K. Yokoyama, E. Tamiya, I. Karube: J. Electroanal. Chem. **273**, 107 (1989)
19. B.A. Gregg, A. Heller: Anal. Chem. **62**, 258 (1990)
20. B.A. Gregg, A. Heller: J. Phys. Chem. **95**, 5976 (1991)
21. M. Elmgren, S.-E. Lindquist: J. Electroanal. Chem. **362**, 227 (1993)
22. T.J. Ohara, R. Rajagopalan, A. Heller: Anal. Chem. **65**, 3512 (1993)
23. T.J. Ohara, R. Rajagopalan, A. Heller: Anal. Chem. **66**, 2451 (1994)
24. T. de Lumley-Woodyear, P. Rocca, J. Lindsay, Y. Dror, A. Freeman, A. Heller: Anal. Chem. **67**, 1332 (1995)
25. C. Taylar, G. Kenausis, I. Katakis, A. Heller: J. Electroanal. Chem. **396**, 511 (1995)
26. T. Tatsuma, K. Saito, N. Oyama: Anal. Chem. **66**, 1002 (1994)
27. E.J. Calvo, C. Danilowicz, L.J. Diaz: J. Chem. Soc. Faraday Trans. **89**, 377 (1993)
28. E.J. Calvo, C. Danilowicz, L. Diaz: J. Electroanal. Chem. **369**, 279 (1994)
29. E.J. Calvo, R. Etchenique, C. Danilowicz, L. Diaz: Anal. Chem. **68**, 4186 (1996)
30. H. Nagasaka, T. Saito, H. Hatakeyama, M. Watanabe: Electrochem. **63**, 1088 (1995)
31. H. Bu, S.R. Mikkelsen, A.M. English: Anal. Chem. **67**, 4071 (1995)
32. K. Yokoyama, T. Shibasaki, Y. Murakami: Electrochem. **64**, 1221 (1996)
33. S. Koide, K. Yokoyama: J. Electroanal. Chem. **468**, 193 (1999)
34. K. Yokoyama, Y. Kayanuma: Anal. Chem. **70**, 3368 (1998)
35. L.A. Coury Jr., B.N. Oliver, J.O. Egekeze, C.S. Sosnoff, J.C. Brumfield, R.P. Buck, R.W. Murray: Anal. Chem. **62**, 452 (1990)
36. P.D. Barker, H.A.O. Hill, N.J. Walton: J. Electroanal. Chem. **260**, 303 (1989)
37. C.P. Andrieux, C. Blocman, J.M. Dumas-Bouchiat, F.M. Halla, J.M. Saveant: J. Electroanal. Chem. **113**, 19 (1980)
38. http://www.glucometer.com/
39. http://www.medisense.co.uk/
40. http://www.abbottdiagnostics.com/
41. http://www.arkray.co.jp/
42. http://www.accu-chek.com/
43. http://www.lifescan.com/
44. Japanese Industrial Standard Committee, JIS K 0102, Testing methods for industrial waste water. Japanese Standards Association, Tokyo, 1974
45. D.G. Hafeman, J.W. Parce, H.M. McConnell: "Light-addressable potentiometric sensor for biochemical systems", Science **240**, 1182 (1988)
46. J.C. Owicki, J.W. Parce: "Biosensors based on the energy metabolism of living cells: The physical chemistry and cell biology of extracellular acidification", Biosens. Bioelect. **7**, 255 (1992)

47. M. Adami, M. Sartore, E. Baldini et al.: "New measuring principle for LAPS devices", Sensors and Actuators **B 9**, 25 (1992)
48. M. Sartore, M. Adami, C. Nicolini et al.: "Minority carrier diffusion length effects on light-addressable potentiometric sensor (LAPS) devices", Sensors and Actuators **A32**, 431 (1992)
49. J.M. Libby, H.G. Wada: "Detection of *Neisseria meningitidis* and *Yersinia pestis* with a novel silicon-based sensor", J. Clin. Microbiol. **27**, 1456 (1989)
50. J. Briggs, P.R. Panfili: "Quantitation of DNA and protein impurities in biopharmaceuticals", Anal. Chem. **63**, 850 (1991)
51. Y. Sasaki, Y. Kanai, H. Uchida et al.: "Highly sensitive taste sensor with a new differential LAPS method", Sensors and Actuators **B24–25**, 819 (1995)
52. A. Pecora, G. Fortunato, R. Carluccio et al.: "Hydrogenated amorphous silicon based light-addressable potentiometric sensor (LAPS) for hydrogen detection", Non-Crystal. Solids **164–166**, 793 (1993)
53. J.W. Parce, J.C. Owicki, K.M. Kercso et al.: "Detection of cell-affecting agents with a silicon biosensor", Science **246**, 243 (1989)
54. J.C. Owicki, J.W. Parce, K.M. Kercso et al.: "Continuous monitoring of receptor-mediated changes in the metabolic rates of living cells", Proc. Natl. Acad. Sci. USA **87**, 4007 (1990)
55. H.M. McConnell, J.C. Owicki, J.W. Parce et al.: "The cytosensor microphysiometer: Biological applications of silicon technology", Science **257**, 1906 (1992)
56. H.G. Wada, S.R. Indelicato, L. Meyer et al.: "GM-CSF triggers a rapid, glucose dependent extracellular acidification by TF-1 cells: Evidence for sodium/proton antiporter and PKC mediated activation of acid production", J. Cell. Physiol. **154**, 129 (1993)
57. S.D.H. Chan, D.M. Antoniucci, K.S. Fok et al.: "Heregulin activation of extracellular acidification in mammary carcinoma cells is associated with expression of HER2 and HER3", J. Biol. Chem. **270**, 22608 (1995)
58. Japanese Industrial Standard Committee: *JIS K 3602, Apparatus for the Estimation of Biochemical Oxygen Demand (BOD$_S$) with Microbial Sensor* (Japanese Standards Association, Tokyo, 1990)
59. I. Karube, T. Matsunaga, S. Mitsuda et al.: "Microbial electrode BOD sensors", Biotechnol. Bioeng. **19**, 1535 (1977)
60. I. Karube, S. Mitsuda, T. Matsunaga et al.: "A rapid method for estimation of BOD by using immobilized microbial cells", J. Ferment. Technol. **55**, 243 (1979)
61. M. Hikuma, H. Suzuki, T. Yasuda et al.: "Amperometric estimation of BOD by using living immobilized yeasts", Euro. J. Appl. Microbiol. Biotechnol. **8**, 289 (1979)
62. T.C. Tan, F. Li, K.G. Neoh: "Measurement of BOD by initial rate of response of a microbial sensor", Sensors and Actuators **B10**, 137 (1993)
63. H. Tanaka, E. Nakamura, Y. Minamiyama et al.: "BOD biosensor for secondary effluent from wastewater treatment plants", Wat. Sci. Technol. **30**, 215 (1994)
64. A. Ohki, K. Shinohara, O. Ito et al.: "A BOD sensor using *Klebsiella oxytoca* AS1", Intern. J. Environ. Anal. Chem. **56**, 261 (1994)
65. F. Li, T.C. Tan, Y.K. Lee: "Effects of pre-conditioning and microbial composition on the sensing efficacy of a BOD biosensor", Biosens. Bioelect. **9**, 197 (1994)

66. F. Li, T.C. Tan: "Monitoring BOD in the presence of heavy metal ions using a poly(4-vinylpyridine)-coated microbial sensor", Biosens. Bioelect. **9**, 445 (1994)
67. R. Iranpour, B. Straub, T.J. Jugo: "Real time BOD monitoring for wastewater process control", Environ. Eng. **123**, 154 (1997)
68. C.K. Hyun, E. Tamiya, T. Takeuchi et al.: "A novel BOD sensor based on bacterial luminescence", Biotech. & Bioeng. **41**, 1107 (1993)
69. X.M. Li, F.C. Ruan, W.Y. Ng et al.: "Scanning optical sensor for the measurement of dissolved oxygen and BOD", Sensors and Actuators **B21**, 143 (1994)
70. S.D.W. Comber, M.J. Gardner, A.M. Gunn: "Measurement of absorbance and fluorescence as potential alternatives to BOD", Environ. Technol. **17**, 771 (1996)
71. Y. Murakami, T. Kikuchi, A. Yamamura, T. Sakaguchi, K. Yokoyama, Y. Ito, M. Takiue, H. Uchida, T. Katsube, E. Tamiya: "An organic pollution sensor based on surface photovoltage", Sensors and Actuators B–Chemical **53**, 163 (1999)
72. T. Chiyo, K. Matsui, Y. Murakami, K. Yokoyama, E. Tamiya: "Yeast-immobilized SPV device for koji quality control in sake brewing process", Biosens. Bioelect., in press
73. J.C. Eccles: J. Neurosci. **10**, 3769 (1990)
74. P.T. Kissinger: in: *Microdialysis in Neuroscience*, T.E. Robinson and J.B. Justice, Jr., eds., pp. 103–115 (Elsevier, Amsterdam, 1991)
75. O. Niwa, K. Torimitsu, M. Morita, P. Osborne, K. Yamamoto: Anal. Chem. **68**, 1865 (1996)
76. M. Suzuki, T. Kumagai, Y. Nakashima: Kagaku Kogaku Ronbunshu **25**, 177 (1999)
77. A. Akiyama, T. Kato, Y. Ishii, E. Yasuda: Anal. Chem. **57**, 1518 (1985)
78. M. Suzuki, H. Suzuki, I. Karube, R.D. Schmid: Anal. Lett. **22**, 2915 (1989)
79. E. Tamiya, Y. Sugiura, A. Akiyama, I. Karube: Ann. N.Y. Acad. Sci. **613**, 396 (1990)
80. E. Tamiya, Y. Sugiura, N.E. Navera, S. Mizoshita, K. Nakajima, A. Akiyama, I. Karube: Anal. Chim. Acta **251**, 129 (1991)
81. E. Tamiya, Y. Sugiura, Y. Amou, I. Karube, A. Ajima, R.T. Kado, M. Ito: Sensors and Materials **7**, 249 (1995)
82. Z. Markus, A. Ursula, Q.D. Kim, S. Peter, C. Michel: J. Neurochem. **52**, 1919 (1989)
83. E. Tamiya, Y. Sugiura, T. Takeuchi, M. Suzuki, I. Karube, A. Akiyama: Sensors and Actuators **B10**, 179 (1993)
84. M. Koudelka: Sensors and Actuators **9**, 249 (1986)
85. G. Jobst, G. Urban, A. Jachimowicz, F. Kohl, O. Tilado, I. Lettenbichler, G. Nauer: Biosens. Bioelect. **8**, 123 (1993)
86. P. Arquint, A. van den Berg, B.H. van der Schoot, N.F. de Rooij, H. Bühler, W.E. Morf, L.F.J. Dürselen: Sensors and Actuators **B13–14**, 340 (1993)
87. P. Arquint, M. Koudelka-Hep, B.H. van der Schoot, P. van der Wal, N.F. de Rooij: Clin. Chem. **40**, 1805 (1994)
88. K. Tsukada, Y. Miyahara, Y. Shibata, H. Miyagi: Sensors and Actuators **B2**, 291 (1990)
89. W. Gumbrecht, D. Peters, W. Schelter, W. Erhardt, J. Henke, J. Steil, U. Sykora: Sensors and Actuators **B18–19**, 704 (1994)

90. H. Suzuki, E. Tamiya, I. Karube: Anal. Chem. **60**, 1078 (1988)
91. H. Suzuki, A. Sugama, N. Kojima, F. Takei, K. Ikegami: Biosens. Bioelect. **6**, 395 (1991)
92. K. Shimada, M. Yano, K. Shibatani, Y. Komoto, M. Esashi, T. Matsuo: Med. Biol. Eng. Comput. **18**, 741 (1980)
93. T. Sekiguchi, Y. Nagai, T. Makino, K. Ohno, M. Nakamura, H. Hosaka, H. Sakio, S. Ohtsu, H. Takahashi: Sensors and Actuators **B49**, 171 (1998)
94. H. Suzuki, H. Arakawa, S. Sasaki, I. Karube: Anal. Chem. **71**, 1737 (1999)
95. J. Janata, R.J. Huber: Ion-Select. Electrode Rev. **1**, 31 (1979)
96. F. Keplinger, R. Glatz, A. Jachimowicz, G. Urban, F. Kohl, F. Olcaytug, O.J. Prohaska: Sensors and Actuators **B1**, 272 (1990)
97. C. Dumschat, R. Frömer, H. Rautschek, H. Müller, H.J. Timpe: Anal. Chim. Acta **243**, 179 (1991)
98. V.V. Cosofret, M. Erdösy, T.A. Johnson, R.P. Buck, R.B. Ash, M.R. Neumann: Anal. Chem. **67**, 1647 (1995)
99. A. Uhlig, E. Lindner, C. Teutloff, U. Schnakenberg, R. Hintsche: Anal. Chem. **69**, 4032 (1997)
100. U. Lemke, K. Cammann, C. Kötter, C. Sundermeier, M. Knoll: Sensors and Actuators **B7**, 488 (1992)
101. H. Meyer, H. Drewer, J. Krause, K. Cammann, R. Kakerow, Y. Manoli, W. Mokwa, M. Rospert: Sensors and Actuators **B18–19**, 229 (1994)
102. B.K. Oh, C.Y. Kim, H.J. Lee, K.L. Rho, G.S. Cha, H. Nam: Anal. Chem. **68**, 503 (1996)
103. V.V. Cosofret, M. Erdosy, T.A. Johnson, D.A. Bellinger, R.P. Buck, R.B. Ash, M.R. Neuman: Anal. Chim. Acta **314**, 1 (1995)
104. J. Janata: *Principles of Chemical Sensors*, p. 106 (Plenum Press, New York, 1989)
105. Y. Hanazato, M. Nakao, S. Shiono: IEEE Trans. Electron. Devices **ED-30**, 47 (1986)
106. S. Nakamoto, N. Ito, T. Kuriyama, J. Kimura: Sensors and Actuators **13**, 165 (1988)
107. H.J. Lee, U.S. Hong, D.K. Lee, J.H. Shin, H. Nam, G.S. Cha: Anal. Chem. **70**, 3377 (1998)
108. M.A. Nolan, S.H. Tan, S.P. Kounaves: Anal. Chem. **69**, 1244 (1997)
109. R.L. Smith, D.C. Scott: IEEE Trans. Biomed. Eng. **BME-33**, 83 (1986)
110. A. van den Berg, A. Grisel, H.H. van den Vlekkert, N.F. de Rooij: Sensors and Actuators **B1**, 425 (1990)
111. S. Yee, H. Jin, L.K.C. Lam: Sensors and Actuators **15**, 337 (1987)
112. H. Suzuki, H. Shiroishi, S. Sasaki, I. Karube: Anal. Chem. **71**, 5069 (1999)
113. D.R. Matthews, R.R. Holman, E. Bown, J. Steemson, A. Watson, S. Hughes, D. Scott: Lancet **i**, 778 (1987)
114. G.B. Broun: in: *Methods in Enzymology*, ed. by K. Mosbach, Vol. 44, pp. 263–280 (Academic Press, New York, 1976)
115. P. Bergveld: Biosensors **2**, 15 (1986)
116. B.H. van der Schoot, P. Bergveld: Biosensors **3**, 161 (1987/88)
117. A.M.N. Hendji, N. Jaffrezic-Renault, C. Martelet, A.A. Shul'ga, S.V. Dzydevich, A.P. Soldatkin, A.V. El'skaya: Sensors and Actuators **B21**, 123 (1994)
118. A.A. Shul'gaß, A.P. Soldatkin, A.V. El'skaya, S.V. Dzyadevich, S.V. Patskovsky, V.I. Strikha: Biosens. Bioelect. **9**, 217 (1994)

119. D.C. Cullen, R.S. Sethi, C.R. Lowe: Anal. Chim. Acta **231**, 33 (1990)
120. A. Manz, N. Graber, H. Widmer: Sensors and Actuators **B1**, 244 (1990)
121. J.C. Fettinger, A. Manz, H. Lüdi, H.M. Widmer: Sensors and Actuators **B17**, 19 (1993)
122. G. Jobst, I. Moser, P. Svasek, M. Varahram, Z. Trajanoski, P. Wach, P. Kotanko, F. Skrabal, G. Urban: Sensors and Actuators **B43**, 121 (1997)
123. H.H. van den Vlekkert, N.F. de Rooij, A. van den Berg, A. Grisel: Sensors and Actuators **B1**, 395 (1990)
124. B.H. van der Schoot, H.H. van den Vlekkert, N.F. de Rooij, A. van den Berg, A. Grisel: Sensors and Actuators **B4**, 239 (1991)
125. S. Shoji, M. Esashi: Sensors and Actuators **B8**, 205 (1992)
126. S. Shoji, M. Esashi, T. Matsuo: Sensors and Actuators **14**, 101 (1988)
127. W. Gumbrecht, W. Schelter, B. Montag, M. Rasinski, U. Pfeiffer: Sensors and Actuators **B1**, 477 (1990)
128. Y. Murakami, T. Takeuchi, K. Yokoyama, E. Tamiya, I. Karube, M. Suda: Anal. Chem. **65**, 2731 (1993)
129. O. Niwa, R. Kurita, T. Horiuchi, K. Torimitsu: Electroanalysis **11**, 356 (1999)
130. B.H. van der Schoot, S. Jeanneret, A. van den Berg, N.F. de Rooij: Sensors and Actuators **B15-16**, 211 (1993)
131. W. Hoffmann, R. Rapp: Sensors and Actuators **B34**, 471 (1996)
132. B.H. van der Schoot, S. Jeanneret, A. van den Berg, N.F. de Rooij: Sensors and Actuators **B6**, 57 (1992)
133. F. van Steenkiste, K. Baert, D. Debruyker, V. Spiering, B. van der Schoot, P. Arquint, R. Born, K. Schumann: Sensors and Actuators **B44**, 409 (1997)
134. T. Laurell: Sensors and Actuators **B13-14**, 323 (1993)
135. R. Steinkuhl, C. Sundermeier, H. Hinkers, C. Dumschat, K. Cammann, M. Knoll: Sensors and Actuators **B33**, 19 (1996)
136. E. Dempsey, D. Diamond, M.R. Smyth, G. Urban, G. Jobst, I. Moser, E.M.J. Verpoorte, A. Manz, H.M. Widmer, K. Labenstein, R. Freaney: Anal. Chim. Acta **346**, 341 (1997)
137. K. Kobayashi, T. Tokuda, H. Suzuki: Trans. MRS-J **26**, 131 (2001)
138. K. Mabuchi et al.: *Proceedings of 21th Annual International Conference of the IEEE Engineering in Medicine and Biology Society*, CD-ROM version (Atlanta, 1999)
139. K. Mabuchi et al.: *Proceedings of 20th Annual International Conference of the IEEE Engineering in Medicine and Biology Society*, Part 1/6, pp. 458–461 (Hong Kong, 1998)
140. K.-E. Hagbarth, Å B. Vallbo: Acta Physiol. Scand. **69**, 121 (1967)
141. Å B. Vallbo: Acta Physiol. Scand. **80**, 552 (1970)
142. Å B. Vallbo et al.: Physiol. Rev. **59**, 919 (1979)
143. M. Knibestol: J. Physiol. **232**, 427 (1973)
144. S. Iwase et al.: "Microneurographic study on sympathetic control of sweating and skin blood flow", in: *Advanced Techniques and Clinical Applications in Biomedical Thermology*, ed. by K. Mabuchi (Harwood Academic Publishers, Chur, 1994)
145. J.L. Ochoa, H.E. Torebjörk: J. Physiol. **342**, 633 (1983)
146. M. Kunimoto et al.: J. Physiol. **442**, 391 (1991)
147. K. Mabuchi et al., Biomed. Thermol. **17**, No. 4, 24 (1997)
148. P. Heiduschka, S. Thanos: Progr. Neurobiol. **55**, 433 (1998)

149. T. Stieglitz, J.-U. Meyer: "Microtechnical interfaces to neurons", in: *Microsystem Technology in Chemistry and Life Sciences*, ed. by A. Manz and H. Becker (Springer, Berlin, 1998)
150. D. Banks: IEE Eng. Sci. Educ. J. **7**, No. 3, 135 (1998)
151. K. Najafi: Eng. Med. Biol. **13**, No. 3, 375 (1994)
152. E.M. Schmidt: "Electrodes for many single neuron recordings", in: *Methods for Neural Ensemble Recordings*, ed. by M.A.L. Nicolelis (CRC Press, Boca Raton, 1999)
153. K. Miki, Y. Hayashida, K. Shiraki: Am. J. Physiol. **264**, R369 (1993)
154. R.B. Stein et al.: Brain Res. **128**, 21 (1977)
155. I. Ninomiya, Y. Yonezawa, M. F. Wilson: J. Appl. Phys. **41**, No. 1, 111 (1976)
156. T. Lefurge et al.: Ann. Biomed. Eng. **19**, 197 (1991)
157. K.-E. Hagbarth, Å.B. Vallbo: Acta Physiol. Scand. **69**, 121 (1967)
158. Å.B. Vallbo, K.-E. Hagbarth: Exp. Neurol. **21**, 270 (1968)
159. T. Suzuki et al.: *Proceedings of 21st International Conference of the IEEE EMBS* (CD-ROM) pp. 457 (Atlanta, 1999)
160. A.C. Hoogerwerf, K.D. Wise: IEEE Trans. BME **41**, No. 12, 1136 (1994)
161. K.D. Wise, K. Najafi: Science **254**, 1335 (1991)
162. P.K. Campbell et al.: IEEE Trans. BME **38**, No. 8, 758 (1991)
163. A. Branner, R.B. Stein, R.A. Normann: *Proceedings of 21st International Conference of the IEEE EMBS* (CD-ROM) p. 377 (Atlanta, 1999)
164. G.T.A. Kovacs et al.: IEEE Trans. BME **41**, No. 6, 567 (1994)
165. T. Suzuki et al.: *Proceedings of 19th International Conference of the IEEE EMBS* (CD-ROM) (Chicago, 1997)
166. K.W. Hiller, W. Seidal, W.J.A. Kolff: Trans. Am. Soc. Intern. Organs **8**, 125 (1962)
167. K. Imachi et al.: Jap. J. Artificial Organs **5**, No. 6, 321 (1976), in Japanese
168. W.S. Pierce et al.: Trans. Am. Soc. Intern. Organs **8**, 125 (1962)
169. Y. Abe et al.: "A new automatic control method of total artificial heart: 1/R control method". in: *Artificial Heart*, ed. by Y. Sezai (Harwood Academic Pulishers, Chur, 1993)
170. P.T. Kissinger, J.B. Hart, R.N. Adams: Brain Res. **55**, 209 (1973)
171. M. Buda et al.: Brain Res. **273**, 197 (1983)
172. P. Bergveld: IEEE Trans. Biomed. Eng. BME **17**, 70 (1970)
173. Y. Miyahara, T. Moriizumi, K. Ichimura: Sensors and Actuators **7**, 1 (1985)
174. K. Maeda et al.: Jap. J. Artificial Organs **17**, No. 3, 1035 (1988)
175. R.M. Wightman, L.J. May, A.C. Michael: Anal. Chem. **60**, 769G (1988)
176. U. Ungerstedt et al.: Neurosci. Lett. **10**, 493 (1982)
177. L. Bousse: Sensors and Actuators B **B34**, No. 1–3, 270 (1996)
178. B.D. Debusschere et al.: *Proceedings of Transducers '99* (CD-ROM version) (Sendai, 1999)
179. W.S. Pierce et al.: Trans. Am. Soc. Artif. Intern. Organs **22**, 347 (1976)
180. Y. Abe et al.: Trans. Am. Soc. Artif. Intern. Organs **40**, No. 3, M506 (1994)
181. S. Iwase et al.: "Microneurographic study on sympathetic control of sweating and skin blood flow", in: *Advanced Techniques and Clinical Applications in Biomedical Thermology*, ed. by K. Mabuchi (Harwood Academic Publishers, Chur, 1994)
182. Å.B. Vallbo et al.: Physiol. Rev. **59**, 919 (1979)
183. K.-E. Hagbarth, Å.B. Vallbo: Acta Physiol. Scand. **69**, 121 (1967)

184. D.R. Kenshalo, ed.: *Sensory Functions of the Skin of Humans* (Plenum Press, New York, 1979)
185. A. Iggo: "Electrophysiological and histological studies of cutaneous mechanoreceptors", in: *The Skin Senses*, ed. by D.R. Kensalo (Thomas, Springfield, IL, 1968)
186. I. Darian-Smith: "The sense of touch: Performance and peripheral neural process", in: *Handbook of Physiology*, ed. by J.M. Brookhart and V.B. Mountcastle, Sect. 1, The Nervous System, Vol. III (American Physiological Society, Bethesda, 1984)
187. A. Iggo: Ann Rev. Neurosci. **5**, 1 (1982)
188. J. Ochoa, E. Torebjörk: J. Physiol. **342**, 633 (1983)
189. G.A. Gescheider et al.: Somatosens. Motor Res. **16**, No. 3, 229 (1999)
190. W.R. Uttal: J. Comp. Physiol. Psychol. **53**, 47 (1960)
191. E. Schmid: J. Exp. Psychol. **61**, 400 (1961)
192. D.S. Alles: IEEE Trans. Man–Machine Systems **MMS-11**, 85 (1970)
193. J.H. Kirman: Perception and Psychophysics **15**, 1 (1974)
194. C.E. Sherrick: "Studies of apparent tactual movement", in: *The Skin Senses*, ed. by D.R. Kensalo (Thomas, Springfield, IL, 1968)
195. K. Mabuchi et al.: *IFESS 2000*, pp. 403–406 (Aalborg, 2000)
196. K. Mabuchi et al.: *Proceedings of 21st International Conference of the IEEE EMBS*, CD-ROM version (Atlanta, 1999)
197. M. Shimojyo: *IEEE Int. Conf. Robotics and Automation*, Vol. 1, pp. 831–836 (Nagoya, 1995)
198. J.K. Chapin et al.: Nature Neurosci. **2**, 664 (1999)
199. J. Wessberg et al.: Nature **408**, 361 (2000)
200. K. Mabuchi et al: The Int. J. Artif. Organs **20**, No. 1, 37 (1997)
201. S. Schtz, J. Czynski, H.E. Hummel: Proceedings of Transducers '99 (CD-ROM version) (Sendai, 1999)

Index

a-g linkage 250
absorber 96
absorbing particle 31
acoustic impedance 55, 102
active shielding 146
adaptation 250
additional positive feedback 135
adverse effects of diagnostic ultrasound 109
affinity 39
aliasing 139
angiologic thermatome 41
anisotropic etching 243
anti-aliasing filter 139
apparent tactile movement 281
Ar ion laser 37
artifact 83
artificial heart system 270
artificial soamtic sensation 275
atherosclerosis 52
atomic force microscope 32, 35
attractive force 35
automated diagnosis 84
autonomic nervous system 263
avalanche photodiode 102
axetylcholine 240
axial cross-sectional signal 97

B-mode echo 70
backscattered light 102
Beer-Lambert law 100
biological application 37
biological structure 101
biomolecule 36
bionic medicine 257
Biot-Savert law 124
blood vessel 101
BOD 231
BOD sensor 231
Brownian motion 38
buoyant force 34

calibration 147
cantilever 32
carbon dioxide electrode 243
carbon-fiber electrode 239
catecholamine 264
CCD camera 38
cell membrane 36
cell-based biosensor 269
chaotic (signal) 64
cold-water immersion test 49
collagen electrode 258
complex refractive index 94
computer processing for 3D ultrasound 79
condenser microphone 62
conjugate material 36
contour map 170
conventional microscope 38
convolution of pulse shape 100
coronary artery stenosis 52
cross-sectional imaging 94
cuff electrode 258
current dipole 149
current element distribution 163
current elements 120
cutaneous saltation 281

dc SQUID 125
decay length 34
deferential amplifier 38
defocusing lens method 81
Delauney triangulation 172
dermatomal thermatome 43
dewar 142

dichroic mirror 37
dielectric particle 31
diffraction limit of light 36
direct offset integration technique 134
displacement signal 38
DNA 33, 39
Doppler technique 71
drug delivery system 102

electrics current dipole 120
electrochemical measurement 264
electromagnetic field 31
enhancement factor 35
equivalent current dipole 154
evanescent illumination 34
event-related field 189
extraction of head and brain from MRI 174

FA I 278
far-infrared (FIR) imaging 40
feedback stabilization 38
field observation 195
field-assisted bonding 243
FISH 37
fluorescence 35
fluorescence beads 35
fluorescence emission 39
fluorescence image 35
fluorescence in situ hybridization 37
fluorescence intensity 35
flux modulation 133
flux-locked loop 133
flux-to-voltage transfer coefficient 131
fluxgate magnetometer 122
forward light scattering 102
functional electrical stimulation (FES) 286
functionality on particle 36

galvanomirror 38
Gaussian beam 30
Gaussian distribution 100
Geselowitz's formula 152
glass particle 35
glass substrate 35
glutamate 239
gold colloidal particle 35
gradient force 30
gravity 34

Harmonic frequencies 72
humoral system 263

I-V characteristics of SQUIDs 129
Ia afferent nerve fiber 250
imaging mechanism 102
in vivo biosensor 239
intensity profile 30
intrafascicular electrode 259
ISFET 246

knife edge 95
knockout (mouse) 269

LAPS 232
laser 30
laser displacement gauge 56
laser intensity feedback 33
laser phonocardiograph 52
laser trapping 30
laser trapping NSOM 35
laser-induced cell modification 28
laser-trapping NSOM 39
lateral resolution 36, 100
left ventricular assist device 275
lift-off 242
light absorption 31
linearity of trapping force 33
liquid helium 142

M-mode tracing 70
magnetic dipole 118
magnetic flux 119
magnetically shielded room 146
magnetoencephalography 138, 177
magnitude estimation method 278
mammography 102
manipulation of living body 29
Markov chain Monte Carlo methods 158
masking effects 281
mathematical model for skin temperature 45
maximum entropy method (MEM) 52, 61
maximum resolving power 38

mechanical contact 30
mechanoreceptor 250, 276
membrane permeability 109
metabolic thermatome 43
metallic particle 31, 35, 37
Metropolis algorithm 158
micro-stimulation 278
microbiosensor 247
microcapsule collapse 105
microcapsules 103
microdailysis probe 239
microdialysis 267
microelectrode 257
microglutamate sensor 239
microneurography 250, 272, 278
microscope 29
microscope objective 32, 36
microstimulation 250
molecular self-assembly 36
momentum change 30
momentum conservation theorem 30
mosquito 249
motion artifact 82
murmur 54
μTAS 248
myotomal thermatome 43

nanometric-chemical sensor 36
Nd:YLF laser 37
near-field imaging 35
near-field scanning optical microscope 35
near-infrared power supply 21
near-infrared transcutaneous transmission 23
needle array electrode 260
needle-electrode 256
nerve electrode 257
nerve-regenerating electrode 286
neural interface 257
neurotransmitter 239
Newton cooling equations 49
NSOM 35
Nyquist limit 72
Nyquist rate 139

optical levitation 31
optoacoustic tomography 94
oxygen electrode 243

parallel receiving processing 84
parallel tempering 159
parasympathetic nervous system 271
particle displacement 33
PC (Pacinian type) 278
periodic (signal) 64
pH 35
phantom sensation 281
Phi-V characteristics 129
photocrosslinking polyvinyl alcohol 240
photodetector 102
photolithography 242
photomultiplier 102
photomultiplier tube 37
photon pressure 30
photosentitive polymers 243
piezo tube 38
PIN photo-detector 95
pinhole 36
platinized carbon-fiber 240
polymethylmethacrylate 104
polystyrene particle 33
post-stenotic turbulence 52
power spectral density 63
pressure-conductive rubber 282
probabilistic approach 158
probe electrode 260
probe particle 33
Protein kinase C 110
proton magnetometer 122
pulse shape 100, 102
pulsed laser beam 57
pump 249

quadrant detector 38
quantization error 141
quasi-periodic (cycle) 64

RA (rapidly adapting type) 278
radiation force 31
radiation pressure 30
ray 30
real-time 3D ultrasound 83
real-time ultrasonic beam tracing 81
receptive field 254
rectangular wave 96
red blood cell 32

reference electrode 246
reflection measurement 102
reflection system 102
refraction 30
refractive index 31
regeneration electrode 261
repulsive component 35
reversible jump 160
RNA polymerase 33
rubber damper 96

SA I (slowly adapting type I) 278
SA II (slowly adapting type II) 278
safety standard 102
scanning microscope 35
scattering coefficient 100
scattering efficiency 35, 94
scattering force 30
scattering image 35
scattering medium 100
scattering theory 102
screen-printing 242
search coil 121
section reconstruction 79
sensory dermatome 42
side lobe 100
silicone rubber 96
simulated annealing 161
skin sympathetic nerve activity (SSNA) 271
skin tissue 101
somatic sensation 257
sono-poration 110
sonoluminescence 94
spherical conductor model 152
spinal cord evoked magnetic field 192
spot position 38
spot size 31
spring constant 32
SPV 232
stabilized condition 168
sudomotor 271
surface force 33
surface modification 36
surface rendering 79
surfactant 100
sympathetic nervous system 271

2-photon laser ablation 25
2D array probe 84
3D data acquisition 79
3D data set construction 79
3D probe 79
the American Institute of Ultrasound in Medicine (AIUM) 109
the resonant frequency of a microbubble 104
The World Federation for Ultrasound in Medicine and Biology (WFUMB) 109
thermal coronary angiography 51
thermal energy 38
thermogram 40
three orthogonal sectional images 79
three-dimensional laser trapping 31
three-dimensional ultrasound 79
time delay 98
time-sequential image processing 47
tissue characterization 74, 84
total internal reflection 34
trajectory 64
transmission system 101
tungsten microelectrode 251, 259
turbulent flow 53
two-dimensional moving target indication 105
two-dimensional ultrasound 76

ultrasonic backscatter 74
ultrasonography 69
ultrasound field 102
ultrasound transducer 95
ultrasound velocity 99

vagal nerve 275
vasomotor 271
vector map 174
victorial blue 96
vinylidenedichloride 104
visible wavelength 36
volume current 150
volume rendering 79

weak force 33
whole-cell sensor 269
wire electrode 258, 275

YOYO-1 iodide 39

Printing: Mercedes-Druck, Berlin
Binding: Stein + Lehmann, Berlin